KU-535-435

WITHDRAWN

WS 1001689 9

CARDIOVASCULAR PHYSIOLOGY

WEST SUSSEX INSTITUTE OF HIGHER EDUCATION LIBRARY	
AUTHOR	ACCESSION No
BERNE, R.M.	BO 40551
TITLE	CLASS No

STANDARD LOAN

UNLESS RECALLED BY ANOTHER READER
THIS ITEM MAY BE BORROWED FOR

FOUR WEEKS

To renew, telephone:
816089 (Bishop Otter)
01243 812099 (Bognor Regis)

1 5 JAN 2007

2 4 MAR 2007

1 6 APR 2007   1 3 DEC 2010

2 1 MAY 2007   1 1 MAR 2011
1 0 OCT 2007
2 1 JAN 2008   2 8 MAR 2011

1 2 MAY 2009   2 9 MAR 2011

1 / 5 / 10

CARDIOVASCULAR PHYSIOLOGY

ROBERT M. BERNE, M.D., DSc. (Hon.)

Chairman and Charles Slaughter Professor of Physiology,
Department of Physiology, University of Virginia School of Medicine,
Charlottesville, Virginia

MATTHEW N. LEVY, M.D.

Chief of Investigative Medicine, Mount Sinai Hospital of Cleveland;
Professor of Physiology, Medicine, and Biomedical Engineering,
Case Western Reserve University,
Cleveland, Ohio

W. SUSSEX INSTITUTE
OF
HIGHER EDUCATION
LIBRARY

FOURTH EDITION

With **218** illustrations

The C. V. Mosby Company

ST. LOUIS • TORONTO • LONDON   1981

MOSBY

1906 **75** 1981
YEARS

A TRADITION OF PUBLISHING EXCELLENCE

Editor: John E. Lotz
Manuscript editor: Patricia J. Milstein
Design: Susan Trail
Production: Barbara Merritt

FOURTH EDITION

Copyright © 1981 by The C.V. Mosby Company

All rights reserved. No part of this book may be reproduced
in any manner without written permission of the publisher.

Previous editions copyrighted 1967, 1972, 1977

Printed in the United States of America

The C.V. Mosby Company
11830 Westline Industrial Drive, St. Louis, Missouri 63141

Library of Congress Cataloging in Publication Data

Berne, Robert M.
  Cardiovascular physiology.

  Includes bibliographies and index.
  1. Cardiovascular system.   I. Levy, Matthew N.
II. Title.
QP102.B47   1981   612'.1   81-2039
ISBN 0-8016-0655-1   AACR2

C/VH/VH   9   8   7   6   5   4   3   2   1   01/B/010

To

Carl J. Wiggers

who introduced us to cardiovascular physiology

and to each other

PREFACE

This book is designed primarily for medical and graduate students. With this fundamental purpose in mind, an attempt has been made to emphasize general concepts and to ignore isolated facts, except where they are deemed to be essential. In accordance with this principle, little documentation is afforded for many of the assertions made, and only a few references have been included in the bibliographies at the end of each chapter. Review articles have been given preference over scientific papers, and articles have been selected primarily for their appropriateness for the beginning student, for the depth of the interpretation included in their discussion sections, and for the comprehensiveness of their bibliographies.

Because many of the broad principles of cardiovascular physiology are complex and confusing to the student, simplified models have been employed throughout the book. Unquestionably, there are advantages and disadvantages to this pedagogical device. In formulating a model, the instructor retains those elements and properties of the system under consideration that are deemed germane and discards those other components of the system that are deemed trivial. Furthermore, for those elements to be included in the model, the behavior of certain of these components is assumed to be less complex than is actually the case. One justification for such simplification is that this behavior is reasonably accurate over a certain limited range of values. However, once the underlying basis of the system is understood, the complicating details can then be added to approximate more closely the true system. Although models can serve an extremely useful function when employed properly, they can also lead to erroneous conclusions when misused. Therefore the reader must constantly be aware of the assumptions inherent in a given model and must decide whether a more detailed model is necessary at times to understand the specific problem under consideration.

In a sense normal physiology serves as a framework that students of medicine must comprehend before they can understand the derangements caused by disease or toxic agents. There is no intensive consideration of pathological physiology in this text. However, many examples of abnormal function are provided to illustrate more lucidly the behavior of the system under consideration and to indicate the direction in which students will be proceeding in their continuing efforts to understand the effect on the body of the multitude of disease processes that afflict humans.

In revising and updating this book, we have profited greatly from the many helpful criticisms and suggestions received from the readers, as well as from our own experience with the book in teaching cardiovascular physiology to medical and graduate students. We particularly want to thank Drs. B.R. Duling, R.A. Murphy, R. Rubio, and D.G. Ward. Approaches that have proved to be useful in development of some subjects have been retained, whereas others that have not met our expectations or have proved to be too complex have been deleted. As a result some sections have undergone little change, but others have been considerably revised. Furthermore, the chapter on special circulations now contains sections on the pulmonary, renal, and splanchnic circulations in addition to those on skin, muscle, brain, and fetal circulation, which were present in previous editions of the book. Throughout the book attempts have been made to incorporate the most recent information, and where the subject is still controversial, this has been indicated. Emphasis has been placed on control mechanisms; thus, in order to present the clearest view of the various mechanisms involved in the regulation of the circulatory system, the component parts of the system are discussed individually. However, since the body functions as a whole, in the last chapter we have tried to show how the cardiovascular system operates in an integrated fashion in response to a physiological stimulus (exercise) and a pathophysiological stimulus (hemorrhage).

We wish to express our appreciation to our readers for their constructive comments and hope that they will continue to provide the input necessary for us to make further improvements in future editions. We also wish to thank the numerous investigators and publishers who have granted us permission to use illustrations from their publications. In most cases these illustrations have been altered somewhat to increase their didactic utility. In some cases unpublished data from our own studies have been presented. These investigations were supported by grants HL-10384 and HL-15758 from the U.S. Public Health Service, to which we are indebted.

Robert M. Berne
Matthew N. Levy

CONTENTS

CARDIOVASCULAR PHYSIOLOGY

CHAPTER 1

THE CIRCUITRY

The circulatory, endocrine, and nervous systems constitute the principal coordinating and integrating systems of the body. Whereas the nervous system is primarily concerned with communications and the endocrine glands with regulation of certain body functions, the circulatory system serves to transport and distribute essential substances to the tissues and to remove by-products of metabolism. The circulatory system also shares in such homeostatic mechanisms as regulation of body temperature, humoral communication throughout the body, and adjustments of oxygen and nutrient supply in different physiological states.

The cardiovascular system that accomplishes these chores is made up of a pump, a series of distributing and collecting tubes, and an extensive system of thin vessels that permit rapid exchange between the tissues and the vascular channels. The primary purpose of this text is to discuss the function of the components of the vascular system and the control mechanisms (with their checks and balances) that are responsible for alteration of blood distribution necessary to meet the changing requirements of different tissues in response to a wide spectrum of physiological and pathological conditions.

Before considering the function of the parts of the circulatory system in detail, it is useful to consider it as a whole in a purely descriptive sense. The heart consists of two pumps in series: one to propel blood through the lungs for exchange of oxygen and carbon dioxide (the *pulmonary circulation*) and the other to propel blood to all other tissues of the body (the *systemic circulation*). Unidirectional flow through the heart is achieved by the appropriate arrangement of effective flap valves. Although the cardiac output is intermittent in character, continuous flow to the periphery occurs by virtue of distension of the aorta and its branches during ventricular contraction *(systole)* and elastic recoil of the walls of the large arteries with forward propulsion of the blood during ventricular relaxation *(diastole)*. Blood moves rapidly through the aorta and its arterial branches, which become narrower and whose walls become thinner and change histologically toward the periphery. From a predominantly elastic structure, the aorta, the peripheral arteries become more muscular in character until at the arterioles the muscular layers predominates (Fig. 1-1). As far out as the beginning of the arterioles, frictional resistance to blood flow is relatively small, and, despite a rapid flow in the arteries, the pressure drop from the root of the aorta to the point of origin of the arterioles is relatively small (Fig. 1-2). The arterioles, the stopcocks of the vascular tree, are the principal points of resistance to blood flow in the circulatory system. The large resistance offered by the arterioles is reflected by the considerable fall in pressure from arterioles to capillaries. Adjustment in the degree of contraction of the circular muscle of these small vessels permits regulation of tissue blood flow and aids in the control of arterial blood pressure.

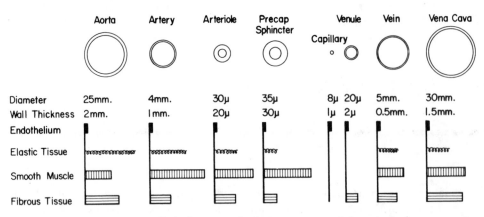

Fig. 1-1. Internal diameter, wall thickness, and relative amounts of the principal components of the vessel walls of the various blood vessels that compose the circulatory system. Cross sections of the vessels are not drawn to scale because of the huge range from aorta and venae cavae to capillary. (Redrawn from Burton, A. C.: Physiol. Rev. **34**:619, 1954.)

In addition to a sharp reduction in pressure across the arterioles, there is also a change from pulsatile to steady flow. The pulsatile character of arterial blood flow, caused by the intermittency of cardiac ejection, is damped at the capillary level by the combination of distensibility of the large arteries and frictional resistance in the arterioles. Many capillaries arise from each arteriole so that the total cross-sectional area of the capillary bed is very large, despite the fact that the cross-sectional area of each individual capillary is less than that of each arteriole. As a result, blood flow becomes quite slow in the capillaries, analogous to the decrease in flow rate seen at the wide regions of a river. Since the capillaries consist of short tubes whose walls are only one cell thick and since flow rate is slow, conditions in the capillaries are ideally suited for the exchange of diffusible substances between blood and tissue.

On its return to the heart from the capillaries, blood passes through venules and then through veins of increasing size. As the heart is

approached, the number of veins decreases, the thickness and composition of the vein walls change (Fig. 1-1), the total cross-sectional area of the venous channels is progressively reduced, and velocity of blood flow increases (Fig. 1-2). Also note that the greatest proportion of the circulating blood is located in the venous vessels (Fig. 1-2). The cross-sectional area of the venae cavae is larger than that of the aorta (although not evident from Fig. 1-2 because cross-sectional areas of venae cavae and aorta are so close to zero with a scale that includes the capillaries), and hence the flow is slower than that in the aorta. Blood entering the right ventricle via the right atrium is then pumped through the pulmonary arterial system at mean pressures about one-seventh those developed in the systemic arteries. The blood then passes through the lung capillaries, where carbon dioxide is released and oxygen taken up. The oxygen-rich blood returns via the pulmonary veins to the left atrium and ventricle to complete the cycle. Thus in the normal intact circulation the total volume of blood is constant

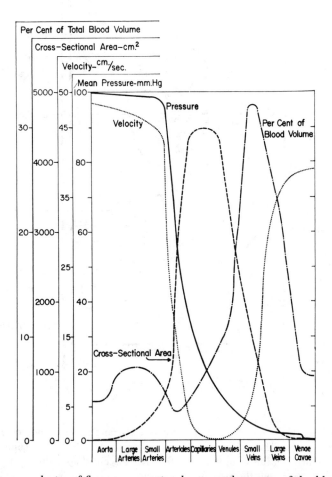

Fig. 1-2. Pressure, velocity of flow, cross-sectional area, and capacity of the blood vessels of the systemic circulation. The important features are the inverse relationship between velocity and cross-sectional area, the major pressure drop across the arterioles, the maximal cross-sectional area and minimal flow rate in the capillaries, and the large capacity of the venous system. The small but abrupt drop in pressure in the venae cavae indicates the point of entrance of these vessels into the thoracic cavity and reflects the effect of the negative intra-thoracic pressure. To permit schematic representation of velocity and cross-sectional area on a single linear scale, only approximations are possible at the lower values.

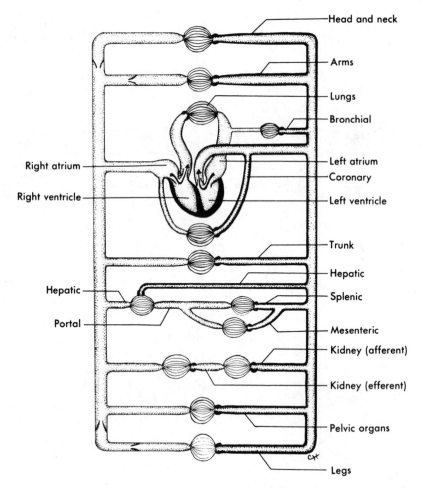

Fig. 1-3. Schematic diagram of the parallel and series arrangement of the vessels composing the circulatory system. The capillary beds are represented by thin lines connecting the arteries (on the right) with the veins (on the left). The crescent-shaped thickenings proximal to the capillary beds represent the arterioles (resistance vessels). (Redrawn from Green, H. D.: In Glasser, O., editor: Medical physics, vol. 1, Chicago, 1944, Year Book Medical Publishers, Inc.)

and an increase in the volume of blood in one area must be accompanied by a decrease in another. However, the velocity at which the blood circulates through the different regions of the body is determined by the output of the left ventricle and by the contractile state of the arterioles (resistance vessels) of these regions. The circulatory system is composed of conduits arranged in series and in parallel, as schematized in Fig. 1-3.

CHAPTER 2

ELECTRICAL ACTIVITY OF THE HEART

The experiments on "animal electricity" conducted by Galvani and Volta in the last half of the eighteenth and early nineteenth centuries prepared the stage for the discovery that electrical phenomena were involved in the spontaneous contractions of the heart. In 1855 Kölliker and Müller observed that when they placed the nerve of a nerve-muscle preparation in contact with the surface of a frog's heart, the muscle twitched with each cardiac contraction. This phenomenon may be observed in the laboratory by allowing the phrenic nerve of an anesthetized dog to lie across the exposed surface of the heart; the diaphragm will contract with each heartbeat. Precise measurement of this electrical activity was not feasible until the end of the past century, when the construction of sensitive, high-fidelity galvanometers permitted registration of the changes in electrical potential during the various phases of the cardiac cycle, which led to the science of electrocardiography. Furthermore, progress in electronics and the acquisition of knowledge concerning the cyclic changes in cardiac excitability have paved the way for the development of devices for converting various abnormal cardiac rhythms to normal rhythms and for maintaining normal heart rates in patients with severe conduction blocks (artificial pacemakers).

TRANSMEMBRANE POTENTIALS

The electrical behavior of single cardiac muscle cells has been investigated by inserting microelectrodes into the interior of cells from various regions of the heart. The potential changes recorded from a typical ventricular muscle fiber are illustrated schematically in Fig. 2-1. When two electrodes are situated in an electrolyte solution near a strip of quiescent cardiac muscle, there will be no potential difference measurable between the two electrodes (from point A to point B, Fig. 2-1). At point B one of the electrodes, a microelectrode with a tip diameter less than 1 μ, was inserted into the interior of a cardiac muscle fiber. Immediately the galvanometer recorded a potential difference across the cell membrane, indicating that the potential of the interior of the cell was

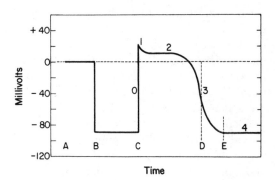

Fig. 2-1. Changes in potential recorded by an intracellular microelectrode. From time A to B the microelectrode was outside the fiber; at B the fiber was impaled by the electrode. At time C an action potential began in the impaled fiber. Time C to D represents the effective refractory period, and D to E represents the relative refractory period.

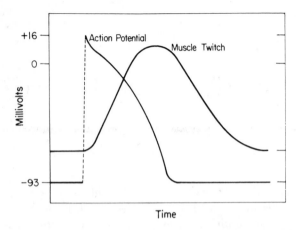

Fig. 2-2. Time relationships between the mechanical tension developed by a thin strip of ventricular muscle and the changes in transmembrane potential. (Redrawn from Kavaler, F., Fisher, V. J., and Stuckey, J. H.: Bull. N.Y. Acad. Med. **41:**592, 1965.)

about 90 mV. lower than that of the surrounding medium. Such electronegativity of the interior of the resting cell with respect to the exterior is also characteristic of skeletal and smooth muscle, of nerve, and indeed of most cells within the body. At point *C* a propagated action potential was transmitted to the cell impaled with the microelectrode. Very rapidly the cell membrane became depolarized; actually, the potential difference was reversed (positive over-shoot), so that the potential of the interior of the cell exceeded that of the exterior by about 20 mV. The rapid upstroke of the action potential is designated phase 0. Immediately after the upstroke, there was a brief period of partial repolarization (phase 1), followed by a *plateau* (phase 2) that persisted for about 0.1 to 0.2 second. The potential then became progressively more negative (phase 3), until the resting state of polarization was again attained (at point *E*). The process of rapid repolarization (phase 3) proceeds at a much slower rate of change than does the process of depolarization (phase 0). The interval from completion of repolarization until the beginning of the next action potential is designated phase 4.

The time relationships between the electrical events and the actual mechanical contraction are shown in Fig. 2-2. It can be seen that rapid depolarization (phase 0) precedes tension development and that completion of repolarization coincides approximately with peak tension development. The duration of contraction tends to parallel the duration of the action potential. Also as the frequency of cardiac contraction is increased, there is a progressive reduction in the duration of both the action potential and the mechanical contraction.

Principal types of cardiac action potentials

Two main types of action potentials are observed in the heart, as shown in Fig. 2-3. One type, the so-called *fast response*, occurs in the normal myocardial fibers in the atria and ventricles and in the specialized conducting fibers *(Purkinje fibers)* in these chambers. The action potentials shown in Figs. 2-1 and 2-2 are also typical fast responses. The other type of action potential, the so-called *slow response*, is found

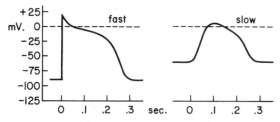

Fig. 2-3. A fast and a slow response action potential recorded from the same canine Purkinje fiber. In the left panel the Purkinje fiber bundle was perfused with a solution containing K^+ at a concentration of 4 mM. In the right panel epinephrine was added and the K^+ concentration was raised to 16 mM. (Redrawn from Wit, A. L., Rosen, M. R., and Hoffman, B. F.: Am. Heart J. **88:**515, 1974.)

in the *sinoatrial* (S-A) *node*, the natural pacemaker region of the heart, and in the *atrioventricular* (A-V) *node*, the specialized tissue involved in conducting the cardiac impulse from atria to ventricles. Furthermore, fast responses may be converted to slow responses either spontaneously or under certain experimental conditions. For example, in a myocardial fiber, a gradual shift of the resting membrane potential from its normal level of about -80 to -90 mV. to a value of about -60 mV. will cause a conversion of subsequent action potentials to the slow response. Such conversions may occur spontaneously in patients with severe coronary artery disease, in those regions of the heart in which the blood supply has been severely curtailed.

As shown in Fig. 2-3, not only is the resting membrane potential of the fast response considerably more negative than that of the slow response, but also the slope of the upstroke (phase 0), the amplitude of the action potential, and the extent of the overshoot of the fast response are greater than in the slow response. It will be explained later that the magnitude of the resting potential is largely responsible for these other distinctions between the fast and

slow responses. Furthermore, the amplitude of the action potential and the rate of rise of the upstroke are important determinants of propagation velocity. Hence, in cardiac tissue characterized by the slow response, conduction velocity is very much slower and there is a much greater tendency for impulses to be blocked than in tissues displaying the fast response. Slow conduction and tendency toward block are conditions that increase the likelihood of certain rhythm disturbances in the heart.

Ionic basis of the resting potential

The various phases of the cardiac action potential are associated with changes in the permeability of the cell membrane, mainly to Na, K, and Ca ions. These changes in permeability produce alterations in the rate of passage of these ions across the membrane. Just as with all other cells in the body, the concentration of potassium ions inside a cardiac muscle cell, $[K^+]_i$, greatly exceeds the concentration outside the cell, $[K^+]_o$, as shown in Fig. 2-4. The reverse concentration gradient exists for Na ions and for unbound Ca ions. Furthermore, the resting cell membrane is relatively permeable to K^+, but much less so to Na^+ and Ca^{++}. Because of the high permeability to K^+, there tends to be a net diffusion of K^+ from the inside to the outside of the cell, in the direction of the concentration gradient, as shown on the right side of the cell in Fig. 2-4. Many of the anions (labeled A^-) inside the cell, such as the proteins, are not free to diffuse out with the K^+. Therefore, as the K^+ diffuses out of the cell and leaves the A^- behind, the deficiency of cations causes the interior of the cell to become electronegative, as shown on the left side of the cell in Fig. 2-4.

Therefore, two opposing forces are involved in the movement of K^+ across the cell membrane. A chemical force, based on the concentration gradient, results in the net outward diffusion of K^+. The counter force is an electrostatic one; the positively charged K ions

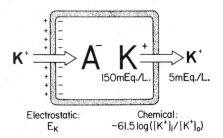

Electrostatic: Chemical:
E_K $-61.5 \log ([K^+]_i / [K^+]_o)$

Fig. 2-4. The balance of chemical and electrostatic forces acting on a resting cardiac cell membrane, based on a 30:1 ratio of the intracellular to extracellular K^+ concentrations, and the existence of a nondiffusible anion (A^-) inside but not outside the cell.

are attracted to the interior of the cell by the negative potential that exists there. If the system came into equilibrium, the chemical and the electrostatic forces would be equal. This equilibrium is expressed by the Nernst equation for potassium:

$$E_K = -61.5 \log ([K^+]_i / [K^+]_o)$$

The right-hand term represents the chemical potential difference at the body temperature of 37° C. The left-hand term, E_K, represents the electrostatic potential difference that would exist across the cell membrane if K^+ were the only diffusible ion. E_K is called the *potassium equilibrium potential*. When the measured concentrations of $[K^+]_i$ and $[K^+]_o$ for mammalian myocardial cells are substituted into the Nernst equation, the calculated value of E_K equals about -90 to -100 mV. This value is close to, but slightly more negative than, the resting potential actually measured in myocardial cells. Therefore there is a small potential of about 10 to 15 mV. tending to drive K^+ out of the resting cell.

The balance of forces acting on the Na ions is entirely different in resting cardiac cells. The intracellular Na$^+$ concentration, $[Na^+]_i$, is much lower than the extracellular concentration, $[Na^+]_o$. At 37° C, the *sodium equilibrium*

potential, E_{Na}, expressed by the Nernst equation is $-61.5 \log ([Na^+]_i / [Na^+]_o)$. For cardiac cells, E_{Na} is about $+40$ to $+60$ mV. At equilibrium, therefore, an electrostatic force of 40 to 60 mV., oriented with the inside of the cell more positive than the outside, would be necessary to counterbalance the chemical potential for Na$^+$. However, the actual polarization of the resting cell membrane is just the opposite. The resting membrane potential of myocardial fibers is about -80 to -90 mV. Hence both chemical and electrostatic forces act to pull extracellular Na$^+$ into the cell. The influx of Na$^+$ through the cell membrane is small, however, because the permeability of the resting membrane to Na$^+$ is very low. Nevertheless, it is mainly this small inward current of positively charged Na ions that causes the potential on the inside of the resting cell membrane to be slightly less negative than the value predicted by the Nernst equation for K^+.

The steady inward leak of Na$^+$ would cause a progressive depolarization of the resting cell membrane were it not for the metabolic pump that continuously extrudes Na$^+$ from the cell interior and pumps in K^+. The metabolic pump involves the enzyme, Na$^+$, K$^+$-activated ATPase, which is located in the cell membrane itself. Because the pump must move Na$^+$ against both a chemical and an electrostatic gradient, operation of the pump requires the expenditure of metabolic energy. Increases in $[Na^+]_i$ or in $[K^+]_o$ accelerate the activity of the pump. The quantity of Na$^+$ extruded by the pump slightly exceeds the quantity of K^+ transferred into the cell. Therefore, the pump itself tends to create a potential difference across the cell membrane, and thus it is termed an *electrogenic pump*. If the pump is partially inhibited, as by large doses of digitalis, the concentration gradients for Na$^+$ and K^+ are partially dissipated, and the resting membrane potential becomes less negative than normal.

The dependence of the transmembrane potential, V_m, on the intracellular and extracellular concentrations of K^+ and Na^+ and on the permeabilities (P_K and P_{Na}) to these ions is described by the Goldman constant-field equation for Na^+ and K^+:

$$V_m = -61.5 \log\frac{[K^+]_i + (P_{Na}/P_K)[Na^+]_i}{[K^+]_o + (P_{Na}/P_K)[Na^+]_o}$$

For a given ion, the permeability, P, is defined as the net quantity of the ion that diffuses across each unit area of membrane per unit concentration gradient and per unit membrane thickness. It is apparent from the constant-field equation that it is the relative permeabilities to Na^+ and K^+, and not the absolute magnitude of each permeability, that determine the resting potential. In the resting cardiac cell, because P_{Na} is so much less than P_K (that is, $P_{Na} / P_K \cong 0.01$), the Goldman equation reduces essentially to the Nernst equation for K^+. When the ratio $[K^+]_i/[K^+]_o$ is decreased experimentally by raising $[K^+]_o$, the measured value of V_m (dashed line, Fig. 2-5) approximates that predicted by the Nernst equation for K^+ (continuous line). For extracellular K^+ concentrations of about 5 mM. and above, the measured values correspond closely with the predicted values. The measured levels are slightly less than those predicted by the Nernst equation because of the small but finite value of P_{Na}. For values of $[K^+]_o$ below about 5 mM., it has been found that the membrane properties become altered, such that there is a progressive reduction in P_K as $[K^+]_o$ is diminished. As P_K is decreased, the effect of the Na^+ gradient on the transmembrane potential becomes relatively more important, as predicted by the constant-field equation. This change in P_K accounts for the greater deviation of the measured V_m from that predicted by the Nernst equation for K^+ at low levels of $[K^+]_o$ (Fig. 2-5). Also, in accordance with the constant-field equation, changes in $[Na^+]_o$ have relatively little effect on resting V_m (Fig. 2-6), because of the low value of P_{Na}.

Ionic basis of the fast response

Any process that abruptly changes the resting membrane potential to a critical value (called the *threshold*) will result in a propagated action potential. The characteristics of fast response action potentials resemble those shown on the left side of Fig. 2-3. The rapid depolarization that takes place during phase 0 is related almost exclusively to the inrush of Na^+ by virtue of a sudden increase in the permeability of the cell membrane to Na^+. The amplitude of the action potential (the magnitude of the potential change during phase 0) varies linearly with the logarithm of $[Na^+]_o$, as shown in Fig. 2-6. When $[Na^+]_o$ is reduced from its normal value of about 140 mM. to about 10 to 30 mM., the cell is no longer excitable.

The physical and chemical forces responsible for these transmembrane movements of Na^+ are explained in Fig. 2-7. When the resting membrane potential, V_m, is suddenly changed to the threshold level of about -60 to -70 mV., there is a dramatic change in the properties of the cell membrane. It is believed that *fast channels* for Na^+ exist in the membrane, and that the flux of Na^+ through these channels is controlled by two polar components, referred to as "gates." The opening and closing of these gates are governed principally by the electrostatic charge, V_m, across the membrane. One of these gates, the *m* gate, tends to open the channel as V_m becomes less negative and is therefore called an *activation gate*. The other, the *h* gate, tends to close the channel as V_m becomes less negative and hence is called an *inactivation gate*. The *m* and *h* designations were originally employed by Hodgkin and Huxley in their mathematical model of conduction in nerve fibers.

With the cell at rest, V_m is about -80 to -90 mV. At this level, the *m* gates are closed and the *h* gates are wide open, as shown in Fig. 2-7, *A*. The concentration of Na^+ is much

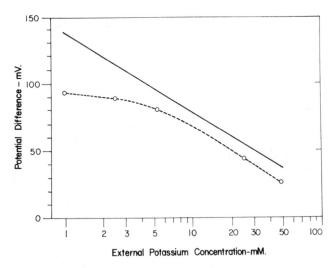

Fig. 2-5. Transmembrane potential of a cardiac muscle fiber varies inversely with the potassium concentration of the external medium (dashed curve). The continuous line represents the change in transmembrane potential predicted by the Nernst equation for E_K. (Redrawn from Page, E.: Circulation **26**:582, 1962.)

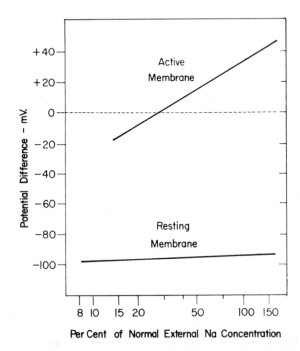

Fig. 2-6. Concentration of sodium in the external medium is a critical determinant of the amplitude of the action potential in cardiac muscle (upper curve) but has relatively little influence on the resting potential (lower curve). (Redrawn from Weidmann, S.: Elektrophysiologie der Herzmuskelfaser, Bern, 1956, Verlag Hans Huber.)

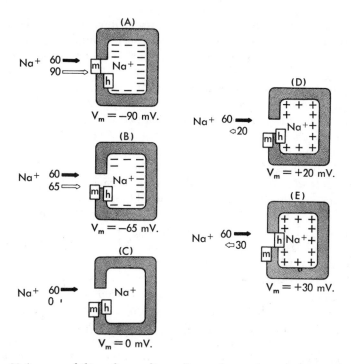

Fig. 2-7. The Na^+ permeability of a cardiac cell membrane during phase 4 (panel A) and during various stages of phase 0, when the transmembrane potential (V_m) had attained values of -65 mV. (panel B), 0 mV. (panel C), $+20$ mV. (panel D), and $+30$ mV. (panel E). The positions of the m and h electrostatic gates in the fast Na^+ channels are shown at the various levels of V_m. The electrostatic forces are represented by the white arrows, and the chemical (diffusional) forces by the black arrows.

greater outside than inside the cell, and the interior of the cell is electrically negative with respect to the exterior. Hence, both chemical and electrostatic forces are oriented to draw Na^+ into the cell. The electrostatic force in Fig. 2-7, A, is a potential difference of 90 mV., and it is represented by the white arrow. The chemical force, based on the difference in Na^+ concentration between the outside and inside of the cell, is represented by the black arrow. For a Na^+ concentration difference of about 130 mM., a potential difference of 60 mV. (inside more positive than the outside) would be necessary to counterbalance the chemical, or diffusional, force, according to the Nernst

equation for Na^+. Therefore, we may represent the net chemical force favoring the inward movement of Na^+ in Fig. 2-7 (black arrows) as being equivalent to a potential of 60 mV. With the cell at rest, therefore, the total electrochemical force favoring the inward movement of Na^+ is 150 mV. (panel A). The m gates are closed, however, and therefore the permeability of the resting cell membrane to Na^+ is very low. Hence, virtually no Na^+ moves into the cell; that is, the *inward Na^+ current* is negligible.

Any process that tends to make V_m less negative tends to open the m gates, thereby "activating" the fast Na^+ channels. The activation of the fast channels is therefore called a *voltage-dependent* phenomenon. The potential at which the *m* gates swing open varies somewhat from channel to channel in the cell membrane. As V_m becomes progressively less negative, therefore, more and more *m* gates will open. As the *m* gates open, Na^+ enters the cell (Fig. 2-7, *B*), by virtue of the chemical and electrostatic forces referred to before.

The entry of positively charged Na^+ into the interior of the cell tends to neutralize some of the negative charges inside the cell and thereby tends to diminish further the transmembrane potential, V_m. The resultant reduction in V_m, in turn, tends to open more *m* gates, thereby producing a still greater increase in the inward Na^+ current. Hence, this is called a *regenerative process*.

As V_m approaches the threshold value of about -65 mV., the remaining *m* gates rapidly swing open in the fast Na^+ channels, until virtually all of the *m* gates are open (Fig. 2-7, *B*). There is a rapid inrush of Na^+, which produces an abrupt reduction of V_m. This accounts for the rapid rate of change of V_m during phase 0 of the action potential (Fig. 2-1). The maximum rate of change of V_m (that is, the maximum dV_m/dt) has been found to be from 100 to 200 V./sec. in myocardial cells and from 500 to 1000 V./sec. in Purkinje fibers. Although the quantity of Na^+ that enters the cell during one action potential alters V_m by over 100 mV., it is too small to change the intracellular Na^+ concentration measurably. Therefore, the chemical force remains virtually constant, and only the electrostatic force changes throughout the action potential. Hence, the lengths of the black arrows in Fig. 2-7 remain constant at 60 mV., whereas the white arrows change in magnitude and direction.

As the Na^+ rushes into the cardiac cell during phase 0, the negative charges inside the cell are neutralized, and V_m becomes progressively less negative. When V_m becomes zero (Fig. 2-7, *C*), there is no longer an electrostatic

force pulling Na^+ into the cell. As long as the fast Na^+ channels are open, however, Na^+ continues to enter the cell because of the large concentration gradient that still exists. This continuation of the inward Na^+ current causes the inside of the cell to become positively charged with respect to the exterior of the cell (Fig. 2-7, *D*). This reversal of the membrane polarity is the so-called overshoot of the cardiac action potential, which is evident in Fig. 2-1. Such a reversal of the electrostatic gradient would, of course, tend to repel the entry of Na^+ (Fig. 2-7, *D*). However, as long as the inwardly directed chemical forces exceed these outwardly directed electrostatic forces, the net flux of Na^+ will still be inward, although the rate of influx will be diminished. The inward Na^+ current finally ceases when the *h* (inactivation) gates close (Fig. 2-7, *E*).

The activity of the *h* gates is governed by the value of V_m just as is that of the *m* gates. However, whereas the *m* gates tend to open as V_m becomes less negative, the *h* gates tend to close under this same influence. Furthermore, the opening of the *m* gates occurs very rapidly (in about 0.1 to 0.2 msec.), whereas the closure of the *h* gates is a relatively slow process, requiring 1 msec. or more. Phase 0 is finally terminated when the *h* gates have closed and have thereby "inactivated" the fast Na^+ channels.

The *h* gates then remain closed until the cell has partially repolarized during phase 3 (at about point *D* in Fig. 2-1). Until these gates do reopen partially, the cell is refractory to further excitation. This mechanism therefore prevents a sustained, tetanic contraction of cardiac muscle, which would of course be inimical to the intermittent pumping action of the heart.

In cardiac cells that have a prominent plateau and especially in Purkinje fibers, phase 1 constitutes an early, brief period of limited repolarization between the end of the upstroke and the beginning of the plateau. This early phase of repolarization has been ascribed to a transient, inward Cl^- current, although recent work has tended to discredit this explanation.

During the plateau (phase 2) of the action potential, there is a weak flow of Ca^{++} and Na^+ into the cell through *slow channels*, which appear to be entirely different from the fast Na^+ channels that are operating during phase 0. The activation, inactivation, and recovery processes are much slower for the slow than for the fast channels. The fast channels may be blocked by tetrodotoxin, whereas the slow channels may be blocked by Mn^{++} or verapamil, agents that are known to impede the movement of Ca^{++} into the cell. The slow channels are activated when V_m reaches their threshold voltage of about -30 to -40 mV. Activation probably represents the opening of electrostatic gates in the slow channels, similar to the process that occurs in the fast channels. Other gates in the slow channels then begin to close, thereby initiating the process of inactivation, which helps terminate the plateau.

The slow inward current of Ca^{++} and Na^+ is not affected appreciably by the resting level of V_m, in contrast to the fast inward Na^+ current during phase 0. Also, the slow inward current is increased by catecholamines, such as epinephrine and norepinephrine. This is probably one of the principal mechanisms whereby the catecholamines enhance the contractile process in cardiac muscle. The Ca^{++} that enters the myocardial cell during the plateau is involved in excitation-contraction coupling, as described in Chapter 4. The Ca^{++} that enters during a given action potential probably contributes toward activating the contractile proteins during the contraction that results from that excitation. It probably also triggers the release of larger quantities of Ca^{++} from intracellular stores, such as those located in the sarcoplasmic reticulum.

During the plateau of the action potential, the concentration gradient for K^+ between the inside and outside of the cell is virtually the same as it is during phase 4, but V_m is close to 0 mV. Therefore, the chemical forces acting on K^+ greatly exceed the electrostatic forces during the plateau, and K^+ tends to diffuse out of the cell. The efflux of this positively charged ion would tend to make the interior of the cell membrane more negative; that is, it would tend to repolarize the cell membrane, thereby terminating the plateau. In nerve fibers, P_K increases when the neuron is depolarized, and the resultant outward current of K^+ promotes rapid repolarization. In cardiac cells, conversely, the permeability of the cell membrane to K^+ in the outward direction diminishes during phase 2, although the permeability in the inward direction is considerably greater. This unidirectional decrease in P_K during the plateau has been called *anomalous rectification*. As a consequence of the reduction in P_K in the outward direction, there is only a small outward current of K^+ during the plateau. It tends to balance the slow inward currents of Ca^{++} and Na^+, and thereby contributes to the maintenance of a prolonged plateau at a level of V_m close to 0 mV. The roles of both the slow inward Ca^{++} and Na^+ currents and of the reduction in P_K in the production of the plateau have been demonstrated by the administration of verapamil. If the slow inward currents are completely blocked with this agent, a plateau still exists, but it occurs at more negative voltages than when the slow channels are not blocked.

The process of final repolarization (phase 3) appears to depend on two principal processes; namely: (1) an increase in P_K, and (2) inactivation of the slow inward Ca^{++} and Na^+ currents. The increase in P_K may be induced by the elevation in intracellular Ca^{++}, consequent to the inward Ca^{++} current during the plateau. The enhancement of P_K leads to an efflux of K^+ from the cell. The outward current of K^+ is no longer balanced by the slow inward currents of Ca^{++} and Na^+. This efflux of positive K ions therefore causes the charge on the inside of the cell membrane to become progressively more negative. The increase in P_K is voltage depen-

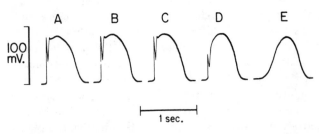

Fig. 2-8. Effect of tetrodotoxin on the action potential recorded in a calf Purkinje fiber perfused with a solution containing epinephrine and 10.8 mM K^+. The concentration of tetrodotoxin was 0 M in A, 3×10^{-8} M in B, 3×10^{-7} M in C, and 3×10^{-6} M in D and E; E was recorded later than D. (Redrawn from Carmeliet, E., and Vereecke, J.: Pflügers Arch. **313:**300, 1969.)

dent; as the inside of the cell becomes more negative, P_K increases and the outward flux of K^+ is accelerated. Hence, this rapid phase of repolarization (phase 3) can be considered to be a *regenerative process*, just as is the inward current of Na^+ during phase 0. The efflux of K^+ during phase 3 rapidly restores the resting level of membrane potential. During the subsequent rest period (phase 4), and probably throughout the action potential as well, the active membrane pump eliminates the excess Na^+ that had entered the cell principally during phases 0 and 2, in exchange for the K^+ that had exited chiefly during phases 2 and 3.

Ionic basis of the slow response

Fast response action potentials may be considered to consist of two principal components, a spike (phases 0 and 1) and a plateau (phase 2). In the slow response, the first component is absent or inoperative, and the second component accounts for the entire action potential. In the fast response, the spike is produced by the inward Na^+ current through the fast channels. These channels can be blocked by certain interventions, such as the administration of tetrodotoxin. When the fast Na^+ channels are blocked, slow responses may be generated in the same fibers under appropriate conditions.

The Purkinje fiber action potentials shown in

Fig. 2-8 clearly exhibit the two components. In the control tracing (panel A), a prominent notch separates the spike from the plateau. In panels B to E, progressively larger quantities of tetrodotoxin were added to the bathing solution in order to produce a graded blockade of a larger and larger fraction of the fast Na^+ channels. It is evident that the spike becomes progressively less prominent in panels B to D, and it disappears entirely in E. Thus, the tetrodotoxin had a pronounced effect on the spike, and only a negligible influence on the plateau. With elimination of the spike (panel E), the action potential resembles a typical slow response.

Certain cells in the heart, notably those in the S-A and A-V nodes, are normally slow response fibers. In such fibers, depolarization is achieved by the inward currents of Ca^{++} and Na^+ through the slow channels. These ionic events closely resemble those that occur during the second component of fast response action potentials. The slow channels in nodal cells can be blocked by Mn^{++} or verapamil, just as in fast response fibers.

CONDUCTION IN CARDIAC FIBERS

An action potential traveling down a cardiac muscle fiber is propagated by local circuit currents, similar to the process that occurs in

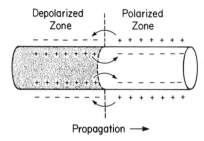

Depolarized Zone | Polarized Zone

Propagation ⟶

Fig. 2-9. The role of local currents in the propagation of a wave of excitation down a cardiac fiber.

nerve and skeletal muscle fibers. In Fig. 2-9, consider that the left half of the cardiac fiber has already been depolarized, whereas the right half is still in the resting state. The fluids normally in contact with the external and internal surfaces of the membrane are essentially solutions of electrolytes and thus are good conductors of electricity. Hence, current (in the abstract sense) will flow from regions of higher to those of lower potential, denoted by the plus and minus signs, respectively. In the external fluid, current will flow from right to left between the active and resting zones, and it will flow in the reverse direction intracellularly. In electrolyte solutions, the true current is carried by a movement of cations in one direction and anions in the opposite direction. At the cell exterior, for example, cations will flow from right to left, and anions from left to right (Fig. 2-9). In the cell interior, the opposite migrations will occur. These local currents at the border between the depolarized and polarized sections of the fiber will tend to depolarize the region of the resting fiber adjacent to the border.

Conduction of the fast response

In the fast response, the fast Na^+ channels will be activated when the transmembrane potential is suddenly brought to the threshold value of about -70 mV. The inward Na^+ current will then depolarize the cell very rapidly at that site. This portion of the fiber will become part of the depolarized zone, and the border will be displaced accordingly (to the

right in Fig. 2-9). The same process will then begin at the new border. This process will be repeated over and over, and the border will move continuously down the fiber as a wave of depolarization.

At any given point on the fiber, the greater the amplitude of the action potential and the greater the rate of change of potential (dV_m/dt) during phase 0, the more rapid the conduction down the fiber. The amplitude of the action potential equals the difference in potential between the fully depolarized and the fully polarized regions of the cell interior (Fig. 2-9). The magnitude of the local currents is proportional to this potential difference. Since these local currents shift the potential of the resting zone toward the threshold value, they are the local stimuli that depolarize the adjacent resting portion of the fiber to its threshold potential. The greater the potential difference between the depolarized and polarized regions (that is, the greater the amplitude of the action potential), the more efficacious the local stimuli, and the more rapidly the wave of depolarization is propagated down the fiber.

The rate of change of potential (dV_m/dt) during phase 0 is also an important determinant of the conduction velocity. The reason can be appreciated by referring again to Fig. 2-9. If the active portion of the fiber depolarizes very gradually, the local currents across the border between the depolarized and polarized regions would be very small. Thus, the resting region adjacent to the active zone would be depolarized very slowly, and consequently each new section of the fiber would require more time to reach threshold.

The level of the resting membrane potential is also an important determinant of conduction velocity. This factor operates through its influence on the amplitude and maximum slope of the action potential. The level of the resting potential may vary for a variety of reasons; (1) it can be altered experimentally by varying $[K^+]_o$ (Fig. 2-5); (2) in cardiac fibers that are intrinsically automatic, V_m becomes progressively less negative during phase 4 (Fig. 2-12, *B*); (3) during a premature contraction, repolarization may not have been completed before the beginning of the next excitation (Fig. 2-11). In general, the less negative the level of V_m, the less the velocity of impulse propagation, regardless of the reason for the alteration of the level of V_m.

The reason for the effect of the level of V_m on conduction velocity resides in the fact that the inactivation, or *h*, gates (Fig. 2-7) in the fast Na^+ channels are voltage dependent. The less negative the value of V_m, the greater the number of *h* gates that tend to close. During the normal process of excitation, depolarization proceeds so rapidly during phase 0 that the comparatively slow *h* gates do not close until the end of that phase. If partial depolarization is produced by a more gradual process, however, such as by elevating the level of external K^+, then the *h* gates do have ample time to close. When the cell is in a partial state of depolarization, therefore, many of the fast Na^+ channels will already be inactivated, and only a fraction of these channels will be available to conduct the inward Na^+ current during phase 0.

The results of an experiment in which the resting V_m of a bundle of Purkinje fibers was varied by altering the value of $[K^+]_o$ are shown in Fig. 2-10. When $[K^+]_o$ was 3 mM. (panels *A* and *F*), the resting V_m was -82 mV., and the slope of phase 0 was great. At the end of phase 0, the overshoot attained a value of $+30$ mV. Hence, the amplitude of the action potential was 112 mV. The tissue was stimulated at some

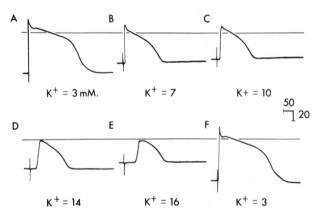

Fig. 2-10. The effect of changes in external potassium concentration on the transmembrane action potentials recorded from a Purkinje fiber. The stimulus artifact appears as a biphasic spike to the left of the upstroke of the action potential. Horizontal calibration, 50 msec.; vertical calibration, 20 mV. The horizontal lines near the peaks of the action potentials denote 0 mV. (From Myerburg, R. J., and Lazzara, R. In Fisch, E., editor: Complex electrocardiography, Philadelphia, 1973, F. A. Davis Co.)

distance from the impaled cell, and the stimulus artifact appears as a diphasic deflection just before phase 0. The time separating this artifact from the beginning of phase 0 is inversely proportional to the conduction velocity.

When $[K^+]_o$ was increased to 7 and 10 mM. (panels B and C), resting V_m became -63 and -58 mV., respectively. There were concomitant reductions in the amplitude and duration of the action potential, and in the slope of phase 0. Also, the stimulus artifact appears at a greater and greater distance in front of phase 0, demonstrating that conduction velocity was reduced accordingly. As resting V_m was made less and less negative by elevating $[K^+]_o$, there was a resetting of the resting position of the h gates toward the closed position, thereby reducing the number of available fast Na^+ channels. When the wave of excitation reached the impaled cell in panels B and C, the m gates rapidly swung open as soon as the threshold voltage for activating the fast Na^+ channels was reached in that cell. However, because the h gates were already partially closed at rest by virtue of the elevated $[K^+]_o$, some of the fast Na^+ channels were no longer available. Therefore the magnitude of the inward Na^+ current was considerably curtailed, and so dV_m/dt was diminished. It has been estimated that, on the average, about half of the fast Na^+ channels are activated when the resting V_m is about -70 mV., and all channels are inactivated when resting V_m is about -50 to -55 mV.

When $[K^+]_o$ was raised to levels of 14 and 16 mM. (panels D and E), the resting V_m became less negative than the critical value of about -50 to -55 mV., and therefore the fast Na^+ channels were completely inactivated. The action potentials in panels D and E are characteristic slow responses, presumably mediated by inward Ca^{++} and Na^+ currents through the slow channels exclusively. The great prolongations of conduction time are reflected in the tracings by the long time intervals between the stimulus artifacts and the upstrokes of the action potentials. In some preparations, it has been found that conduction velocity may actually increase slightly when $[K^+]_o$ is elevated to levels of about 10 or 12 mM. However, higher K^+ concentrations invariably retard conduction.

Conduction of the slow response

Local circuits are also responsible for propagation of the slow response (Fig. 2-9). However, the characteristics of the conduction process are entirely different from those of the fast response. The threshold potential is between -45 and -35 mV. for the slow response, and the conduction velocity is of course very much less than for the fast response. The velocities of conduction of the slow responses in the S-A and A-V nodes are about 0.02 to 0.1 m./sec. The conduction velocities of the fast responses are about 0.3 to 1.0 m./sec. for myocardial cells, and 1 to 4 m./sec. for the specialized conducting fibers in the atria and ventricles. Slow responses are more likely to be blocked than are fast responses. Also, the former are not able to be conducted at as rapid repetition rates. Fast responses are easily conducted in either an antegrade or a retrograde direction. The velocity of conduction is virtually the same in both directions. The velocity of conduction of the slow response in one direction may be much greater than in the opposite direction. Not uncommonly, the slow response will be conducted in one direction only, but it will be blocked in the opposite direction. This condition of *undirectional* block is a sine qua non for reentry. This is the basis for many arrhythmias, as explained on pp. 28-30.

CARDIAC EXCITABILITY

Currently, considerable attention is being focused on obtaining more detailed knowledge of cardiac excitability because of the rapid development of artificial pacemakers and other electrical devices for correcting certain serious disturbances of rhythm. The excitability characteristics of cardiac cells differ considerably, depending on whether the action potentials are fast or slow responses.

Fast response

Once the fast response has been initiated, the depolarized cell will no longer be excitable until about the middle of the period of final repolarization (phase 3). The interval from the beginning of the action potential until the fiber is able to conduct another action potential is called the *effective refractory period*. In the fast response, this period extends from the beginning of phase 0 to a point in phase 3 where repolarization has proceeded to a value of about −50 mV. (period C to D in Fig. 2-1). It is at about this value of V_m that the electrochemical gates (*m* and *h*) for many of the fast Na⁺ channels have been reset.

Full excitability is not regained until the cardiac fiber has been fully repolarized (point *E* in Fig. 2-1). During period *D* to *E* in the figure, an action potential may be evoked, but only with a stimulus that is stronger than that which would be capable of eliciting a response during phase 4. Period *D* to *E* is called the *relative refractory period*.

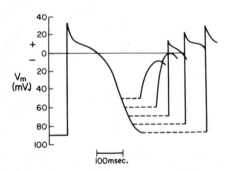

Fig. 2-11. The changes in action potential amplitude and slope of the upstroke as action potentials are initiated at different stages of the relative refractory period of the preceding excitation. (Redrawn from Rosen, M. R., Wit, A. L., and Hoffman, B. F.: Am. Heart J. **88:**380, 1974.)

The characteristics of a fast response evoked during the relative refractory period of an antecedent excitation vary with the membrane potential that exists at the time of stimulation. The nature of this voltage dependency is illustrated in Fig. 2-11. It is evident that as the fiber is stimulated later and later in the relative refractory period, the amplitude of the response and the rate of rise of the upstroke increase progressively. Presumably, the number of fast Na⁺ channels that have recovered from inactivation increase as repolarization proceeds during phase 3. As a consequence of the increase in amplitude and slope of the evoked response, the propagation velocity becomes greater the later the cell is stimulated in the relative refractory period. Once the fiber is fully repolarized, the response is constant at no matter what time in phase 4 the stimulus is applied. By the end of phase 3, the *m* and *h* gates of all channels are in their final positions, and therefore there is no further change in excitability with time.

Slow response

The effective refractory period during the slow response frequently extends well beyond phase 3. Even after the cell has completely repolarized, it may not be possible to evoke a propagated response for some time. The relative refractory period then extends into phase 4, during which V_m is virtually constant in nonpacemaker cells. During the relative refractory period, there is a progressive recovery of excitability despite the presence of this constant level of V_m. The recovery of full excitability usually requires considerably more time than for the fast response. The process might involve a few seconds, as compared to a few tenths of a second for recovery in the fast response. Until full recovery of excitability is achieved, conduction velocity varies with excitability. Impulses arriving at a slow response fiber early in its relative refractory period are conducted much more slowly than those arriving late in that period. The lengthy refractory periods also account for the observed tendency toward conduction blocks. Even when slow responses recur at a relatively low repetition rate, the fiber may be able to conduct only a fraction of those impulses.

NATURAL EXCITATION OF THE HEART

The nervous system exercises control over various aspects of the behavior of the heart, including the frequency at which it beats and the vigor of each contraction. However, cardiac function is certainly not completely dependent on intact nervous pathways. Indeed, in subsequent chapters it will be shown that with a completely denervated heart the animal appears to function normally and adapts surprisingly well to stressful situations.

The properties of *automaticity* (the ability to initiate its own beat) and of *rhythmicity* (the regularity of such pacemaking activity) are intrinsic to cardiac tissue. The heart will continue to beat even when it is completely removed from the body. If the coronary vasculature is artificially perfused, rhythmic cardiac contraction will persist for considerable periods of time. Apparently, at least some cells in the walls of all four cardiac chambers are capable of initiating beats; such cells probably reside in the nodal tissues or specialized conducting fibers of the heart. The region of the mammalian heart that ordinarily displays the highest order of rhythmicity is the *sinoatrial*, or S-A, node; it is called the *natural pacemaker* of the heart. Other regions of the heart that initiate beats under special circumstances are called *ectopic foci* or *ectopic pacemakers*. Ectopic foci may become pacemakers when (1) their own rhythmicity becomes enhanced, (2) the rhythmicity of the higher order pacemakers becomes depressed, or (3) all conduction pathways between the ectopic focus and those regions with a higher degree of rhythmicity become blocked.

When the S-A node is suddenly excised or destroyed, pacemaker cells in the A-V junction usually have the next highest order of rhythmicity, and they become the pacemakers for the entire heart. After some time, which may vary from minutes to days, automatic cells in the atria usually become dominant. In the dog, the most common site for the dominant atrial pacemaking region is at the junction between the inferior vena cava and the right atrium. Purkinje fibers in the specialized conduction system of the ventricles also possess the property of automaticity. Characteristically, they fire at a very slow rate. When the A-V junction is unable to conduct the cardiac impulse from the atria to the ventricles, such *idioventricular pacemakers* in the Purkinje fiber network generate the impulses that initiate the ventricular contractions. Such ventricular contractions occur at a frequency of only 30 to 40 beats per minute.

Sinoatrial node

The S-A node, which is the phylogenetic remnant of the sinus venosus of lower vertebrate hearts, was first described for mammalian hearts by Keith and Flack in 1906. In humans, it is a crescent-shaped structure, approximately 15 mm. long, 5 mm. wide, and 2 mm. thick. It lies in the sulcus terminalis on the posterior aspect of the heart, where the superior vena cava joins the right atrium. The S-A node contains two principal types of cells: (1) small, round cells, which have few organelles and myofibrils, and (2) slender, elongated cells, which are intermediate in appearance between the round and the ordinary myocardial cells. The round cells are probably the pacemaker cells, whereas the transitional cells probably conduct the impulse within the node and to the nodal margins.

A typical transmembrane action potential recorded from a cell in the S-A node is depicted in Fig. 2-12, *B*. In comparison with the action potential recorded from a ventricular myocardial cell (Fig. 2-12, *A*), the resting potential of

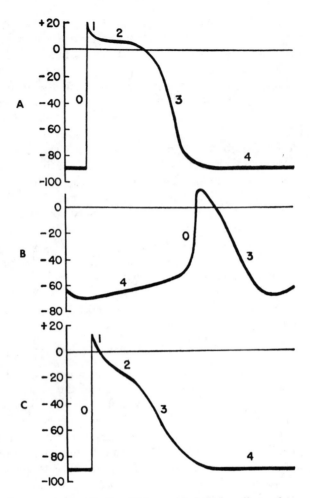

Fig. 2-12. Typical action potentials (in mV.) recorded from cells in the ventricle, **A,** sino-atrial node, **B,** and atrium, **C.** Sweep velocity in **B** is one-half that in **A** or **C.** (From Hoffman, B. F., and Cranefield, P. F.: Electrophysiology of the heart, New York, 1960, McGraw-Hill Book Co., Inc.)

the S-A nodal cell is usually less, the upstroke of the action potential (phase 0) has a much slower velocity, a plateau is absent, and repolarization (phase 3) is more gradual. These are all characteristic of the slow response. Under ordinary conditions, tetrodotoxin has no influence on the S-A nodal action potential. This indicates that the upstroke of the action potential is not produced by an inward current of Na^+

through the fast channels. By special techniques, it has been shown that such fast Na^+ channels do exist in the nodal pacemaker cells. They are much more sparse in nodal cells than in myocardial or Purkinje fibers, however. Also, those fast Na^+ channels are ordinarily inactivated (that is, the h gates are closed [Fig. 2-7]) because of the low threshold value (-40 to -50 mV.) for the action potential upstroke

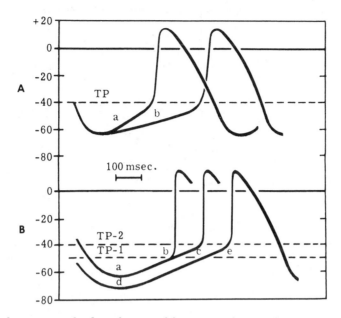

Fig. 2-13. Mechanisms involved in changes of frequency of pacemaker firing. In section **A** a reduction in the slope of the pacemaker potential from *a* to *b* will diminish the frequency. In section **B** an increase in the threshold (from *TP-1* to *TP-2*) or an increase in the magnitude of the resting potential (from *a* to *d*) will also diminish the frequency. (Redrawn from Hoffman, B. F., and Cranefield, P. F.: Electrophysiology of the heart, New York, 1960, McGraw-Hill Book Co., Inc.)

that prevails in S-A nodal cells. However, the principal distinguishing feature of a pacemaker fiber resides in phase 4. In nonautomatic cells the potential remains constant during this phase, whereas in a pacemaker fiber there is a slow, steady depolarization, called the *pacemaker potential*. Depolarization proceeds at a steady rate during phase 4 until a threshold is attained, and then an action potential is triggered.

The frequency of discharge of pacemaker cells may be varied by a change in (1) the rate of depolarization during phase 4, (2) the threshold potential, or (3) the resting potential (Fig. 2-13). With an increase in the rate of depolarization (*b* to *a* in Fig. 2-13, *A*) the threshold potential will be attained earlier, and heart rate will increase. A rise in the threshold potential

(from *TP-1* to *TP-2* in Fig. 2-13, *B*) will delay the onset of phase 0 (from time *b* to time *c*), and heart rate will be reduced accordingly. Similarly, when the magnitude of the resting potential is increased (from *a* to *d*), more time will be required to reach threshold *TP-2* when the slope of phase 4 remains unchanged, and the heart rate will diminish. However, the slope of phase 4 usually does not remain constant, but decreases during hyperpolarization; this exaggerates the extent of the reduction in discharge frequency.

Ordinarily, the frequency of pacemaker firing is controlled by the activity of both divisions of the autonomic nervous system. Increased sympathetic nervous activity, through the release of norepinephrine, raises the heart rate principally by increasing the slope of the

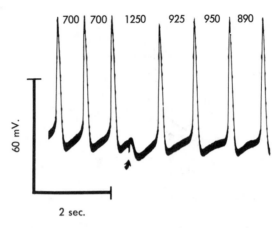

Fig. 2-14. Effect of a brief vagal stimulus (arrow) on the transmembrane potential recorded from an S-A nodal pacemaker cell in an isolated cat atrium preparation. The cardiac cycle lengths, in msec., are denoted by the numbers at the top of the figure. (Modified from Jalife, J., and Moe., G. K.: Circ. Res. **45**:595, 1979.)

pacemaker potential. Increased vagal activity, through the release of acetylcholine, diminishes the heart rate by hyperpolarizing the pacemaker cell membrane and by reducing the slope of the pacemaker potential. This hyperpolarization is ascribable to a significant increase in P_K evoked by the action of acetylcholine on the pacemaker cell membrane. Fig. 2-14 shows the changes in transmembrane potential in an S-A nodal cell in response to a brief vagal stimulus (at the arrow). The immediately ensuing hyperpolarization caused the cardiac cycle length to increase to 1250 msec. from a basic cycle length of 700 msec. The next several cycles were also lengthened. However, these longer cycles were associated with a decreased slope of the pacemaker potential, but no significant hyperpolarization. It is likely that the acetylcholine released at the vagal nerve endings persists only as long as the hyperpolarization. The mechanism responsible for the later phase of bradycardia, associated with the reduced rate of diastolic depolarization, is not known. Recent studies with potassium-sensitive microelectrodes have disclosed that, par-

alleling this secondary phase of bradycardia, there is an elevation in the extracellular concentration of K^+. This undoubtedly reflects the K^+ that leaks out the pacemaker cells as a result of the acetylcholine-induced increase in membrane permeability to K^+. Vagal stimulation frequently also evokes a *pacemaker shift,* wherein the true pacemaker cells are inhibited more than some of the latent pacemakers within the node, and these then become the true pacemakers.

Ionic basis of automaticity

On the basis of the Goldman constant-field equation (p. 9), the gradual phase 4 depolarization characteristic of pacemaker cells could be accounted for either by a progressive increase in P_{Na} or by a progressive reduction in P_K. Measurement of the membrane resistance of automatic Purkinje fibers and certain other types of pacemaker cells has disclosed a gradual increase in resistance during phase 4. This suggests that a progressive reduction in P_K is more likely than an increase in P_{Na} as the mechanism of the pacemaker potential in such cells. As a

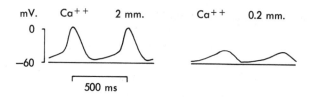

Fig. 2-15. Transmembrane action potentials recorded from an S-A nodal pacemaker cell in an isolated rabbit atrium preparation. The concentration of Ca^{++} in the bath was changed from 2 mM. to 0.2 mM. (Modified from Kohlhardt, M., Figulla, H. -R., and Tripathi, O.: Basic Res. Cardiol. **71:**17, 1976.)

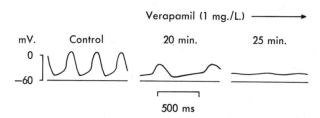

Fig. 2-16. Transmembrane action potentials recorded from an S-A nodal pacemaker cell in an isolated rabbit atrium preparation. After a control record was obtained, verapamil (1 mg/l.) was added to the tissue bath, and records were taken 20 and 25 min. later. (Modified from Kohlhardt, M., Figulla, H.-R, and Tripathi, O.: Basic Res. Cardiol. **71:**17, 1976.)

consequence of the changing P_K, there is a progressive reduction in outward K^+ current during phase 4. Also, there is a small, but steady, inward Na^+ current, reflecting the concentration and electrostatic gradients for this ion (Fig. 2-7, *A*). The imbalance between the steady inward Na^+ current and the gradually diminishing outward K^+ current produces the slow diastolic depolarization that is characteristic of such automatic cells. Because of these two opposing currents during phase 4, raising $[K^+]_o$ or lowering $[Na^+]_o$ diminishes the firing frequency of automatic Purkinje fibers.

Considerably less is known about the ionic mechanisms of automaticity in nodal cells, including the pacemaker cells in the S-A node itself. There is a progressive reduction in K^+ permeability that accompanies the slow dia-

stolic depolarization during phase 4, just as in automatic Purkinje fibers. However, lowering $[Na^+]_o$ or raising $[K^+]_o$ has relatively little influence on such cells, in contrast to the pronounced effects of such changes on automatic Purkinje fibers. Instead, changes in the external Ca^{++} concentration (Fig. 2-15) and the addition of slow channel blocking agents, such as verapamil (Fig. 2-16), decrease the firing rate and alter the characteristics of the action potentials of S-A nodal pacemaker cells. Hence, a steady inward Ca^{++} current is probably responsible for the pacemaker potential in these cells.

Overdrive suppression

The automaticity of pacemaker cells becomes depressed after a period of excitation at a frequency greater than their intrinsic firing rate. This phenomenon is known as *overdrive suppression*. Because of the greater intrinsic rhythmicity of the S-A node than of the other latent pacemaking sites in the heart, the firing of the S-A node tends to suppress the automaticity in the other loci. If an ectopic focus in one of the atria suddenly began to fire at a rate of 150 impulses per minute in an individual with a normal heart rate of 70 beats per minute, the ectopic center would become the pacemaker for the entire heart. When that rapid ectopic focus suddenly stopped firing, the S-A node might remain quiescent briefly, by virtue of overdrive suppression. The interval from the end of the period of overdrive till the S-A node resumes firing is called the *sinus node recovery time*. In patients with the so-called *sick sinus syndrome*, the sinus node recovery time may be markedly prolonged. The resultant period of asystole might result is syncope because of the cessation of cardiac pumping.

The mechanism responsible for overdrive suppression is not known for certain. A reasonable hypothesis is based on the activity of the membrane pump that actively extrudes Na^+ from the cell, in partial exchange for K^+. During depolarization, a certain quantity of Na^+ enters a given cell. The more frequently it is depolarized, therefore, the more Na^+ that enters the cell per minute. Under conditions of overdrive, the Na^+ pump becomes more active in the process of extruding this larger quantity of Na^+ from the cell interior. The quantity of Na^+ extruded by the pump exceeds the quantity of K^+ that enters the cell. This enhanced activity of the pump therefore results in some hyperpolarization of the cell, since there is a net loss of cations from the cell interior. Because of the hyperpolarization, the pacemaker potential requires more time to reach the threshold, as shown in Fig. 2-13, *B*. It has also been postulated that, when the overdrive suddenly ceases, the Na^+ pump does not decelerate instantaneously, but continues to operate at

an accelerated rate for some time after the cessation of overdrive. This excessive extrusion of Na^+ tends to oppose the gradual depolarization of the pacemaker cell during phase 4, thereby suppressing its intrinsic automaticity temporarily.

Atrial conduction

From the S-A node, the cardiac impulse spreads radially throughout the right atrium (Fig. 2-17) along ordinary atrial myocardial fibers, at a conduction velocity of approximately 1 m./sec. There is a special pathway, the *anterior interatrial myocardial band* (or *Bachmann's bundle*), which conducts the impulse from the S-A node directly to the left atrium. There are also three tracts, the *anterior, middle,* and *posterior internodal pathways*, which conduct the cardiac impulse directly from the S-A to the A-V node; of these, the anterior pathway is probably the most important. These pathways consist of a mixture of ordinary myocardial cells and of specialized conducting fibers similar to those that exist in the ventricles.

The configuration of the atrial transmembrane potential is depicted in Fig. 2-12, *C*. In comparison with the potential recorded from a typical ventricular fiber (Fig. 2-12, *A*), it is evident that the plateau (phase 2) is not as well developed and that repolarization (phase 3) occurs at a slower rate.

Atrioventricular conduction

The cardiac action potential proceeds along the internodal pathways in the atrium and ultimately reaches the A-V node. This node (described by Tawara in 1906) is approximately 22 mm. long, 10 mm. wide, and 3 mm. thick. It is situated posteriorly on the right side of the interatrial septum near the ostium of the coronary sinus. The A-V node contains the same two cell types as the S-A node, but the round cells are more sparse and the elongated cells preponderate. The internodal pathways join

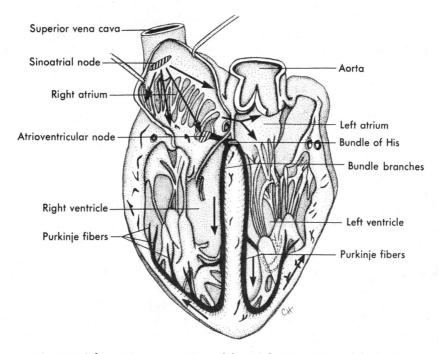

Superior vena cava

Sinoatrial node

Right atrium

Atrioventricular node

Right ventricle

Purkinje fibers

Aorta

Left atrium

Bundle of His

Bundle branches

Left ventricle

Purkinje fibers

Fig. 2-17. Schematic representation of the conduction system of the heart.

the A-V node at its crest and along the convex surface facing the right atrium.

The A-V node has been divided into 3 functional regions: (1) the A-N region, the transitional zone between the atrium and the remainder of the node; (2) the N region, the midportion of the A-V node; and (3) the N-H region, the zone in which the nodal fibers gradually merge with the *bundle of His*, which is the upper portion of the specialized conducting system for the ventricles. Normally, the A-V node and bundle of His constitute the only pathways for conduction from atria to ventricles.

Several features of atrioventricular conduction are of physiological and clinical significance. It is in the AN region of the A-V node that the principal delay occurs in the passage of the impulse from the atria to the ventricular myocardial cells. The conduction velocity is actually less in the N region that in the AN re-

gion. However, the path length is substantially greater in the latter than in the former zone, which accounts for the difference in the total conduction time through the two regions. The conduction times through the AN and N zones account for a considerable fraction of the time interval between the onsets of the atrial and ventricular systoles and between the onsets of the *P wave* (the electrical manifestation of the spread of atrial excitation) and the *QRS complex* (spread of ventricular excitation) in the electrocardiogram. The period of time between the initiation of the P wave and the QRS complex is called the *P-R interval* (Fig. 2-29). Functionally, this delay between atrial and ventricular excitation permits optimal ventricular filling during atrial contraction.

In the N region, the action potentials display many of the characteristics of the slow response. The resting potential is about −50 to −60 mV., the upstroke velocity is very low (about 5 V./sec.), and the conduction velocity is about 0.05 m./sec. Tetrodotoxin, which blocks the fast Na^+ channels, has virtually no effect on the action potentials in this region. Conversely, Mn^{++} and verapamil, which are slow channel blocking agents, have a depressant effect on A-V conduction. The action potentials in the A-N region are intermediate in configuration between those in the N region and atria. Similarly, the action potentials in the N-H region are transitional between those in the N region and His bundle.

The relative refractory period of the cells in the N region extends well beyond the period of complete repolarization. Such refractoriness is said to be *time-dependent*, in contrast to the *voltage-dependent* refractoriness that characterizes most of the other types of cells in the heart (Fig. 2-11). As the repetition rate of atrial depolarizations is increased, conduction through the A-V junction tends to be prolonged. Most of that prolongation takes place in the N region of the A-V node. Impulses tend to be blocked at stimulus repetition rates that are easily conducted in other regions of the heart. If the atria are depolarized at a high frequency, only one-half or one-third of the impulses might be conducted through the A-V junction to the ventricles. This tends to protect the ventricles from excessive contraction frequencies, wherein the filling time between contractions might be inadequate. Retrograde conduction can usually occur through the A-V node. However, the propagation time is significantly longer and the impulse is blocked at lower repetition rates in the retrograde than in the antegrade direction. Finally, the A-V node is a common site for reentry; the underlying mechanisms will be explained later (p. 28).

The autonomic nervous system plays an important role in the regulation of A-V conduction. Weak vagal activity may simply prolong A-V conduction time. Stronger vagal activity may cause some or all of the impulses arriving from the atria to be blocked in the node. The delayed conduction or block tends to occur largely in the N region of the node. The cardiac sympathetic nerves, on the other hand, have a facilitatory effect. They act to decrease the A-V conduction time and to enhance the rhythmicity of the latent pacemakers in the A-V junction. The norepinephrine released at the sympathetic nerve terminals increases the amplitude and rate of rise of the upstroke of the A-V nodal action potentials, principally in the N region of the node.

Ventricular conduction

The bundle of His (described by His in 1893) passes subendocardially down the right side of the interventricular septum for approximately 12 mm. and then divides into the right and left *bundle branches* (Figs. 2-17 and 2-18). The right bundle branch is a direct continuation of the bundle of His and proceeds down the right side of the interventricular septum. The left bundle branch, which is considerably thicker than the right, arises almost perpendicularly from the bundle of His and perforates the interventricular septum. On the subendocardial surface of the left side of the interventricular septum, the main left bundle branch splits into a thin *anterior division* and a thick *posterior division*. Clinically, impulse conduction in the right bundle branch, the main left bundle branch, or either division of the left bundle branch may be impaired. Conduction blocks in one or more of these pathways give rise to characteristic electrocardiographic patterns. Block of either of the main bundle branches is known as right or left *bundle branch block*. Block of either division of the left bundle branch is called *left anterior hemiblock* or *left posterior hemiblock*.

The right bundle branch and the two divisions of the left bundle branch ultimately subdivide into a complex network of conducting fibers called *Purkinje fibers* (described by Purkinje in 1839), which ramify over the subendocardial surfaces of both ventricles. In cer-

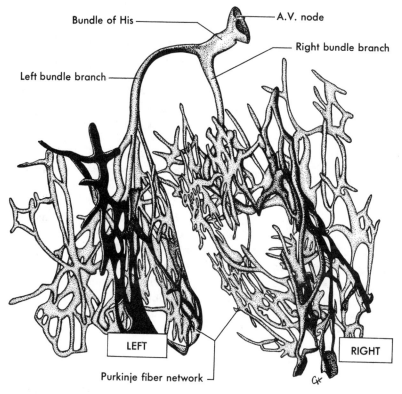

Fig. 2-18. Atrioventricular and ventricular conduction system of the calf heart. (Redrawn from DeHaan, R. L.: Circulation **24**:458, 1961.)

tain mammalian species, such as cattle, the Purkinje fiber network is arranged in discrete, encapsulated bundles (Fig. 2-18).

The Purkinje fibers are the broadest cells in the heart, 70 to 80 μ in diameter, as against 10 to 15 μ for ventricular myocardial cells. The large diameter accounts in part for the greater conduction velocity in Purkinje than in myocardial fibers. Purkinje cells have abundant, linearly arranged sarcomeres, just as do myocardial cells. However, the T-tubular system is absent in the Purkinje cells of many species, but well developed in the myocardial cells.

The conduction velocity for propagation of the action potential over the Purkinje fiber system is the fastest of any tissue within the heart;

estimates vary from 1 to 4 m./sec. This permits a rapid activation of all regions of the endocardial surface of the ventricles.

The configuration of the action potentials recorded from Purkinje fibers is quite similar to that from ordinary ventricular myocardial fibers (Fig. 2-12, A). In general, phase 1 is more prominent in Purkinje fiber action potentials than in those recorded from ventricular fibers, and the duration of the plateau (phase 2) is longer. In certain regions of the Purkinje fiber network, there is a pronounced increase of the action potential duration, and hence in the effective refractory period. The Purkinje fibers in these regions are called *gate cells;* the comparative durations of the action potentials of the

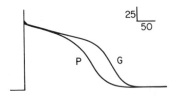

Fig. 2-19. Superimposed action potentials recorded from a regular Purkinje cell *(P)* and from a "gate" cell *(G)*. Horizontal calibration, 50 msec.; vertical calibration, 25 mV. (Redrawn from Kus, T., and Sasyniuk, B. I.: Circ. Res. **37**:844, 1975. By permission of the American Heart Association, Inc.)

ordinary Purkinje cells *(P)* and of the gate cells *(G)* are shown in Fig. 2-19. The actual location of the gate cells is controversial at present. In studies conducted in vitro, the cells with the longest action potentials were located distally in the Purkinje fiber system, near the junction of these specialized fibers with the ordinary myocardial fibers. However, in more recent in vivo studies, the gate cells appeared to be located more proximally, in the main bundle branches.

Because of the long refractory period of the gate cells, many premature activations of the atria are conducted through the A-V junction but are blocked by the gate cells. Therefore, they fail to evoke a premature contraction of the ventricles. This function of protecting the ventricles against the effects of premature atrial depolarizations is especially pronounced at slow heart rates, because the action potential duration and hence the effective refractory period of the gate cells vary inversely with the heart rate. At slow heart rates, the effective refractory period of the gate cells is especially prolonged; as the heart rate increases, the refractory period diminishes. Similar directional changes in refractory period occur in most of the other cells in the heart with changes in rate. However, in the A-V node, the effective refractory period does not change appreciably over the normal range of heart rates, and it ac-

tually increases at very rapid heart rates. Therefore, at high rates, it is the A-V node that protects the ventricles from the arrival of impulses at excessive repetition rates.

The intimate details of the spread of the action potential over the ventricles are of major concern in clinical cardiology. Numerous studies have been conducted to determine the precise course of the wave of excitation under normal and various abnormal conditions. Such knowledge serves as a basis for the interpretation of the electrocardiogram. However, only the elementary, salient features of ventricular activation will be considered here. The first portions of the ventricles to be excited are the interventricular septum (except the basal portion) and the papillary muscles. The wave of activation spreads into the substance of the septum from both its left and its right endocardial surfaces. Early contraction of the septum tends to make it more rigid and allows it to serve as an anchor point for the contraction of the remaining ventricular myocardium. Also, early contraction of the papillary muscles serves to prevent eversion of the A-V valves during ventricular systole.

The endocardial surfaces of both ventricles are activated rapidly, but the wave of excitation spreads from endocardium to epicardium at a slower velocity (about 0.3 to 0.4 m./sec). Because the right ventricular wall is appreciably thinner than the left, the epicardial surface of the right ventricle is activated earlier than that of the left ventricle. Also, apical and central epicardial regions of both ventricles are activated somewhat earlier than their respective basal regions. The last portions of the ventricles to be excited are the posterior basal epicardial regions and a small zone in the basal portion of the interventricular septum.

REENTRY

Under appropriate conditions, a cardiac impulse may reexcite some region through which it has passed previously. This phenomenon,

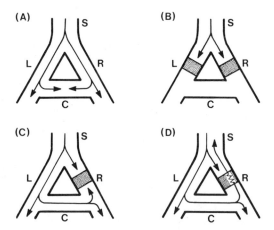

Fig. 2-20. The role of unidirectional block in reentry. In panel A, an excitation wave traveling down a single bundle (S) of fibers continues down the left (L) and right (R) branches. The depolarization wave enters the connecting branch (C) from both ends and is extinguished at the zone of collision. In panel B, the wave is blocked in the L and R branches. In panel C, bidirectional block exists in branch R. In panel D, unidirectional block exists in branch R. The antegrade impulse is blocked, but the retrograde impulse is conducted through, and reenters bundle S.

known as *reentry,* is responsible for many clinical disturbances of cardiac rhythm. The conditions necessary for reentry are illustrated in Fig. 2-20. In each of the four panels, a single bundle (S) of cardiac fibers is seen to divide into a left (L) and a right (R) branch. A connecting bundle (C) runs between the two branches.

Normally, the impulse coming down bundle S is conducted along the L and R branches (panel A). As the impulse reaches connecting link C, it enters from both sides and becomes extinguished at the point of collision. The impulse from the left side cannot proceed further because the tissue beyond is absolutely refractory since it had just undergone depolarization from the other direction. The impulse cannot

pass through bundle C from the right either, for the same reason.

It is obvious from panel B that the impulse cannot make a complete circuit if antegrade block exists in the two branches (L and R) of the fiber bundle. Furthermore, if bidirectional block exists at any point in the loop (for example, branch R in panel C), the impulse will not be able to reenter.

A necessary condition for reentry is that at some point in the loop, the impulse is able to pass in one direction but not in the other. This phenomenon is called *unidirectional block.* As shown in panel D, the impulse may travel down branch L normally and may be blocked in the antegrade direction in branch R. The impulse that had been conducted down branch L and through the connecting branch C may be able to penetrate the depressed region in branch R from the retrograde direction, even though the antegrade impulse had been blocked previously at this same site. Such unidirectional block is commonly observed with slow response action potentials and it is frequently a temporal phenomenon. It is evident from panel D that the antegrade impulse will arrive at the depressed region in branch R earlier than the impulse coming from the opposite direction. The antegrade impulse may be blocked simply because it happens to arrive at the depressed region during its effective refractory period. If the retrograde impulse is delayed sufficiently, the refractory period may have ended, and the impulse will be conducted back into bundle S.

Unidirectional block is a necessary condition for reentry, but not a sufficient one. It is also essential that the effective refractory period of the reentered region be less than the propagation time around the loop. In panel D, if the retrograde impulse is conducted through the depressed zone in branch R and if the tissue just beyond is still refractory from the antegrade depolarization, branch S will not be reexcited. Therefore, the conditions that promote reentry are those that prolong conduction time or shorten the effective refractory period.

The functional components of reentry loops responsible for specific arrhythmias are protean. Some loops are very large, and involve entire specialized conduction bundles. Others are microscopic in size. The loop may include myocardial fibers, specialized conducting fibers, nodal cells, and junctional tissues, in almost any conceivable arrangement. Also, the cardiac cells in the loop may be normal or deranged. Some of the important arrhythmias that occur on the basis of reentry will be described at the end of this chapter.

TRIGGERED ACTIVITY

Considerable attention has been directed recently toward the possible role played by *triggered activity* in the genesis of certain types of arrhythmias. Such a mechanism probably accounts for some of the occurrences of extrasystoles and paroxysmal tachycardias that previously had been ascribed to reentry. Many of the characteristics of arrhythmias evoked by triggered activity closely resemble those induced by reentry.

A triggered depolarization of cardiac tissue is always coupled to a preceding beat. Furthermore, the preceding depolarization is always followed by an afterpotential. Such afterpotentials may be evoked by stretch, hypoxia, ionic changes, catecholamines, and various drugs, such as aconitine and especially toxic concentrations of digitalis.

Certain characteristics of these afterpotentials are shown in Fig. 2-21. The transmembrane potentials were recorded from Purkinje fibers that were exposed to a high concentration of acetylstrophanthidin, a rapidly acting digitalis derivative. In the absence of driving stimuli, these fibers were quiescent. In each panel, there was a sequence of six driven depolarizations at a specific basic cycle length (BCL). When the cycle length was 800 msec.

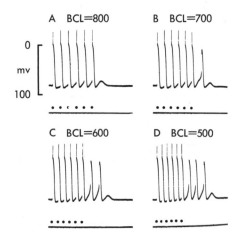

Fig. 2-21. Transmembrane action potentials recorded from isolated canine Purkinje fibers. Acetylstrophanthidin was added to the bath, and sequences of 6 driven beats (denoted by the dots) were produced at basic cycle lengths (BCL) of 800, 700, 600, and 500 msec. Note that afterpotentials occurred after the driven beats, and that these afterpotentials reached threshold at all but the longest BCL. (From Ferrier, G. R., Saunders, J. H., and Mendez, C.: Circ. Res. **32:**600, 1973. By permission of the American Heart Association, Inc.)

(panel *A*), the last driven depolarization was followed by a brief afterdepolarization that did not reach threshold. Once the afterpotential had subsided, the transmembrane potential remained constant until another driving stimulus was given. The upstroke of an afterdepolarization can be observed after each of the first five driven depolarizations. It is evident that the slope of this afterdepolarization increased progressively after each of the first three or four driven depolarizations in the sequence.

When the basic cycle length was diminished to 700 msec. (panel *B*), the afterdepolarization that followed the last driven beat did reach threshold, and a nondriven depolarization (or *extrasystole*) ensued. This extrasystole was itself followed by an afterpotential that was subthreshold. Diminution of the basic cycle length to 600 msec. (panel *C*) also evoked an extrasystole after the last driven depolarization. The afterpotential that followed the extrasystole did reach threshold, however, and a second extrasystole occurred. A sequence of three extrasystoles followed the six driven depolarizations that were separated by intervals of 500 msec. (panel *D*). Slightly shorter basic cycle lengths or slightly greater concentrations of acetylstrophanthidin were found to evoke a continuous sequence of nondriven beats, resembling a *paroxysmal tachycardia* (described on p. 48).

The ionic mechanism responsible for the afterpotentials appears to involve Ca^{++}. The amplitude of the afterpotential in cardiac cells *in vitro* is directly related to the concentration of Ca^{++} in the bath. Also, the afterpotentials may be attenuated or abolished by slow channel blocking agents, such as verapamil or Mn^{++}. It is still not known, however, whether the afterpotentials are ascribable directly to an inward Ca^{++} current, or whether a change in intracellular Ca^{++} may then alter the membrane permeability to some other ion, such as K^{+}. The afterpotential may actually be produced by the influx of that other ion.

BASIS OF ELECTROCARDIOGRAPHY

The electrocardiograph is a valuable instrument, for it enables the physician to infer the course of the cardiac impulse simply by recording the variations in electrical potential at various loci on the surface of the body. Occasionally, cardiac potentials are registered from the thoracic portions of the lumen of the esophagus, from within the chambers of the heart during cardiac catheterization, and from the surface of the heart during thoracic surgery. Ordinarily, however, differences of potential are recorded between pairs of points on the external surface of the body or between specific skin loci and a so-called indifferent, or reference, electrode. By analyzing the details of these potential fluctuations, the physician gains valuable insight concerning (1) the anatomical orientation of the heart, (2) the relative sizes of its chambers, (3) a variety of disturbances of rhythm and of conduction, (4) the extent, location, and progress of ischemic damage to the myocardium, (5) the effects of altered electrolyte concentrations, and (6) the influence of certain drugs (notably digitalis and its derivatives). It should be emphasized, however, that the electrocardiogram gives no direct information concerning the mechanical performance of the heart as a pump. The science of electrocardiography is extensive and complex, but only the elementary basis of electrocardiography will be considered here.

Direct leads

Electrocardiograms are usually recorded from *indirect leads* (recording electrodes located on the skin), which are at some distance from the heart, the source of potential. However, *direct leads* have been applied to the surface of the heart in experimental animals and in humans during thoracic surgical procedures. Electrocardiograms recorded by direct leads will be discussed first because they are simpler to comprehend.

The principles will be illustrated by describ-

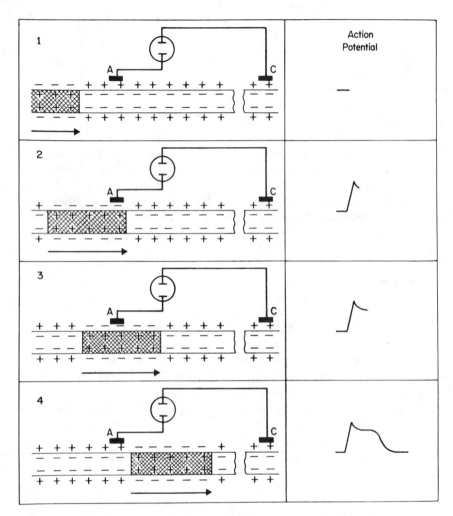

Fig. 2-22. Sequential changes of potential recorded from an external electrode, A, as an action potential travels along a strip of cardiac muscle. Electrode C is located far to the right of A; the entire cycle of depolarization and repolarization is completed under A before the action potential reaches C.

ing the potential changes recorded directly from the surface of a single, long myocardial fiber. In Fig. 2-22 the changes in potential in one localized region (in contact with electrode A) will first be considered. Electrode A is connected to the lower vertical deflecting plate of a cathode ray oscilloscope; electrode C, far to the right of A, is connected to the upper de-

flecting plate. If the strip of muscle is stimulated at its left end, the action potential travels from left to right along the strip. Before excitation reaches region A, however, the external surface of the fiber at A is at the same potential as the surface at C, and no difference in potential is recorded between A and C (section 1). When the action potential reaches region A,

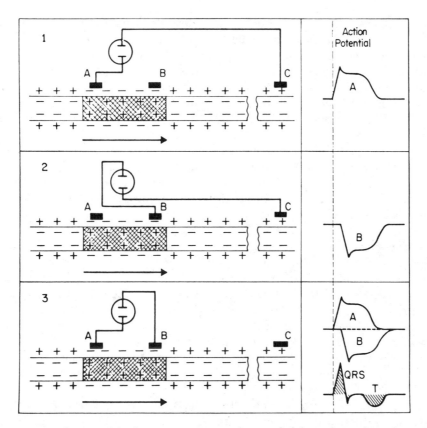

Fig. 2-23. Monophasic and biphasic action potentials recorded from the surface of a strip of cardiac muscle. In section **1**, electrodes *A* and *C* are connected to the oscilloscope as in the preceding figure, and the recorded action potential is therefore identical with that shown in Section **4** of Fig. 2-22. In section **2**, electrodes *B* and *C* are connected to the oscilloscope as shown, and the resulting deflection, *B*, will be inverted and displaced in time relative to deflection *A* in section **1**. In section **3**, electrodes *A* and *B* are connected to the oscilloscope, and the resulting QRS and T waves represent the algebraic sum of the individual deflections, *A* and *B*.

there is a rapid reversal of the transmembrane potential in that region, and the external surface at *A* becomes electronegative with respect to the surface at *C* (section 2). This difference in potential persists as long as region *A* remains depolarized (section 3). As region *A* repolarizes, the potential at *A* again becomes equal to the potential at *C* (section 4). Repolarization

proceeds much more slowly than depolarization; hence the slope of phase 3 is more gradual than that of phase 0.

Fig. 2-23 depicts a similar situation, but illustrates the potential differences that would be recorded when the wave of activation is recorded simultaneously from two regions of the same fiber. In section *1* of Fig. 2-23 the lead locations and oscilloscope connections are identical to those represented in Fig. 2-22. Hence, the recorded action potential is the same. Section 2 of Fig. 2-23 shows the record that is obtained under identical conditions, except that the potential difference between electrodes *B* and *C* is registered. In this case *B* is connected to the upper and *C* to the lower vertical deflecting plate. Relative to the action potential recorded in section *1* of Fig. 2-23, the record is inverted and is displaced in time, depending on the distance between electrodes *A* and *B* and on the propagation velocity of the action potential.

Finally, if electrodes *A* and *B* are connected to the lower and upper deflecting plates, respectively, then the potential difference recorded between them has the configuration shown at the right of section 3. This action potential represents the algebraic sum of the potentials recorded separately under *A* and *B* (with respect to resting region *C*). The externally recorded action potential consists of an initial upright spike, the *R wave*, followed by a small downward deflection, the *S wave*. The second major wave, termed the *T wave*, is inverted and occurs during repolarization. In electrocardiograms the analogous deflection occurring during ventricular depolarization is often triphasic (see Fig. 2-29) and therefore is designated the *QRS complex*. The T wave is of lesser amplitude but of greater duration than the QRS complex because the rate of repolarization is slower than the rate of depolarization (as is evident in the monophasic tracings shown in sections *1* and *2*). The portion of the tracing between the end of the QRS complex and the start of the T wave is called the *S-T segment*. During this interval the regions under both electrodes are depolarized. Since they are of equal negativity, the S-T segment lies along the *isoelectric line*, or line of zero potential difference.

In most normal electrocardiograms the T wave is deflected in the same direction as the QRS complex from the isoelectric line, contrary to the oppositely directed waves seen in section 3 of Fig. 2-23. The explanation may be illustrated by subjecting a strip of cardiac muscle to a temperature gradient (temperature increasing from left to right), as shown in Fig. 2-24. If a stimulus is applied to the left end, then the wave of depolarization is, of course, propagated from left to right. However, be-

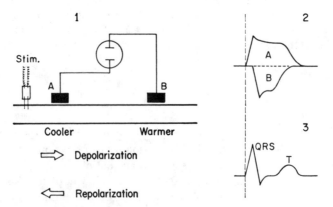

Fig. 2-24. When the wave of depolarization progresses in one direction along a cardiac muscle strip and the wave of repolarization travels in the opposite direction, then the QRS and T waves will be deflected in the same direction relative to the isoelectric line.

cause the right end of the strip is warmer than the left end, the repolarization process may actually proceed from right to left, that is, the reverse of the direction of propagation of the wave of depolarization. Stated in another way, the action potential duration under electrode A (near the cooler end) will exceed that under B (near the warmer end). Therefore, the T wave will be deflected in the same direction as the QRS complex under such conditions, as shown in Fig. 2-24. Thus, when the spread of depolarization and of repolarization occur in the same direction, the T wave is inverted with respect to the QRS complex; when these processes proceed in opposite directions, the deflections occur in the same direction.

As stated on p. 28, the wave of depolarization in the ventricles proceeds from endocardium to epicardium. However, the wave of repolarization normally travels in the opposite direction across the ventricular walls, producing the concordant QRS and T deflections. Stated in another way, the duration of the action potentials in the subendocardial cells is greater than that in the subepicardial cells (analogous to action potentials A and B in section 2 of Fig. 2-24). It has been postulated that the pressure developed in the ventricular chambers during systole retards the repolarization process more in the subendocardial than in the subepicardial region.

Potential distribution in a volume conductor

In recording conventional, or indirect, electrocardiograms the points on the surface of the skin at which potential variations are measured are separated from the source of the potential (that is, the heart itself) by the various tissues of the body. The extracellular and intracellular compartments of these tissues contain electrolytes in aqueous solution and hence are conductors of electric current. Therefore, an understanding of electrocardiography requires an appreciation of the distribution of electric currents and potentials throughout a three-dimen-

sional conductor of electricity, a so-called *volume conductor*. The problem is difficult since (1) the heart is a current source in which, at any moment in time, the distributions of charges are heterogeneous and the pathways of propagation are multidirectional; (2) at any given location the time course of potential change (that is, the cardiac action potential) is complex; (3) the configuration of the action potential varies from region to region in the heart (for example, Fig. 2-12); (4) the anatomical position of the heart continually changes throughout the cardiac and respiratory cycles; (5) the conductivity of the volume conductor is not uniform; and (6) the surface of the body is not simple geometrically.

Recent attempts to simulate this complex condition with mathematical and physical models have more closely approximated the true situation. Some understanding of indirect electrocardiography may be derived from a consideration of the simplest model—that of a *dipole* in a homogeneous volume conductor. A dipole is a pair of point sources of electric charge equal in magnitude but opposite in sign; the terminals of a battery, for example, constitute a dipole. If such a dipole is located in a homogeneous conducting medium of infinite extent, as represented in Fig. 2-25, current will flow from the positive to the negative pole, according to the usual convention. In a solution of electrolytes, anions will actually flow toward the positive pole (anode) and cations toward the negative pole (cathode). The density of current flow will be greatest over the most direct path between the two poles, but some current will flow over less direct paths.

The lines comprised of equal potential, or *isopotential lines*, will form a companion series of curves that intersect the curves of current flow everywhere at right angles, as shown in Fig. 2-25. A graph of isopotential lines is depicted in Fig. 2-26 for a dipole that possesses a potential difference of 8 volts ($+4$ V. and -4 V. for the positive and negative poles, respec-

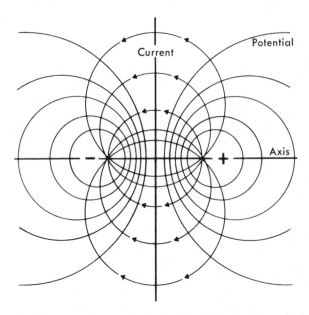

Fig. 2-25. Current flow between the terminals of a dipole in a homogeneous volume conductor. Lines of constant potential cross the lines of current flow everywhere at right angles.

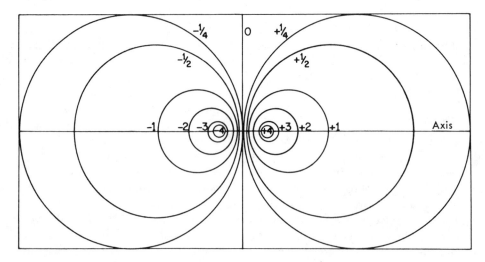

Fig. 2-26. Isopotential lines in a homogeneous volume conductor containing a dipole of +4 and −4 volts.

tively). The *axis of the dipole* is defined as the line passing through the poles. The voltage varies linearly along the axis between the poles. Hence, at the halfway point the potential is midway between that of the two poles; that is, it is zero, in accordance with the convention of assigning potentials of equal magnitudes but of opposite signs to the poles. Furthermore, since the volume conductor in the model is homogeneous, potential will change linearly along all paths of current flow. Therefore, along every path, zero potential will be at a point equidistant from the two poles; the locus of such points will be the line equidistant from the poles.

To illustrate the quantitative changes in potential at points relatively distant from the dipole, another example of a dipole immersed in a homogeneous volume conductor is presented in Fig. 2-27. The potentials at all points to the right of the zero potential line are positive; to the left they are negative. Proceeding to the right from the positive pole, the potential progressively diminishes. At P_1 the measured potential is found to be $+2$ V. At P_2, which is exactly twice as far as P_1 from the center of the dipole, the potential is found to be $+0.5$ V. It can be shown that for any point, P, lying along the axis of the dipole but not between the poles, the potential, V_p, is inversely proportional to the square of the distance, r, from the center of the dipole. This applies quantitatively for values of r that are relatively great compared to the distance between the poles. Also, for any given distance from the center of the dipole the potential varies with the cosine of

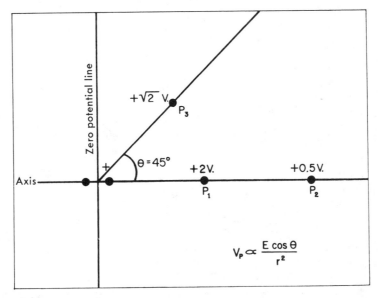

$$V_p \propto \frac{E \cos \theta}{r^2}$$

Fig. 2-27. Example of a dipole in a homogeneous volume conductor, to show that the potential at a given point is proportional to cos θ (where θ is the angle between the axis of the dipole and the line from the given point to the center of the dipole) and inversely proportional to r^2 (where r is the distance from the given point to the center of the dipole). If the potential at P_1 is $+2$ V. and if P_2 is twice as far as P_1 from the center of the dipole, then the potential at P_2 is $+0.5$ V. If θ = 45 degrees and if P_3 and P_1 are equidistant from the center of the dipole, then the potential at $P_3 = +2 \cos 45$ degrees V. $= +\sqrt{2}$ V.

the angular displacement from the axis of the dipole. For example, P_3 and P_1 are equidistant from the center of the dipole, and the line connecting P_3 and the center intersects the axis at an angle, θ, of 45 degrees; the potential at P_3 is found to be $+\ 2 \cos 45$ degrees $=\ +\ \sqrt{2}$ V., since the potential at P_1 is $+2$ V. It is obvious that similar relationships will exist if the electrode position remains fixed but the dipole itself rotates. For example, the potential at P_1 will become $+\ \sqrt{2}$ V. if the dipole rotates 45 degrees in a clockwise direction. Finally, the potential at any point will vary with the strength, E, of the dipole itself. In the example depicted in Fig. 2-27, if the strength of the dipole is doubled, then V_p would become $+4$ V. at P_1; in summary, $V_p \propto \dfrac{E \cos \theta}{r^2}$ in a homogeneous volume conductor. It must be emphasized that this expression is an approximation and applies only at points in the volume conductor that are at great distances from the center of the dipole relative to the interpolar distance.

Vectorcardiography

From the previous consideration of conduction of a charge in a volume conductor, it is evident that the potential at any point in the conducting medium is dependent on both the strength and the orientation of the dipole. Since magnitude and direction are involved, the potential distribution in a volume conductor is a problem in vector analysis. When two dipoles of different strengths and orientations exist in a volume conductor, their effects are

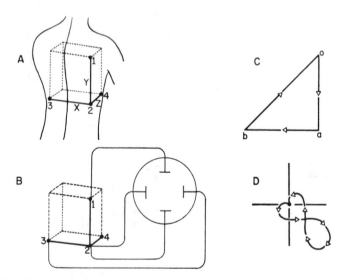

Fig. 2-28. Illustration of some of the basic principles of vectorcardiography. **A,** Location of electrodes in the cube system. **B,** Connections of electrodes to the deflecting plates of the oscilloscope for recording the projection of the cardiac vector on the frontal plane. **C,** Loop recorded when electrode *1* first becomes negative with respect to the other electrodes (beam deflects downward, to *a*), then the potential at *3* becomes equal to that at *1* (beam deflects to the left, to *b*), and finally the potentials at *1* and *3* return to their original values at the same rate of change of potential (beam deflects diagonally back to *0*). **D,** Typical QRS loop in the frontal plane recorded from a patient.

additive *vectorially*. During the passage of the impulse through the myocardium, the potential at any point on the surface of the skin depends on the resultant of an infinite number of dipoles of diverse strengths and orientations.

By means of a cathode-ray oscilloscope it is possible to record the continuous changes of the actual cardiac vector, called a *vectorcardiogram*, by applying recording electrodes to certain sites on the skin. Various lead systems have been devised, but only the cube system developed by Grishman will be described to illustrate the basis of vectorcardiography. Electrode placement in this system is illustrated in Fig. 2-28, *A*. Electrode *1* is located on the posterior surface of the right shoulder, and electrode *2* is located directly below it at the level of the first lumbar vertebra. Together these electrodes constitute the *Y* axis or vertical lead. Electrode 3 is placed on the left side of the back, at the same level as *2* and equidistant from the midline. Electrodes *2* and *3* constitute the *X* axis, or transverse lead. Electrode *4* also is located at the same level as *2*, but on the right anterior surface of the body. Electrodes *2* and *4* constitute the *Z* axis, or sagittal lead. The *X* and *Y* axes together define the frontal plane; the *X* and *Z* axes, the horizontal plane; and the *Y* and *Z* axes, the right sagittal plane.

To record the projection of the cardiac vector on the frontal plane, for example, the electrodes are connected to the deflecting plates of the oscilloscope as shown in Fig. 2-28, *B*. To understand the nature of the vectorcardiogram, consider the following sequence of events, which is illustrated in Fig. 2-28, *C*. When all deflecting plates are at the same potential, the electron beam (negatively charged) strikes the center of the oscilloscope screen, which is labelled *0* in the figure. Let electrode *1* suddenly become negative with respect to the other electrodes. The beam will then be repelled straight downward, to point *a*. Next, let electrode *3* suddenly attain the same potential as electrode *1*. The beam will then be deflected to the left from *a*, to point *b*, equal to the distance from *0* to *a*. If the electrodes *1* and *3* return at the same rate to the same potential as *2*, the electron beam will follow a diagonal path from *b* back to point *0*.

At those times when the isoelectric line is being inscribed in both the vertical and transverse leads, the beam of the vectorcardiographic oscilloscope will be quiescent at the origin, *0*. However, during inscription of the P, QRS, and T waves, the electron beam will move in some path from the origin. It will return again to the origin on completion of each wave, provided these waves terminate at the isoelectric line. Therefore, actual P, QRS, and T loops will appear on the oscilloscope screen during each cardiac cycle. A line from the origin to any point on one of these loops will represent the magnitude and direction of the vector in that plane at that moment of time. A time base is usually incorporated by interrupting the electron beam at fixed time intervals. A typical QRS loop in the frontal plane is shown in Fig. 2-28, *D*. Similar principles are involved in recording vectorcardiograms in the horizontal and sagittal planes.

Scalar electrocardiography

The systems of leads used to record routine electrocardiograms are oriented in certain planes of the body. A system of leads oriented in a given plane detects only the projection of the three-dimensional vector on that plane. Furthermore, the potential difference between two recording electrodes represents the projection of the vector on the line between the two leads. Components of vectors projected on such lines are not vectors but are *scalar quantities* (having magnitude, but not direction). Hence, a recording of the changes with time of the differences of potential between two points on the surface of the skin is called a *scalar electrocardiogram*.

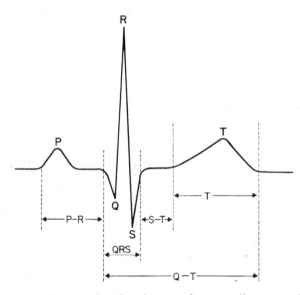

Fig. 2-29. Configuration of a typical scalar electrocardiogram, illustrating the important deflections and intervals.

Configuration of the scalar electrocardiogram. The scalar electrocardiograph detects the changes with time of the electrical potential between some point on the surface of the skin and an indifferent electrode or between pairs of points on the skin surface. The cardiac impulse progresses through the heart in an extremely complex three-dimensional pattern. Hence, the precise configuration of the electrocardiogram varies from individual to individual, and in any given individual the pattern varies with the anatomical location of the leads.

In general, the pattern consists of P, QRS, and T waves (Fig. 2-29). The P-R interval is a measure of the time from the onset of atrial activation to the onset of ventricular activation; it normally ranges from 0.12 to 0.20 second. A considerable fraction of this time involves passage of the impulse through the A-V conduction system. Pathological prolongations of this interval are associated with disturbances of A-V conduction produced by inflammatory, circulatory, pharmacological, or nervous mechanisms.

The configuration and amplitude of the QRS complex vary considerably among individuals. The duration is usually between 0.06 and 0.10 second. Abnormal prolongation may indicate a block in the normal conduction pathways through the ventricles (such as a block of the left or right bundle branch). The duration of the S-T segment is of little clinical significance and is usually not measured in routine analyses of electrocardiograms. During this interval the entire ventricular myocardium is depolarized. Therefore, the S-T segment lies on the isoelectric line under normal conditions. Any appreciable deviation from the isoelectric line is noteworthy and may indicate ischemic damage of the myocardium. The Q-T interval is sometimes referred to as the period of "electrical systole" of the ventricles. Its duration is about 0.4 second, but it varies inversely with the heart rate, in part because the myocardial cell action potential duration varies inversely with the heart rate.

In most leads the T wave is deflected in the

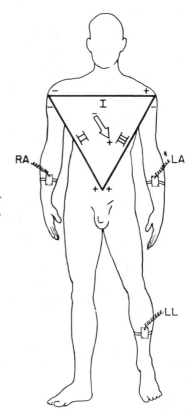

Fig. 2-30. Einthoven triangle, illustrating the galvanometer connections for standard limb Leads I, II, and III.

same direction from the isoelectric line as the major component of the QRS complex, although biphasic or oppositely directed T waves are perfectly normal in certain leads. When the T wave and QRS complex deviate in the same direction from the isoelectric line, it indicates that the repolarization process does not follow the same route as the depolarization process, as explained previously (Fig. 2-24). T waves that are abnormal either in direction or amplitude may indicate myocardial damage, electrolyte disturbances, or other abnormal conditions.

Standard limb leads. The original electrocardiographic lead system was devised by Willem Einthoven (1860 to 1927), who was professor of physiology at the University of Leiden. In his lead system the *resultant cardiac vector* (the vector sum of all electrical activity occurring in the heart at any given moment) was considered to lie in the center of a triangle (assumed to be equilateral) formed by the left and right shoulders and the pubic region (Fig. 2-30). This triangle, called the *Einthoven triangle*, is oriented in the frontal plane of the body. Hence, only the projection of the resultant cardiac vector on the frontal plane will be detected by this system of leads. For convenience, the electrodes are connected to the right and left forearms rather than to the corresponding shoulders, since the arms are considered to represent simple extensions of the leads from the shoulders; this assumption has been validated experimentally. Similarly, the leg (the

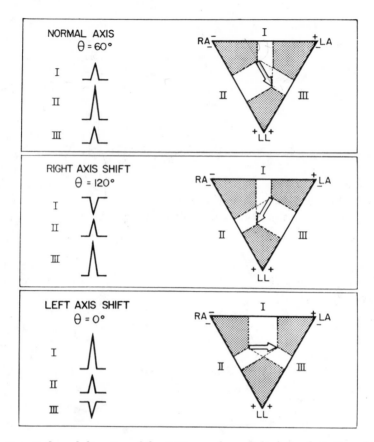

Fig. 2-31. Magnitude and direction of the QRS complexes in limb Leads I, II, and III, when the mean electrical axis (θ) is 60 degrees (top section), 120 degrees (middle section), and 0 degrees (bottom section).

left leg, by convention) is taken as an extension of the lead system from the pubis, and the third electrode is therefore connected to the left leg.

Certain conventions prevail in the manner in which these so-called *standard limb leads* are connected to the galvanometer. Lead I records the potential difference between the left arm (LA) and the right arm (RA). The galvanometer connections are such that when the potential at LA (V_{LA}) exceeds the potential at RA (V_{RA}), the galvanometer will be deflected upward from the isoelectric line. In Figs. 2-30 and

2-31 this arrangement of the galvanometer connections for Lead I is designated by a (+) at LA and a (−) at RA. Lead II records the potential difference between RA and LL (left leg) and yields an upward deflection when L_{LL} exceeds V_{RA}. Finally, Lead III registers the potential difference between LA and LL and yields an upward deflection when V_{LL} exceeds V_{LA}. It will become evident from the following discussion that these galvanometer connections were arbitrarily chosen so that the QRS complexes will be upright in all three standard limb leads in the majority of normal individuals.

Let the frontal projection of the resultant cardiac vector at some moment in time be represented by an arrow (tail negative, head positive), as in Fig. 2-30. Then the potential difference, $V_{LA} - V_{RA}$, recorded in Lead I will be represented by the component of the vector projected along the horizontal line between LA and RA, as shown in Fig. 2-31. If the vector makes an angle, θ, of 60 degrees with the horizontal (as in the top section of Fig. 2-31), the magnitude of the potential recorded by Lead I will equal the vector magnitude times cosine 60 degrees. The deflection recorded in Lead I will be upward, since the positive arrowhead lies closer to LA than to RA. The deflection in Lead II will also be upright, since the arrowhead lies closer to LL than to RA. The magnitude of the Lead II deflection will be greater than that in Lead I, since in this example the direction of the vector parallels that of Lead II; therefore, the magnitude of the projection on Lead II exceeds that on Lead I. Similarly, in Lead III the deflection will be upright, and in this example, where $\theta = 60$ degrees, its magnitude will equal that in Lead I.

If the vector in the top section of Fig. 2-31 happens to represent the resultant of the electrical events occurring during the peak of the QRS complex, then the orientation of this vector is said to represent the *mean electrical axis* of the heart in the frontal plane. The positive direction of this axis is taken in the clockwise direction from the horizontal (contrary to the usual mathematical convention). For normal individuals the average mean electrical axis is approximately + 60 degrees (as in the top section of Fig. 2-31). Therefore the QRS complexes are usually upright in all three leads, and largest in Lead II.

Changes in the mean electrical axis may occur with alterations in the anatomical position of the heart or with changes in the relative preponderance of the right and left ventricles. For example, the axis tends to shift toward the left (more horizontal) in short, stocky individuals

and toward the right (more vertical) in tall, thin persons. Also, with left or right ventricular hypertrophy (increased myocardial mass), the axis will shift toward the hypertrophied side.

With appreciable shift of the mean electrical axis to the right (middle section of Fig. 2-31, where $\theta = 120$ degrees), the displacements of the QRS complexes in the standard leads will change considerably. In this case the largest upright deflection will be in Lead III and the deflection in Lead I will be inverted, since the arrowhead will be closer to RA than to LA. With left axis shift (bottom section of Fig. 2-31, where $\theta = 0$ degrees), the largest upright deflection will be in Lead I, and the QRS complex in Lead III will be inverted.

As is evident from the above discussion, the standard limb leads, I, II, and III, are oriented in the frontal plane at 0, 60, and 120 degrees, respectively, from the horizontal. Other limb leads, which are also oriented in the frontal plane, are usually recorded in addition to the standard leads. These leads (principally the unipolar limb leads of Wilson or the augmented unipolar limb leads of Goldberger) lie along axes at angles of + 90, - 30, and - 150 degrees from the horizontal. Such lead systems are described in all textbooks on electrocardiography and will not be considered further here.

To obtain information concerning the projections of the cardiac vector on the sagittal and transverse planes of the body in scalar electrocardiography, the so-called *precordial leads* are usually recorded. Most commonly, each of six selected points on the anterior and lateral surfaces of the chest in the vicinity of the heart is connected in turn to the galvanometer. The other galvanometer terminal is usually connected to a *central terminal*, which is composed of a junction of three leads from LA, RA, and LL, each in series with a 5,000-ohm resistor. It can be shown that the voltage of this central terminal remains at a theoretical zero potential throughout the cardiac cycle.

ARRHYTHMIAS
Altered sinoatrial rhythms

The frequency of pacemaker discharge varies by the mechanisms described earlier in this chapter (Fig. 2-13). Changes in S-A nodal discharge frequency are usually produced by the cardiac autonomic nerves. Examples of electrocardiograms of sinus tachycardia and sinus bradycardia are shown in Fig. 2-32. The P, QRS, and T deflections are all normal, but the duration of the cardiac cycle (the so-called *P-P interval*) is altered. Characteristically, when sinus bradycardia or tachycardia occurs under natural conditions, the cardiac frequency changes gradually and requires several beats to attain its new steady state value. Electrocardiographic evidence of *respiratory cardiac arrhythmia* is common and is manifested as a rhythmic variation in the P-P interval at the respiratory frequency (p. 152).

Atrioventricular transmission blocks

Various physiological, pharmacological, and pathological processes can impede impulse transmission through the atrioventricular conduction tissue. The site of block can now be localized more precisely by the recent development of a technique for recording the *His bundle electrogram* (Fig. 2-33). To obtain such tracings, an electrode catheter is introduced into a peripheral vein and is threaded centrally until the tip containing the electrodes lies in the junctional region between the right atrium and ventricle. When the electrodes are properly positioned, a distinct deflection (Fig. 2-33, *H*) is registered, which represents the passage of the cardiac action potential down the His bundle. The time intervals required for propagation from the atrium to the His bundle (*A-H interval*) and from the His bundle to the ventricles (*H-V interval*) may be measured accu-

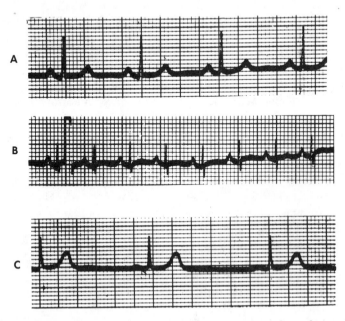

Fig. 2-32. Sinoatrial rhythms. **A,** Normal sinus rhythm. **B,** Sinus tachycardia. **C,** Sinus bradycardia.

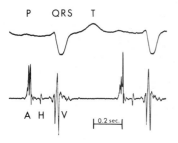

Fig. 2-33. His bundle electrogram (lower tracing, retouched) and Lead II of the scalar electrocardiogram (upper tracing). The deflection, *H*, which represents the impulse conduction over the bundle of His, is clearly visible between the atrial, *A*, and ventricular, *V*, deflections. The conduction time from the atria to the bundle of His is denoted by the *A-H* interval; that from the His bundle to the ventricles, by the *H-V* interval. (Courtesy Dr. J. Edelstein.)

rately. Abnormal prolongation of the former or latter interval indicates block above or below the His bundle, respectively.

Three degrees of A-V block can be distinguished, as shown in Fig. 2-34. *First-degree A-V block* is characterized by a prolonged P-R interval. In Fig. 2-34, *A*, the P-R interval is 0.28 second; an interval greater than 0.2 second is usually considered to be abnormal. In most cases of first-degree block the A-H interval of the His bundle electrogram is prolonged, whereas the H-V interval is normal. Hence the delay is located above the bundle of His, that is, in the A-V node.

In *second-degree A-V block* all QRS complexes are preceded by P waves, but not all P waves are followed by QRS complexes. The ratio of P waves to QRS complexes is usually the

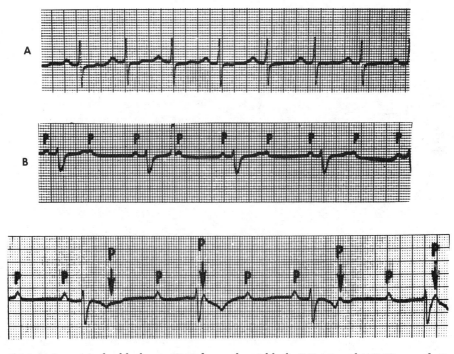

Fig. 2-34. Atrioventricular blocks. **A,** First-degree heart block; P-R interval is 0.28 second. **B,** Second-degree heart block (2:1). **C,** Third-degree heart block; note that there is a dissociation between the P waves and the QRS complexes.

ratio of two small integers (such as, 2:1, 3:1, 3:2). Fig. 2-34, *B*, illustrates a typical 2:1 block. His bundle electrograms have demonstrated that the site of block may be above or below the bundle of His. When the block arises above the bundle, the H deflections are absent after the A waves of the blocked beats. When the block occurs below the bundle, the H deflection is readily apparent after each A wave, but the V wave is absent during the blocked beats. This type of block implies a graver prognosis than when the block exists above the bundle, and an artificial pacemaker is frequently required.

Third-degree A-V block is often referred to as *complete heart block* because the impulse is unable to traverse the A-V conduction pathway from atria to ventricles. His bundle electrograms reveal that the most common sites of block are distal to the His bundle; that is, simultaneous block of the right and left bundle branches, or of the right bundle branch and the two divisions of the left bundle branch. In complete heart block the atrial and ventricular rhythms are entirely independent. A classical example is displayed in Fig. 2-34, *C*, where the QRS complexes bear no fixed relationship to the P waves. Because of the slow ventricular rhythm (32 beats per minute in this example), circulation is often inadequate, especially during muscular activity Third-degree block is often associated with syncope (so-called Stokes-Adams attacks) caused principally by insufficient cerebral blood flow. It is for such a condition that artificial pacemakers are most often installed to ensure a more nearly normal ventricular frequency.

Premature systoles

Premature contractions occur at times in most normal individuals but are more common under certain abnormal conditions. They may originate in the atria, A-V junction, or ventricles. Two principal mechanisms are believed to be responsible for premature beats. One type

of premature beat is coupled to a normally conducted beat. If the normal beat is suppressed in some way (for example, by vagal stimulation), the premature beat will also be abolished. Such premature beats are called *coupled extrasystoles,* or simply *extrasystoles,* and they probably reflect a reentry phenomenon (Fig. 2-20). The other type of premature beat occurs as the result of enhanced automaticity in some ectopic focus. This ectopic center may fire regularly and be protected in some way from depolarization by the normal cardiac impulse. If this premature beat occurs at a regular interval or at a simple multiple of that interval, the disturbance is called *parasystole.*

A *premature atrial systole* is shown in the electrocardiogram in Fig. 2-35, *A*. The normal interval between beats was 0.89 second (heart rate, 68 beats per minute). The premature atrial contraction (second P wave in the figure) followed the preceding P wave by only 0.56 second. The configuration of the premature P wave differs from the configuration of the other, normal P waves because the course of atrial excitation, originating at some ectopic focus in the atrium, is different from the normal spread of excitation originating at the S-A node. The QRS complex of the premature beat is usually normal in configuration because the spread of ventricular excitation occurs over the usual pathways. However, the QRS complex of the premature beat is sometimes bizarre, revealing *aberrant intraventricular conduction,* especially with very premature contractions. This is because the action potential duration of the ventricular Purkinje fibers is so long. Premature atrial impulses, traversing the A-V conduction system, may find some of the Purkinje fibers still refractory. In this event the sequence of ventricular activation may be abnormal, and therefore the configuration of the QRS may be aberrant.

The duration of the interval following a premature atrial systole depends, in part, on the location of the ectopic activity. The impulse is

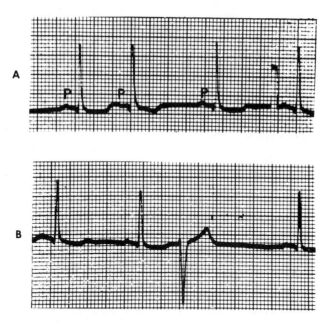

Fig. 2-35. A premature atrial systole, **A,** and a premature ventricular systole, **B,** recorded from the same patient. The premature atrial systole (the second beat in the top tracing) is characterized by an inverted P wave and normal QRS and T waves. The interval following the premature beat is not much longer than the usual interval between beats. The brief rectangular deflection just before the last beat is a standardization signal. The premature ventricular systole, **B,** is characterized by bizarre QRS and T waves and is followed by a compensatory pause.

conducted from the site to all regions of the atria and to the S-A node. If the impulse reaches the S-A node before it has spontaneously generated its next natural impulse, the S-A node is depolarized and then begins a completely new cycle of activity. The interval between a premature atrial beat and the next normal beat is therefore equal to the duration of a normal cardiac cycle plus the time required for the ectopic impulse to be conducted to the S-A node. One additional factor that can also prolong this interval is that a premature depolarization of pacemaker cells may transiently depress their rhythmicity. In Fig. 2-35, A, the interval from the premature depolarization to the succeeding normal P wave is virtually equal to the normal cardiac cycle duration for that patient. Therefore the ectopic focus must have been near the S-A node. If the ectopic focus is sufficiently far from the S-A node, the S-A node will fire naturally before the premature excitation wave will have reached it. In this event the natural cycle of activity of the S-A node will not have been disturbed; the interval from the P wave of the beat just preceding the premature contraction to the P wave of the beat just following it will be equal to the duration of two normal cycles. However, if the S-A node is depolarized by the excitation originating in an ectopic focus, then the interval from the beat preceding the premature contraction to the beat following it will be significantly less than the duration of two normal cycles (such as in the example illustrated in Fig. 2-35, A).

A *premature ventricular systole*, recorded from the same patient, appears as tracing *B* of Fig. 2-35. Since the premature excitation originated at some ectopic focus in the ventricles, the impulse spread was aberrant and the configurations of the QRS and T waves were entirely different from the normal deflections. The premature QRS complex followed the preceding normal QRS complex by only 0.47 second. The interval following the premature excitation was 1.28 seconds, considerably longer than the normal interval between beats (0.89 second). The interval (1.75 seconds) from the QRS complex just before the premature excitation to the QRS complex just after it, was virtually equal to the duration of two normal cardiac cycles (1.78 seconds).

The prolonged interval that usually follows a premature ventricular contraction is called a *compensatory pause*. The reason for the compensatory pause after a premature ventricular contraction is that, contrary to most premature atrial contractions, the ectopic ventricular impulse does not disturb the natural rhythm of the S-A node. Either the ectopic ventricular impulse is not conducted retrograde through the A-V conduction system or, if it is, the time required is such that the S-A node has already fired at its natural interval before the ectopic impulse could have reached it. Likewise, the S-A nodal impulse usually does not affect the ventricle, since the A-V junction and perhaps also the ventricles are still refractory from the premature contraction. In Fig. 2-35, *B*, the P wave originating in the S-A node at the time of the premature systole occurred at the time of the T wave of the premature cycle, and therefore cannot be identified in the tracing.

Ectopic tachycardias

When a tachycardia originates from some ectopic site in the heart, the onset and termination are typically abrupt, as distinguished from the more gradual changes in heart rate in sinus tachycardia. Because of the sudden appearance and abrupt reversion to normal, such ectopic tachycardias are usually referred to as *paroxysmal tachycardias*. Episodes of ectopic tachycardia may persist for only a few beats or for many hours or days, and the episodes are often recurrent. Paroxysmal tachycardias may occur either as the result of (a) the rapid firing of an ectopic pacemaker, (b) triggered activity secondary to afterpotentials that reach threshold, or (c) an impulse circling a reentry loop repetitively.

Paroxysmal tachycardias originating in the atria or in the A-V junctional tissues are usually indistinguishable, and therefore both are included in the term *paroxysmal supraventricular tachycardia*. In many instances, the tachycardia occurs as the result of an impulse repetitively circling a reentry loop that includes atrial tissue and the A-V junction. An electrocardiogram illustrating this arrhythmia is shown in Fig. 2-36, *A*. The QRS complexes are normal since ventricular activation proceeds over the normal pathways. When the supraventricular frequency is excessively rapid, the A-V conduction tissue may be incapable of conducting all impulses, and second-degree A-V blocks (for example, 2:1 block) may be a concomitant of the paroxysmal tachycardia. Not infrequently the supraventricular impulses may be conducted through the junctional tissues at a frequency that is too high for some of the ventricular Purkinje fibers (because of their long refractory periods). In this event the QRS complexes will display the resultant aberrant intraventricular conduction, and the arrhythmia may be very difficult to distinguish from a paroxysmal tachycardia of ventricular origin.

Paroxysmal ventricular tachycardia originates from an ectopic focus in the ventricles. The electrocardiogram is characterized by the rapid repeated, bizarre QRS complexes that reflect the aberrant intraventricular impulse conduction (Fig. 2-36, *B*). Paroxysmal ventricular tachycardia is much more ominous than supraventricular tachycardia because it is fre-

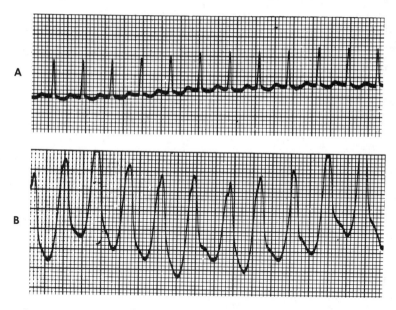

Fig. 2-36. A, Paroxysmal supraventricular and **B,** ventricular tachycardia.

quently a precursor of ventricular fibrillation, a lethal arrhythmia that will be described in the following section. The paroxysmal ventricular tachycardia illustrated in Fig. 2-36, *B*, was recorded from a patient immediately after resuscitation from ventricular fibrillation.

Fibrillation

Under certain conditions cardiac muscle undergoes an extremely irregular type of contraction that is entirely ineffectual in propelling blood. Such an arrhythmia is termed *fibrillation* and may involve either the atria or ventricles. Fibrillation probably represents a reentry phenomenon, in which the reentry loop fragments into multiple, irregular circuits.

The tracing in Fig. 2-37, *A*, illustrates the electrocardiographic changes in *atrial fibrillation*. In this condition, which occurs quite commonly in various types of chronic heart disease, the atria do not contract and relax sequentially during each cardiac cycle and hence do not contribute to ventricular filling. Instead, the atria undergo a continuous, uncoordinated, rippling type of activity. On the electrocardiogram there are no P waves; they are replaced by continuous irregular fluctuations of potential, called *f* waves. The A-V node is activated at intervals that may vary considerably from cycle to cycle. Hence there is no constant interval between QRS complexes and therefore between ventricular contractions. Since the strength of ventricular contraction depends on the interval between beats (because it determines the time available for ventricular filling, and for other reasons as well, as explained on p. 166), the pulse is extremely irregular with regard to both rhythm and force.

Although atrial fibrillation is compatible with life and even with full activity the onset of *ventricular fibrillation* leads to loss of consciousness within a few seconds. The irregular, continuous, uncoordinated twitchings of the ventricular muscle fibers result in no output of blood. Death ensues except when immediate, effective resuscitation is achieved or when ven-

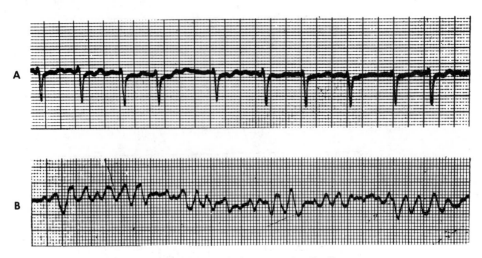

Fig. 2-37. A, Atrial and **B,** ventricular fibrillation.

tricular fibrillation reverts to a more normal rhythm spontaneously, which rarely occurs. Ventricular fibrillation may supervene when the entire ventricle, or some portion of it, is deprived of its normal coronary blood supply. It may also occur as a result of electrocution or in response to certain drugs and anesthetics. In the electrocardiogram (Fig. 2-37, *B*) large, irregular fluctuations of potential are manifest.

Fibrillation is often initiated when a premature impulse arrives during the so-called *vulnerable period*. In the ventricles, this period coincides with the downslope of the T wave. During this period, there is some variability in the excitability of the cardiac cells. Some fibers are still in their effective refractory periods; others have almost fully recovered their excitability; and still others are able to conduct impulses, but only at very slow conduction velocities. As a consequence, the action potentials are propagated over the chambers in multiple wavelets that travel along circuitous paths and at widely varying conduction velocities. As a region of cardiac cells becomes excitable again, it will ultimately be reentered by one of the

wave fronts traveling about the chamber. The process tends to be self-sustaining.

Atrial fibrillation may be reverted to a normal sinus rhythm by means of certain depressant drugs, such as quinidine, which acts in part by prolonging the refractory period. Therefore the cardiac impulse, on retracing its path, may find the myocardial fibers no longer excitable. However, much more dramatic therapy is required in ventricular fibrillation. Conversion to a normal sinus rhythm is accomplished by means of a strong electric current that places the entire myocardium in a refractory state. Originally it was necessary to open the patient's chest and apply the electrodes directly to the walls of the heart. However, techniques have now been developed so that the current can be administered safely through the intact chest wall. In successful cases the S-A node again takes over as the normal pacemaker for the entire heart. Direct current shock has been found to be more effective than alternating current shock, and it is now widely used clinically to treat not only ventricular fibrillation, but atrial fibrillation and certain other arrhythmias as well. Electrical abolition of such disturbances of rhythm is often called *cardioversion*.

BIBLIOGRAPHY
Journal articles

Abildskov, J. A.: The sequence of normal recovery of excitability in the dog heart, Circulation **52**:442, 1975.

Burgess, M. J.: Relation of ventricular repolarization to electrocardiographic T wave-form and arrhythmia vulnerability, Am. J. Physiol. **236**:H391, 1979.

Carmeliet, E.: Repolarization and frequency in cardiac cells, J. Physiol. (Paris) **73**:903, 1977.

Childers, R.: The AV node: normal and abnormal physiology, Prog. Cardiovasc. Dis. **19**:361, 1977.

Coraboeuf, E.: Ionic basis of electrical activity in cardiac tissues, Am. J. Physiol. **234**:H101, 1978.

Cranefield, P. F.: Action potentials, afterpotentials, and arrhythmias, Circ. Res. **41**:415, 1977.

Cranefield, P. F.: Ventricular fibrillation, N. Engl. J. Med. **289**:732, 1973.

Damato, A. N., Lau, S. H., Berkowitz, W. D., Rosen, K. M., and Lisi, K. R.: Recording of specialized conducting fibers (A-V nodal, His bundle, and right bundle branch) in man using an electrode catheter technic, Circulation **39**:435, 1969.

Elharrar, V., and Zipes, D. P.: Cardiac electrophysiologic alterations during myocardial ischemia, Am. J. Physiol. **233**:H329, 1977.

Ferrier, G. R.: Digitalis arrhythmias: role of oscillatory afterpotentials, Prog. Cardiovasc. Dis. **19**:459, 1977.

Glitsch, H. G.: Characteristics of active Na transport in intact cardiac cells, Am. J. Physiol. **236**:H189, 1979.

Hauswirth, O., and Singh, B. N.: Ionic mechanisms in heart muscle in relation to the genesis and the pathological control of cardiac arrhythmias, Pharmacol. Rev. **30**:5, 1979.

Irisawa, H.: Comparative physiology of the cardiac pacemaker mechanism, Physiol. Rev. **58**:461, 1978.

Jalife, J., and Moe, G. K.: Phasic effects of vagal stimulation on pacemaker activity of the isolated sinus node of the young cat, Circ. Res. **45**:595, 1979.

James, T. N., and Sherf, L.: Specialized tissues and preferential conduction in the atria of the heart, Am. J. Cardiol. **28**:414, 1971.

Kohlhardt, M., and Mnich, Z.: Studies on the inhibitory effect of verapamil on the slow inward currents in mammalian ventricular myocardium, J. Mol. Cell. Cardiol. **10**:1037, 1978.

Lazzara, R., El-Sherif, N., Befeler, B., and Scherlag, B. J.: Regional refractoriness within the ventricular conduction system, Circ. Res. **39**:254, 1976.

Martin, P.: The influence of the parasympathetic nervous system on atrioventricular conduction, Circ. Res. **41**:593, 1977.

Paes de Carvalho, A., Hoffman, B. F., and Carvalho, M. deP.: Two components of the cardiac action potential, J. Gen. Physiol. **54**:607, 1969.

Pappano, A. J.: Ontogenetic development of autonomic neuroeffector transmission and transmitter reactivity in embryonic and fetal hearts, Pharmacol. Rev. **29**:3, 1977.

Reuter, H.: Divalent cations as charge carriers in excitable membranes, Progr. Biophys. Mol. Biol. **26**:1, 1973.

Spear, J. F., Kronhaus, K. D., Moore, E. N., and Kline, R. P.: The effect of brief vagal stimulation on the isolated rabbit sinus node, Circ. Res. **44**:75, 1979.

Strauss, H. C., Prystowsky, R. N., and Scheinman, M. M.: Sino-atrial and atrial electrogenesis, Prog. Cardiovasc. Dis. **19**:385, 1977.

Tsien, R. W., and Carpenter, D. O.: Ionic mechanisms of pacemaker activity in cardiac Purkinje fibers, Fed. Proc. **37**:2127, 1978.

Vassalle, M.: Cardiac automaticity and its control, Am. J. Physiol. **233**:H625, 1977.

Wit, A. L., and Cranefield, P. F.: Reentrant excitation as a cause of cardiac arrhythmias, Am. J. Physiol. **235**:H1, 1978.

Books and monographs

Carmeliet, E., and Vereecke, J.: Electrogenesis of the action potential and automaticity. In Handbook of physiology; Section 2: Cardiovascular system, vol. 1, Bethesda, Md., 1979, American Physiological Society, pp. 269-334.

Cranefield, P. F.: The conduction of the cardiac impulse, Mount Kisco, N. Y., 1975, Futura Publishing Co., Inc.

Fozzard, H. A.: Conduction of the action potential. In Handbook of physiology; Section 2: Cardiovascular system, vol. 1, Bethesda, Md., 1979, American Physiological Society, pp. 335-356.

Noble, D.: The initiation of the heartbeat, Oxford, 1975, Oxford University Press.

Scher, A. M., and Spach, M. S.: Cardiac depolarization and repolarization and the electrocardiogram. In Handbook of physiology; Section 2: Cardiovascular system, vol. 1, Bethesda, Md., 1979, American Physiological Society, pp. 357-392.

Sperelakis, N.: Origin of the cardiac resting potential. In Handbook of physiology; Section 2: Cardiovascular system, vol. 1, Bethesda, Md., 1979, American Physiological Society, pp. 187-267.

Vassalle, M.: Cardiac physiology for the clinician, New York, 1976, Academic Press, Inc.

CHAPTER 3

HEMODYNAMICS

The problem of treating the pulsatile flow of blood through the cardiovascular system in precise mathematical terms is virtually insuperable. The heart is an extremely complicated pump, the behavior of which is affected by a large variety of physical and chemical factors. The blood vessels are multibranched, elastic conduits of continuously varying dimensions. The blood itself is not a simple fluid, but is a suspension of red and white corpuscles, platelets, and lipid globules suspended in a colloidal solution of proteins. Despite these complicating factors, considerable insight may be gained from an understanding of the more elementary principles of fluid mechanics as they pertain to simpler physical systems. Such principles will be expounded in the following sections to explain the interrelationships among velocity of blood flow, blood pressure, and the dimensions of the various components of the systemic circulation.

VELOCITY OF THE BLOODSTREAM

In describing the variations in blood flow in different vessels it is first essential to distinguish between the terms *velocity* and *flow*. The former term, sometimes designated as linear velocity, refers to the rate of displacement with respect to time and has the dimensions of distance per unit time, for example, cm./sec. The latter term is frequently designated as volume flow and has the dimensions of volume per unit time, for example, cm.^{3}/sec. In a conduit of varying cross-sectional dimensions, velocity, v, flow, Q, and cross-sectional area, A, are related by the equation:

$$v = Q/A$$

The interrelationships among velocity, flow, and area are portrayed in Fig. 3-1. In the case of an incompressible fluid flowing through rigid tubes, the flow past successive cross sections must be the same. For a given constant flow the velocity varies inversely as the cross-sectional area. Thus for the same volume of fluid per second passing from section *a* into section *b*, where the cross-sectional area is five times greater, the velocity of flow diminishes to one-fifth of its previous value. Conversely, when the fluid proceeds from section *b* to section *c*, where the cross-sectional area is one-tenth as great, the velocity of each particle of fluid must increase tenfold. The velocity at any point in the system is dependent not only on area, but also on the magnitude of the flow, Q. This, in turn, depends on the pressure gradient, the properties of the fluid, and the dimensions of the entire hydraulic system, as discussed in the following section. For any given flow, however, the ratio of the velocity past one cross section relative to that past a second cross section depends only on the inverse ratio of the respective areas; that is,

$$v_1/v_2 = A_2/A_1$$

This rule pertains regardless of whether a given cross-sectional area applies to a single large tube or to several smaller tubes in parallel.

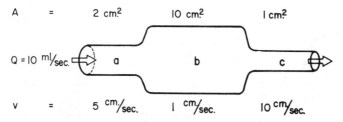

Fig. 3-1. As fluid flows through a tube of variable cross-sectional area, A, the linear velocity, v, varies inversely as the cross-sectional area.

As shown in Fig. 1-2, there is a progressive reduction in velocity as the blood traverses the aorta, its larger primary branches, the smaller secondary branches, and the arterioles. Finally, a minimum value is reached in the capillaries. As the blood then passes through the venules and continues centrally toward the venae cavae, the velocity progressively increases again. The relative velocities in the various components of the circulatory system are not related directly to the pressure gradients or to any other physical factors, but only to the cross-sectional area. For this reason each point on the curve representing the total cross-sectional area is inversely proportional to the corresponding point on the curve representing velocity for any given subdivision of the vascular bed (Fig. 1-2).

RELATIONSHIP BETWEEN VELOCITY AND PRESSURE

In that portion of a hydraulic system in which the total energy remains virtually constant, changes in velocity may be accompanied by appreciable alterations in the measured pressure. Consider three sections (A, B, and C) of such a hydraulic system, as depicted in Fig. 3-2. Six pressure probes, or *pitot tubes*, have been inserted. The openings of three of these (*2, 4,* and *6*) are tangential to the direction of flow and hence measure the *lateral*, or *static*, pressure within the tube. The openings of the remaining three pitot tubes (*1, 3,* and *5*)

face upstream. Therefore, they detect the *total pressure*, which is the lateral pressure plus a pressure component ascribable to the kinetic energy of the flowing fluid. This dynamic component, P_d, of the total pressure may be calculated from the following equation:

$$P_d = \rho v^2/2$$

where ρ is the density of the fluid, and v the velocity. If the midpoints of segments A, B, and C are at the same hydrostatic level, then the corresponding total pressure, P_1, P_3, and P_5, will be virtually equal, if the energy loss from viscosity in these segments is negligible. However, with the changes in cross-sectional area, the consequent alteration in velocity induces variations in the dynamic component.

In sections A and C, let $\rho = 1$ gm./cm.^{3}, and $v = 100$ cm./sec. Therefore,

$$P_d = 5,000 \text{ dynes/cm}^2$$
$$= 3.8 \text{ mm. Hg}$$

since 1,330 dynes/cm^2 = 1 mm. Hg. In the narrow section, B, let the velocity be twice as great as in sections A and C. Therefore,

$$P_d = 20,000 \text{ dynes/cm}^2$$
$$= 15.0 \text{ mm. Hg}$$

Hence, in the wide sections of the conduit, the lateral pressures (P_2 and P_6) will be only 3.8 mm. Hg less than the respective total pressures (P_1 and P_5). However, in the narrow sec-

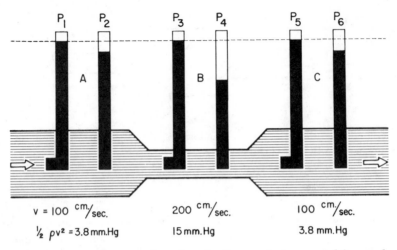

Fig. 3-2. In a narrow section, *B*, of a tube, the linear velocity, *v*, and hence the dynamic component of pressure, $\frac{1}{2}\rho v^2$, are greater than in the wide sections, *A* and *C*, of the same tube. If the total energy is virtually constant throughout the tube (that is, if the energy loss due to viscosity is negligible), the total pressures $(P_1, P_3$ and $P_5)$ will not be detectably different, but the lateral pressure, P_4, in the narrow section will be less than the lateral pressures $(P_2$ and $P_6)$ in the wide sections of the tube.

tion, the lateral pressure (P_4) is 15 mm. Hg less than the total pressure (P_3).

The peak velocity of flow in the ascending aorta of normal dogs has been found to be in the range of 100 to 200 cm./sec. Therefore the measured pressure may vary significantly, depending on the orientation of the pressure probe. In the descending thoracic aorta the peak velocity is only half as great as in the ascending aorta, and lesser magnitudes have been recorded in other major arterial sites. The dynamic component will thus be a negligible fraction of the total pressure, and the orientation of the pressure probe will not materially influence the magnitude of the pressure recorded. However, at the site of a constriction, such as a stenotic heart valve or coarctation of the aorta, the dynamic component may attain substantial values.

In *aortic stenosis*, for example, the entire output of the left ventricle is ejected through a narrow valve orifice. The high flow velocity is

associated with a large kinetic energy, and therefore the lateral pressure is correspondingly reduced. The orifices of the right and left coronary arteries are located in the sinuses of Valsalva, just behind the valve leaflets. Hence the initial segments of these vessels are oriented at right angles to the direction of blood flow through the aortic valves. Therefore the lateral pressure is that component of the total pressure which propels the blood through the two major coronary arteries. During the ejection phase of the cardiac cycle, the lateral pressure is diminished by virtue of the conversion of potential to kinetic energy. This process is grossly exaggerated in aortic stenosis, because of the high flow velocities. Angiographic studies in patients with aortic stenosis have revealed that the direction of flow often reverses in the large coronary arteries toward the end of the ejection phase of systole. The decreased lateral pressure in the aorta in aortic stenosis is undoubtedly an important factor in causing this reversal of flow, because the aortic pressure suddenly drops below that in the coronary arteries. An important aggravating feature in this condition is that the demands of the heart muscle for oxygen are greatly increased. Therefore, the pronounced drop in lateral pressure during cardiac ejection may be a contributory factor in the tendency for patients with severe aortic stenosis to experience *angina pectoris* (anterior chest pain associated with inadequate blood supply to the heart muscle) and to die suddenly.

RELATIONSHIP BETWEEN PRESSURE AND FLOW

The most fundamental law governing the flow of fluids through cylindrical tubes was derived empirically in 1842 by Jean Léonard Marie Poiseuille. This French physician was primarily interested in the physical determinants of blood flow, but substituted simpler liquids for blood in his measurements of flow through glass capillary tubes. His work was precise and of such importance that his observations have been designated *Poiseuille's law*. Subsequently, this same law has been derived on certain theoretical bases.

Poiseuille's law is applicable to the flow of fluids through cylindrical tubes only under special conditions. It applies to the case of steady, laminar flow of Newtonian fluids. The term *steady flow* signifies the absence of variations of flow in time, that is, a nonpulsatile flow. *Laminar flow* is the type of motion in which the fluid moves as a series of individual layers, with each stratum moving at a different velocity from its neighboring layers. In the case of flow through a tube the fluid consists of a series of infinitesimally thin concentric tubes sliding past one another. Laminar flow will be described in greater detail below, where it will be distinguished from turbulent flow. Also a *Newtonian fluid* will be defined more precisely. For the present discussion it will suffice to consider it as a homogeneous fluid, such as air or water, in contradistinction to a suspension, such as blood.

Pressure is one of the principal determinants of the rate of flow. The pressure, P, in dynes/cm.^{2}, at a distance h centimeters below the surface of a liquid is

$$P = h\rho g$$

where ρ is the density of the liquid in gm./cm.^{3} and g is the acceleration of gravity in cm./sec.^{2} For convenience, however, pressure is frequently expressed simply in terms of height, h, of the column of liquid above some arbitrary reference point.

Consider the tube connecting reservoirs A and B in Fig. 3-3. Let reservoir A be filled with liquid to height h_1, and let reservoir B be empty, as in section 1 of Fig. 3-3. The outflow pressure, P_0, is therefore equal to the atmospheric pressure, which shall be designated as the zero, or reference, level. The inflow pressure, P_i, is then equal to the same reference level plus the height, h_1, of the column of liquid in reservoir A. Under these conditions let the flow, Q, through the tube be 5 ml./sec. If reservoir A is filled to height h_2, which is twice h_1, and reservoir B is again empty (as in section 2), the flow will be twice as great, that is, 10 ml./sec. Thus with reservoir B empty, the flow will be directly proportional to the inflow pressure, P_i. If reservoir B is now allowed to fill to height h_1, and the fluid level in A is maintained at h_2 (as in section 3), the flow will again become 5 ml./sec. Thus flow is directly proportional to the difference between inflow and outflow pressures:

$$Q \propto P_i - P_o$$

If the fluid level in B attains the same height as in A, flow will cease (section 4).

For any given pressure difference between the two ends of a tube, the flow will be dependent on the dimensions of the tube. Consider the tube connected to reservoir 1 in Fig. 3-4. With length l_1 and radius r_1, the flow Q_1 is

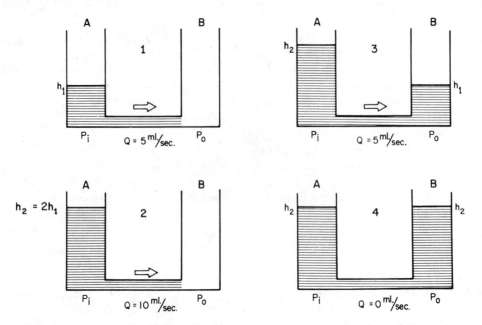

Fig. 3-3. The flow, Q, of fluid through a tube connecting two reservoirs, A and B, is proportional to the difference between the pressure, P_i, at the inflow end and the pressure, P_o, at the outflow end of the tube.

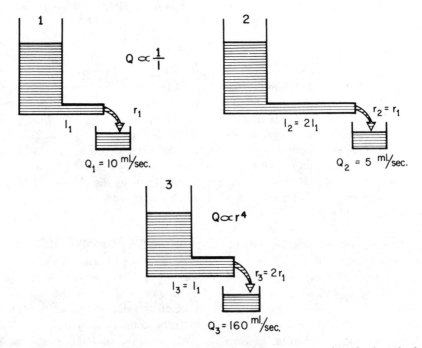

Fig. 3-4. The flow, Q, of fluid through a tube is inversely proportional to the length, l, and directly proportional to the fourth power of the radius, r, of the tube.

observed to be 10 ml./sec. The tube connected to reservoir *2* has the same radius, but is twice as long. Under these conditions the flow Q_2 is found to be 5 ml./sec., or only half as great as Q_1. Conversely, for a tube half as long as l_1 the flow would be twice as great as Q_1. In other words, flow is inversely proportional to the length of the tube:

$$Q \propto 1/l$$

The tube connected to reservoir *3* in Fig. 3-4 is the same length as l_1, but the radius is twice as great. Under these conditions, the flow Q_3 is found to increase to a value of 160 ml./sec., which is sixteen times greater than Q_1. The precise measurements of Poiseuille revealed that flow varies directly as the fourth power of the radius:

$$Q \propto r^4$$

Thus, in the example above, since $r_3 = 2r_1$, Q_3 will be proportional to $(2r_1)^4$, or $16r_1{}^4$; therefore, Q_3 will equal $16Q_1$.

Finally, for a given pressure difference and for a cylindrical tube of given dimensions, the flow will vary, depending on the nature of the fluid itself. This flow-determining property of fluids is termed *viscosity*, η, which has been defined by Newton as the ratio of *shear stress*

to the *rate of strain* of the fluid. These terms may be comprehended most clearly by considering the flow of a homogeneous fluid between parallel plates. In Fig. 3-5 let the bottom plate (the bottom of a large basin) be stationary, and let the upper plate move at a constant velocity along the upper surface of the fluid. The *shear stress*, τ, is defined as the ratio of F:A, where F is the force applied to the upper plate in the direction of its motion along the upper surface of the fluid and A is the area of the upper plate in contact with the fluid. The *rate of strain* is du/dy, where u is the velocity in the direction parallel to the motion of the upper plate for any minute fluid element contained between the plates and y is the distance of that element above the bottom, stationary plate.

For a movable plate traveling with constant velocity across the surface of a homogeneous fluid, the velocity profile of the fluid will be linear. The fluid layer in contact with the upper plate will adhere to it and therefore will move at the same velocity, U, as the plate. Each minute element of fluid between the plates will move at a velocity, u, proportional to its distance, y, from the lower plate. Therefore, the rate of strain will be U/Y, where Y is the total distance between the two plates. Since viscosity, η, is defined as the ratio of

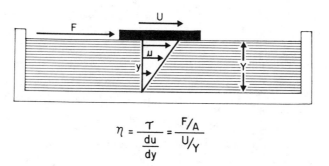

$$\eta = \frac{\tau}{\frac{du}{dy}} = \frac{F/A}{U/Y}$$

Fig. 3-5. For a Newtonian fluid, the viscosity, η, is defined as the ratio of shear stress, τ, to rate of strain, *du/dy*. For a plate of contact area, *A*, moving across the surface of a liquid, τ equals the ratio of the force, *F*, applied in the direction of motion to the contact area, *A*, and *du/dy* equals the ratio of the velocity of the plate, *U*, to the depth of the liquid, *Y*.

shear stress, τ, to the rate of strain, du/dy, in the case illustrated in Fig. 3-5,

$$\eta = (F/A)/(U/Y)$$

Thus the dimensions of viscosity are dynes/cm.^{2} divided by (cm./sec.)/cm., or dyne-sec.-cm.$^{-2}$ In honor of Poiseuille, 1 dyne-sec.-cm.$^{-2}$ has been termed a "poise." The viscosity of water at 20° C. is approximately 0.01 poise, or 1 centipoise. In the case of certain nonhomogeneous fluids, notably suspensions such as blood, the ratio of the shear stress to the rate of strain is not constant; such fluids are said to be *non-Newtonian*.

With regard to the flow of fluids through cylindrical tubes, the flow will vary inversely as the shear stress required to produce a given rate of strain; that is, flow varies inversely as the viscosity. Thus in the example of flow from reservoir *1* in Fig. 3-4, if the viscosity of the fluid in the reservoir were doubled, then the flow would be halved (5 ml./sec. instead of 10 ml./sec.).

In summary, for the steady, laminar flow of a Newtonian fluid through a cylindrical tube, the flow, Q, varies directly as the pressure difference, $P_i - P_o$, and the fourth power of the radius, r, of the tube and varies inversely as the length, l, of the tube and the viscosity, η, of the fluid. The full statement of Poiseuille's law is

$$Q = \frac{\pi(P_i - P_o)r^4}{8 \eta l}$$

where $\pi/8$ is the constant of proportionality.

RESISTANCE TO FLOW

In direct-current electrical circuit theory it has been useful to employ the concept of resistance, R, which is defined as the ratio of voltage dop, E, to current flow, I. Similarly, in fluid mechanics the analogous hydraulic resistance, R, may be defined as the ratio of pressure drop, $P_i - P_o$, to flow, Q. For the steady, laminar flow of a Newtonian fluid through a cylindrical tube the physical components of hydraulic resistance may be perceived readily by rearranging Poiseuille's law to give the hydraulic resistance equation

$$R = \frac{P_i - P_o}{Q} = \frac{8 \eta l}{\pi r^4}$$

Thus, when Poiseuille's law applies, the resistance to flow is dependent only on the dimensions of the tube and on the characteristics of the fluid.

In the circulatory system the length of any given vessel is virtually constant, and ordinarily the viscosity of the blood does not vary appreciably. Thus the major alterations in resistance are produced, physiologically and pathologically, by virtue of variations in radius. From Fig. 1-2 it may be noted that the greatest pressure drop occurs across the arterioles. Since the total flow is the same through the various series components of the circulatory system, it follows that the greatest resistance to flow resides in the arterioles. The arterioles are vested with a thick coat of circularly arranged smooth muscle fibers, by means of which variations in lumen radius may be produced. From the hydraulic resistance equation, wherein R varies inversely as r^4, it is clear that small changes in radius will result in relatively great alterations in resistance.

In the cardiovascular system the various types of vessels listed in Fig. 1-2 lie in series with one another. Furthermore, the individual members of each category of vessels are ordinarily arranged in parallel with one another (Fig. 1-3). For example, the capillaries throughout the body are in most instances parallel elements, with the notable exceptions of the renal vasculature (wherein the peritubular capillaries are in series with the glomerular capillaries) and the splanchnic vasculature (wherein the intestinal and hepatic capillaries are aligned in series). Formulas for the total hydraulic resistance of components arranged in series and in parallel have been derived in a manner analogous to those employed for similar combinations of electrical resistances.

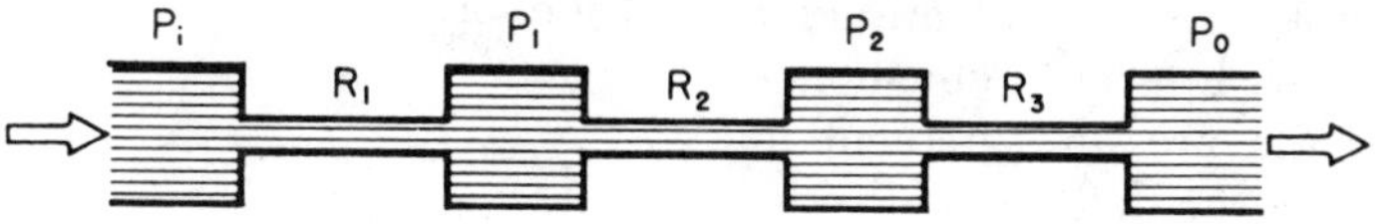

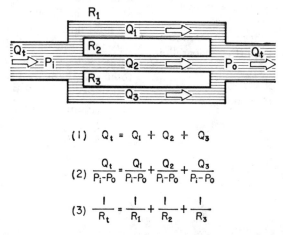

$$(1) \quad P_i - P_o = (P_i - P_1) + (P_1 - P_2) + (P_2 - P_o)$$

$$(2) \quad \frac{P_i - P_o}{Q} = \frac{(P_i - P_1)}{Q} + \frac{(P_1 - P_2)}{Q} + \frac{(P_2 - P_o)}{Q}$$

$$(3) \quad R_t = R_1 + R_2 + R_3$$

Fig. 3-6. For resistances $(R_1, R_2,$ and $R_3)$ arranged in series, the total resistance R_t, equals the sum of the individual resistances.

$$(1) \quad Q_t = Q_1 + Q_2 + Q_3$$

$$(2) \quad \frac{Q_t}{P_i - P_o} = \frac{Q_1}{P_i - P_o} + \frac{Q_2}{P_i - P_o} + \frac{Q_3}{P_i - P_o}$$

$$(3) \quad \frac{1}{R_t} = \frac{1}{R_1} + \frac{1}{R_2} + \frac{1}{R_3}$$

Fig. 3-7. For resistances R_1, R_2, and R_3) arranged in parallel, the reciprocal of the total resistance, R_t, equals the sum of the reciprocals of the individual resistances.

Three hydraulic resistances, R_1, R_2, and R_3, are arranged in series in the schema depicted in Fig. 3-6. The pressure drop across the entire system—that is, the difference between inflow pressure, P_i, and outflow pressure, P_o,—consists of the sum of the pressure drops across each of the individual resistances (equation *1*). Under steady state conditions the flow, Q, through any given cross section must equal the flow through any other cross section. By dividing each component in equation *1* by Q (equation *2*), it becomes evident from the definition

of resistance that the total resistance, R_t, of the entire system equals the sum of the individual resistances, that is,

$$R_t = R_1 + R_2 + R_3$$

For resistances in parallel, as illustrated in Fig. 3-7, the inflow and outflow pressures are the same for all tubes. Under steady state conditions the total flow, Q_t, through the system equals the sum of the flows through the individual parallel elements (equation *1*). Since the pressure gradient $(P_i - P_o)$ is identical for all parallel elements, each term in equation *1* may be divided by the pressure gradient to yield equation *2*. From the definition of resistance, equation *3* may be derived. This states that the reciprocal of the total resistance, R_t, equals the sum of the reciprocals of the individual resistances, that is,

$$\frac{1}{R_t} = \frac{1}{R_1} + \frac{1}{R_2} + \frac{1}{R_3}$$

Stated in another way, if we define hydraulic *conductance* as the reciprocal of resistance, in analogy to the practice in electrical theory, it becomes evident that, for tubes in parallel, the total conductance is the sum of the individual conductances.

By considering a few simple illustrations, some of the fundamental properties of parallel hydraulic systems become apparent. For example, if the resistances of the three parallel elements in Fig. 3-7 were all equal, then

$$R_1 = R_2 = R_3$$

Therefore,

$$1/R_t = 3/R_1$$

and

$$R_t = R_1/3$$

Thus the total resistance is less than any of the individual resistances. After further consideration, it becomes evident that for any parallel arrangement, the total resistance must be less than that of any individual component. For example, consider a system in which a very high-resistance tube is added in parallel to a low-resistance tube. The total resistance must be less than that of the low-resistance component by itself, since the high-resistance component affords an additional pathway, or conductance, for fluid flow.

As a physiological illustration of these principles, consider the relationship between the *total peripheral resistance* (TPR) of the entire systemic vascular bed and the resistance of one of its components, such as the renal vasculature. In an individual with a cardiac output of

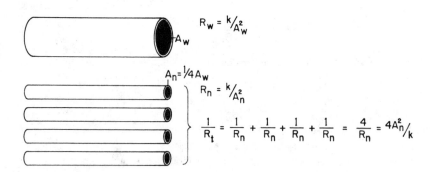

Fig. 3-8. When four narrow tubes, each of area A_n, are connected in parallel, the total cross-sectional area equals the area A_w, of a wide tube of area such that $A_w = 4A_n$. Although the total areas are equal, the total resistance, R_t, to flow through the parallel narrow tubes is four times as great as the resistance, R_w, through the single wide tube.

5,000 ml./min. and an arterial pressure of 100 mm. Hg, the TPR will be 0.02 mm. Hg/ml./min., or 0.02 PRU (peripheral resistance units). Blood flow through one kidney would be approximately 600 ml./min. Renal resistance would therefore be 100 mm. Hg/600 ml./min., or 0.17 PRU, which is 8.5 times as great as the TPR. By similar computations, it is apparent that, since the arterial pressures of children and of adults are approximately the same and since the cardiac output of adults greatly exceeds that of children, the TPR of children must be considerably greater than the TPR of adults.

From Fig. 1-2 it appears to be paradoxical that the resistance to flow through the arterioles (as manifested by the pressure drop between the arterial and capillary ends of these vessels) is considerably greater than that through certain other vascular components, despite the fact that the total cross-sectional area of the arterioles exceeds that for these same vascular components. For example, the total cross-sectional area of the arterioles greatly exceeds that of the large arteries, yet the resistance to flow through the arterioles exceeds that through the large arteries. Consideration of simple models of tubes in parallel will help resolve this apparent paradox. In Fig. 3-8 the resistance to flow through one wide tube of cross-sectional area A_w is compared with that through four narrower tubes in parallel, each of area A_n. The total cross-sectional area of the parallel system of four narrow tubes equals the area of the wide tube; that is,

$$A_w = 4 A_n \tag{1}$$

Since resistance, R, is inversely proportional to the fourth power of the radius, r, and since

$$A = \pi r^2 \tag{2}$$

for cylindrical tubes, it follows that

$$R = k / A^2$$

The proportionality constant, k, is related to tube length and fluid viscosity, both of which will be held constant in the example under consideration. From equation (3), the resistances of the wide tube (R_w) and of a single narrow tube (R_n) are

$$R_w = k/A_w^2 \tag{4}$$

and

$$R_n = k/A_n^2 \tag{5}$$

From equation (3) in Fig. 3-7,

$$\frac{1}{R_t} = \frac{1}{R_n} + \frac{1}{R_n} + \frac{1}{R_n} + \frac{1}{R_n} = \frac{4}{R_n} \tag{6}$$

Substituting the value of R_n in equation (5),

$$1/R_t = 4A_n^2/k \tag{7}$$

Rearranging,

$$R_t = k/4A_n^2 \tag{8}$$

From equations 1 and 4,

$$R_t = 4k/A_w^2 = 4R_w \tag{9}$$

Hence the resistance of four such tubes in parallel is four times as great as that of a single tube of equal total cross-sectional area.

If a similar calculation is made for eight such tubes in parallel, with each tube having one-fourth the cross-sectional area of the single wide tube, it will be found that the total resistance will equal $2R_w$. In this circumstance the resistance to flow through eight such narrow tubes in parallel will still be twice as great as that through the single wide tube, despite the fact that the total cross-sectional area for the eight narrow tubes is twice as great as for the single wide tube. This is analogous to the relationship that exists between resistance and area in the circulatory system when comparing the arterioles with the large arteries. Despite the fact that the total cross-sectional area of all the arterioles greatly exceeds that of all the large arteries, the resistance to flow through the arterioles is considerably greater than that through the large arteries.

If the example is carried still further, it will be found that sixteen such narrow tubes in parallel, now with four times the total cross-sectional area of the single wide tube, will exert a resistance to flow just equal to the resistance through the wide tube. Any number of these narrow tubes in excess of sixteen, then, will possess a lower resistance than that of the single wide tube. This is analogous to the situation for the arterioles and capillaries. The resistance to flow through a single capillary is much greater than that through a single arteriole. Yet the number of capillaries so greatly exceeds the number of arterioles, as reflected by the relative difference in total cross-sectional areas, that the pressure drop across the arterioles is usually considerably greater than the pressure drop across the capillaries.

LAMINAR AND TURBULENT FLOW

Under certain conditions the flow of a fluid in a cylindrical tube will be *laminar* (sometimes called *streamlined*), as illustrated in Fig. 3-9. At the entrance of the tube all the fluid elements will have the same linear velocities,

regardless of their radial positions. In progressing along the tube, however, the thin layer of fluid in contact with the wall of the tube remains adherent to the wall and hence is motionless. The layer of fluid just central to this external lamina must shear against this motionless layer and therefore moves slowly, but with a finite velocity. Similarly, the adjacent, more central layer travels still more rapidly. Close to the tube inlet the fluid layers near the axis of the tube still move with the same velocity and do not shear against one another. However, at a distance from the tube inlet equal to several tube diameters, laminar flow becomes *fully developed*; that is, the velocity profiles do not change with longitudinal distance along the tube. In fully developed laminar flow the longitudinal velocity profile is that of a paraboloid. The velocity of the fluid adjacent to the wall is zero, whereas the velocity at the center of the stream is maximal and equal to twice the mean velocity of flow across the entire cross section of the tube. In laminar flow, fluid elements in one lamina, or streamline, remain in that streamline as the fluid progesses longitudinally along the tube.

Irregular motions of the fluid elements may develop in the flow of fluid through a tube;

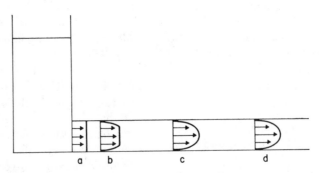

Fig. 3-9. Laminar flow in a cylindrical tube. At the inlet, *a*, the velocities are equal at all radial distances from the center of the tube. Near the inlet, *b*, the velocity profile is flat near the center of the tube, but a velocity gradient is established near the wall. When flow becomes fully developed, *c* and *d*, the velocity profile is parabolic.

such a flow is called *turbulent*. Under such conditions fluid elements do not remain confined to definite streamlines but rapid, radial mixing occurs. A considerably greater pressure difference is required to force a given flow of fluid through the same tube under conditions of turbulence as compared to the pressure difference required for laminar flow. In turbulent flow the pressure drop is approximately proportional to the square of the flow rate, whereas in laminar flow, the pressure drop is proportional to the first power of the flow rate. Hence, to produce a given flow, a pump such as the heart would have to do considerably more work for a given flow if turbulence developed.

Whether turbulent or laminar flow will exist in a tube under given conditions may be predicted on the basis of a dimensionless number called *Reynold's number*, N_R. This number represents the ratio of inertial to viscous forces and equals $\rho D \bar{v}/\eta$, where D is the tube diameter, $\bar{v}$ is the mean velocity, ρ is the density, and η is the viscosity. For $N_R < 2,000$, the flow will usually be laminar; for $N_R > 3,000$, turbulence will usually exist. Various possible conditions may develop in the transition range of N_R between 2,000 and 3,000. Since flow tends to be laminar at low N_R and turbulent at high N_R, it is evident from the definition of N_R that large diameters, high velocities, and low viscosities predispose to the development of turbulence. In addition to these factors, abrupt variations in tube dimensions or irregularities in the tube walls will produce turbulence.

Turbulence is usually accompanied by vibrations in the auditory frequency ranges. When turbulent flow exists within the cardiovascular system, it is usually detected as a *murmur*. The factors listed above that predispose to turbulence may account for murmurs heard clinically. In severe anemia *functional cardiac murmurs* (murmurs not caused by structural abnormalities) are frequently detectable. The physical basis for such murmurs resides in (1) the reduced viscosity of blood caused by the low red cell content and (2) the high flow velocities associated with marked augmentation of cardiac output that usually occurs in anemic patients.

It has recently been found that blood clots, or *thrombi*, are much more likely to develop with turbulent than with laminar flow. One of the problems with the use of artificial valves in the surgical treatment of valvular heart disease is that thrombi tend to occur in association with the prosthetic valve. The thrombi may then be dislodged, often occluding a crucial vessel in the peripheral circulation. It is thus important to design such valves to avert development of turbulent flow through the valve and thereby minimize the danger of thrombosis.

SHEAR STRESS ON THE VESSEL WALL

In Fig. 3-5, an external force was applied to a plate floating on the surface of a volume of liquid contained in a large basin. This force, exerted parallel to the surface, caused a shearing stress on the liquid below, thereby producing a differential motion of each layer of liquid relative to the adjacent layers. At the bottom of the basin, the flowing liquid produced a shearing stress on the surface of the basin in contact with the liquid. By a rearranging the formula for viscosity stated in Fig. 3-5, it is apparent that the shear stress, τ, equals η (du/dy); that is, the shear stress equals the product of the viscosity and the rate of strain. Hence, the greater the rate of flow, the greater the shear stress that the liquid exerts on the walls of the container in which it flows.

For precisely the same reasons, the rapidly flowing blood in a large artery tends to pull the endothelial lining of the artery along with it. This force, or *viscous drag*, is proportional to the rate of strain (du/dy) of the layers of blood near the wall. For Poiseuille flow,

$$\tau = 4\eta Q/\pi r^3$$

The greater the rate of blood flow (Q) in the artery, the greater will be du/dy near the arterial wall, and the greater will be the viscous drag (τ).

In certain types of arterial disease, particularly with hypertension, the subendothelial layers tend to degenerate locally and small regions of the endothelium may lose their normal support. The viscous drag on the arterial wall may cause a tear between a normally supported and unsupported region of the endothelial lining. Blood may then enter the rift in the lining from the vessel lumen and dissect between the various layers of the artery. Such a lesion is called a *dissecting aneurysm*. It occurs most commonly in the proximal portions of the aorta and is extremely serious. One reason for its predilection for this site is the very high velocity of blood flow, with the associated high values of du/dy at the endothelial wall. A major aim of treatment is to depress the heart in order to reduce the velocity of blood flow and hence the viscous drag on the aortic endothelium.

RHEOLOGICAL PROPERTIES OF BLOOD

The viscosity of a Newtonian fluid, such as water, may be determined by measuring the rate of flow of the fluid at a given pressure gradient through a cylindrical tube of known length and radius. As long as the fluid flow is laminar, the viscosity may be computed by substituting these values into the Poiseuille equation. With careful measurement the viscosity of a given Newtonian fluid at a specified temperature will be constant over a wide range of tube dimensions and flow rates. However, for a non-Newtonian fluid, the viscosity calculated by substituting into Poiseuille's equation may vary considerably as a function of tube dimensions and flow rates. Therefore, in considering the rheological properties of a suspension such as blood, the term *viscosity* does not have a unique meaning. The terms *anomalous viscosity* and *apparent viscosity* are frequently applied to the value of viscosity obtained for

blood under the particular conditions of measurement.

Rheologically, blood is a suspension, principally a suspension of erythrocytes in a relatively homogeneous liquid, the blood plasma. For this reason the apparent viscosity of blood varies as a function of the *hematocrit ratio* (ratio of volume of red blood cells to volume of whole blood). In Fig. 3-10 the upper curve represents the ratio of the apparent viscosity of whole blood to that of plasma over a range of hematocrit ratios from 0% to 80%, measured in a tube 1 mm. in diameter and 250 mm. in length. The viscosity of plasma is 1.2 to 1.3 times that of water. From the graph it may be seen that blood, with a normal hematocrit ratio of 45%, has an apparent viscosity 2.4 times that of plasma. With severe anemia, therefore,

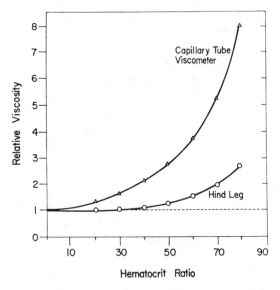

Fig. 3-10. Viscosity of whole blood, relative to that of plasma, increases at a progressively greater rate as hematocrit ratio increases. For any given hematocrit ratio the apparent viscosity of blood is less when measured in a biological viscometer (such as the hind leg of a dog) than in a conventional capillary tube viscometer. (Redrawn from Levy, M. N., and Share, L.: Circ. Res. **1:**247, 1953.)

there is a considerable reduction in blood viscosity. With increasing hematocrit ratios the slope of the curve increases progressively; it is especially steep at the upper range of erythrocyte concentrations. A rise in hematocrit ratio from 45% to 70%, which occurs in *polycythemia*, results in more than a twofold increase in apparent viscosity, with a proportionate effect on the resistance to blood flow. The magnitude of the effect on peripheral resistance may be appreciated when it is recognized that even in the most severe cases of essential hypertension, in which peripheral resistance is augmented by arteriolar constriction rather than by greater blood viscosity, the total peripheral resistance rarely increases by more than a factor of two.

For any given hematocrit ratio the apparent viscosity of blood depends on the dimensions of the tube employed in estimating the viscosity. In 1931 Fåhraeus and Lindqvist demonstrated that the apparent viscosity of blood di-

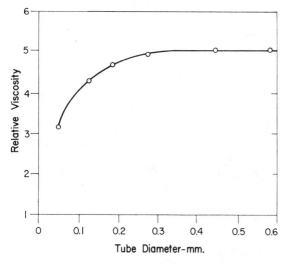

Fig. 3-11. Viscosity of blood, relative to that of water, increases as a function of tube diameter up to a diameter of about 0.3 mm. (Redrawn from Fåhraeus, R., and Lindqvist, T.: Am. J. Physiol. **96:**562, 1931.)

minishes appreciably in glass capillary tubes approaching the radial dimensions of the microscopic blood vessels. Fig. 3-11 illustrates the changes in viscosity of blood relative to that of water as a function of tube diameter. The graph demonstrates that the apparent viscosity of blood diminishes progressively as tube diameter decreases below a value of about 0.3 mm. Since the highest resistance blood vessels, the arterioles, possess diameters considerably less than this critical value, it was pointed out by these investigators that the consequent reduction of apparent viscosity "allows the heart to drive a given volume of blood through the arterioles at a much lower pressure than would be the case if the blood behaved as a [Newtonian] fluid." It was subsequently demonstrated that the apparent viscosity of blood, when measured in living tissues, is considerably less than when measured in a conventional capillary tube viscometer with a diameter greater than 0.3 mm. In the lower curve of Fig. 3-10 apparent relative viscosity of blood was assessed by using the hind leg of a dog as a biological viscometer. It is evident that over the entire range of hematocrit ratios, the apparent viscosity is less as measured in the living tissue than in the capillary tube viscometer (upper curve) and that the disparity becomes greater with higher hematocrit ratios. Undoubtedly, this phenomenon may be ascribed to the so-called Fåhraeus-Lindqvist effect, since the diameter of the high-resistance blood vessels is considerably less than the diameter of the glass-tube viscometer employed in the same study.

The influence of tube diameter on apparent viscosity is ascribable in part to the change in actual composition of the blood as it flows through small tubes. The alteration in composition results from the tendency for the red blood cells to accumulate in the faster axial stream, whereas largely plasma flows in the slower marginal layers. To illustrate this phenomenon, a reservoir such as *A* in Fig. 3-3 could be filled with blood possessing a given

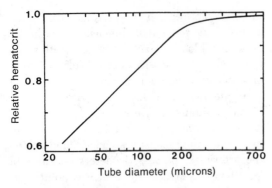

Fig. 3-12. The "relative hematocrit" of blood flowing from a feed reservoir through capillary tubes of various calibers, as a function of the tube diameter. The relative hematocrit is the ratio of the hematocrit of the blood in the tubes to that of the blood in the feed reservoir. (Redrawn from Barbee, J. H., and Cokelet, G. R.: Microvasc. Res. **3:**6, 1971.)

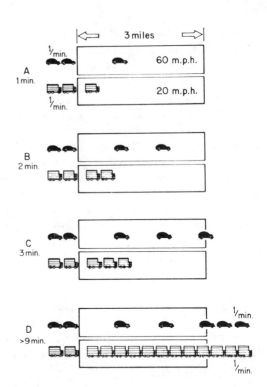

Fig. 3-13. When the car velocity is three times as great as the truck velocity, the ratio of the number of cars to trucks on a bridge will be 1:3, even though one of each type of vehicle enters and leaves the bridge each minute.

hematocrit ratio. If the blood in *A* was constantly agitated to prevent settling and was permitted to flow through a narrow capillary tube into reservoir *B*, the hematocrit ratio of the blood in *B* would not be detectably different from that in *A*. Surprisingly, however, if the capillary tube joining the two reservoirs was suddenly disconnected and the hematocrit ratio of the blood contained within the tube was determined, it would be found to be considerably lower than the hematocrit ratio of the blood in either reservoir. In actual experiments, it has been found that the smaller the tube, the more pronounced the change in hematocrit ratio in the tube. In Fig. 3-12, the relative hematocrit is the ratio of the hematocrit in the tube to that in the reservoir at either end of the tube. For tubes of 500 μ diameter or greater, the relative hematocrit was observed to be close to 1.0. However, with reductions in the tube diameter there was a progressive decrease in the relative hematocrit, such that for a tube diameter of 30 μ, the relative hematocrit was only 0.6.

That this situation results from a disparity in the relative velocities of the red cells and plasma can be appreciated on the basis of the following analogy. Consider the flow of traffic across a bridge that is 3 miles long. Let the cars move in one lane at a speed of 60 miles per hour and the trucks in another lane at 20 miles per hour, as illustrated in Fig. 3-13. If one car and one truck start out across the bridge each minute, then except for the initial few minutes of traffic flow across the bridge, one car and one truck will arrive at the other end each minute. Yet if one counts the actual number of cars and trucks on the bridge at any moment, there will be three times more of the slower moving trucks than of the more rapidly traveling cars.

Since the axial portions of the bloodstream contain a greater proportion of red cells and move with a greater velocity, the red cells tend to traverse the tube in a shorter period of time than the plasma. Therefore, the red cells correspond to the cars in the analogy and the plasma corresponds to the trucks. Measurements of transit times through various organs have shown that red cells do travel with a greater mean velocity than does the plasma. Furthermore, measurements of the hematocrit ratio of the blood contained in various tissues have yielded lower values than have been obtained from blood samples withdrawn from large arteries or veins in the same animal.

The physical forces that are responsible for the drift of the erythrocytes toward the axial stream and away from the vessel walls are not fully understood. One factor is entirely dependent on simple geometrical considerations. Disregarding red cell deformation, it is obvious that it would not be possible for the center of an erythrocyte to be situated closer to the vessel wall than half its thickness. Therefore, even with an otherwise homogeneous distribution of red cells, the thin layer of blood adjacent to the vessel walls would be relatively deficient in erythrocytes. This layer would be of constant dimensions (half the thickness of an erythrocyte); hence the smaller the diameter of the vessel, the greater the relative area occupied by this layer of cell-deficient blood. It is evident, therefore, that the relative packing of erythrocytes in the faster axial stream would be more pronounced in smaller than in larger caliber vessels.

The anomalous rheological behavior of blood also becomes manifest when the flow rate is varied through a capillary tube of given dimensions. During laminar flow the graph relating pressure and flow for a Newtonian fluid is rectilinear and intersects the coordinate system at its origin, as displayed in Fig. 3-14. Thus resistance to flow (which is the reciprocal of the slope of this curve) is constant over the entire

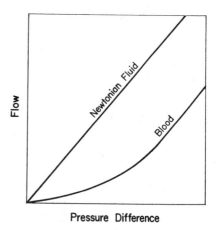

Fig. 3-14. In rigid tubes the pressure-flow curve is rectilinear for Newtonian fluids during laminar flow but is curvilinear for blood. At low flow rates, resistance to flow (apparent viscosity) of blood decreases appreciably as flow increases. At higher flow rates the pressure-flow curve becomes virtually rectilinear, and resistance (apparent viscosity) changes only slightly as a function of flow.

range of pressure and flow during laminar flow. However, for blood, the pressure-flow relationship is curvilinear and convex to the pressure axis. Hence, with increasing flow, the increment of flow for any given increment of pressure difference becomes progressively greater over the curvilinear region of the graph. Consequently, the computed resistance to flow is found to decrease as pressure difference and flow increase. This change in resistance is most pronounced at the lower levels of pressure and flow. At higher flow rates, and over the range that ordinarily prevails in vivo, the curve is virtually rectilinear, and resistance varies only slightly as a function of pressure and flow.

When curvilinear pressure-flow curves for blood such as that depicted in Fig. 3-14 are obtained in rigid tubes, the observed changes in resistance must be ascribable to changes in the viscous characteristics of the blood. Specifically, decreasing resistance at progressively higher flow rates implies that there is a dimi-

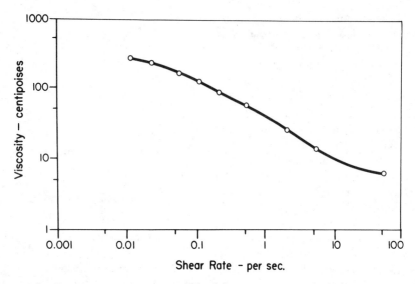

Fig. 3-15. Decrease in apparent viscosity of blood (hematocrit ratio, 51.7%) at increasing rates of shear, both plotted on logarithmic scales. The shear rate refers to the relative velocity of one layer of fluid with respect to that of the adjacent layers, and is directionally related to the rate of flow. (Redrawn from Chien, S.: J. Appl. Physiol. **21**:81, 1966.)

nution in the apparent viscosity of the blood under these conditions; this has been confirmed experimentally many times (for example, Fig. 3-15). The reduction in apparent viscosity with increasing shear rate has been called *shear thinning*. This non-Newtonian behavior has been explained in part on the basis of a greater tendency of the erythrocytes to accumulate in the axial laminae at higher flow rates. However, a more important factor is that at very slow rates of shear there is a distinct tendency for the suspended cells to form aggregates, which would increase viscosity. This tendency toward aggregation becomes progressively less as the rate of flow is augmented, producing the diminution in apparent viscosity that is exhibited in Fig. 3-15. The tendency toward aggregation at low flows is dependent on the concentration in the plasma of the larger protein molecules, such as the globulins and particularly fibrinogen. For this reason, the changes in blood viscosity with shear rate are much more pronounced when the plasma contains a high concentration of fibrinogen than when the content is low. In addition, there is a tendency, at low flow rates, for leukocytes to adhere to the endothelial walls of the microvasculature, thereby increasing the apparent viscosity.

The deformability of the erythrocytes is also a factor in shear thinning, especially at high hematocrit ratios. The mean diameter of human red blood cells is about 8.5 μ, yet they are able to pass through openings with a diameter of only 3.0 μ. As blood, densely packed with erythrocytes, is caused to flow at progressively greater rates, the erythrocytes become more and more deformed. The deformation is such that the apparent viscosity of the blood diminishes. If the red blood cells become hardened, as they are in certain spherocytic anemias, shear thinning may become much less prominent or the apparent viscosity may actually increase slightly at greater rates of shear.

DISTENSIBILITY OF THE HIGH-RESISTANCE BLOOD VESSELS

Under certain conditions the pressure-flow curve for some tissues has a curvilinear configuration similiar to that shown for blood in Fig. 3-14. Thus the computed resistance to flow diminishes as pressure difference and flow increase. Because of the similarity of such curves for the flow of blood through tissues and the flow of non-Newtonian fluids through rigid capillary tubes, it has been suspected that the change in resistance as a function of pressure and flow is related to the anomalous viscous nature of blood. Undoubtedly, this factor is contributory, but apparently it is not the predominant mechanism. The elasticity of the high-resistance vessels (presumably the arterioles) appears to be of far greater importance.

According to Poiseuille's law, flow is proportional to pressure *difference*. For the limited conditions under which Poiseuille's law applies, the flow will be the same for an inflow pressure (P_i) of 100 units and an outflow pressure (P_o) of 0 as it is for a P_i of 200 and a P_o of 100, since the pressure difference is 100 units in both cases. This principle is illustrated in section 3 of Fig. 3-3. However, in the case of distensible tubes or vessels, even at identical pressure differences (P_i-P_o), flow will be greater (hence, resistance less) at higher than at lower absolute levels of P_i and P_o. At the higher absolute pressures the *transmural pressures* (difference between the intraluminal and extravascular pressures) will be greater than at the lower pressures. Thus the vessels will be distended more and resistance will be reduced accordingly. In experiments on perfused vascular beds, it was found that flow did increase significantly as the transmural pressure was raised, even though the pressure difference was held constant. The results were not appreciably different when homogeneous (Newtonian) perfusates were substituted for blood.

BIBLIOGRAPHY
Journal articles

Badeer, H. S., and Rietz, R. R.: Vascular hemodynamics: deep-rooted misconceptions and misnomers, Cardiology **64**:197, 1979.

Barbee, J. H., and Cokelet, G. R.: The Fåhraeus effect, Microvasc. Res. **3**:6, 1971.

Carroll, R. J., and Falsetti, H. L.: Retrograde coronary artery flow in aortic valve disease, Circulation **54**:494, 1976.

Fåhraeus, R., and Lindqvist, T.: The viscosity of blood in narrow capillary tubes, Am. J. Physiol. **96**:562, 1931.

Green, H. D., Lewis, R. N., Nickerson, N. D., and Heller, A. L.: Blood flow, peripheral resistance, and vascular tonus, with observations on the relationship between blood flow and cutaneous temperature, Am. J. Physiol. **141**:518, 1944.

Levy, M. N., and Share, L.: The influence of erythrocyte concentration upon the pressure-flow relationships in the dog's hind limb, Circ. Res. **1**:247, 1953.

Lingard, P. S.: Capillary pore rheology of erythrocytes. IV. Effect of pore diameter and haematocrit, Microvasc. Res. **13**:59, 1977

Lipowsky, H. H., Kovalcheck, S., and Zweifach, B. W.: The distribution of blood rheological parameters in the microvasculature of cat mesentery, Circ. Res. **43**:738, 1978.

Munter, W. A., and Stein, P. D.: Turbulent blood flow and the effects of erythrocytes, Cardiovasc. Res. **8**:338, 1974.

Prokop, E. K., Palmer, R. F., and Wheat, M. W., Jr.: Hydrodynamic forces in dissecting aneurysms, Circ. Res. **27**:121, 1970.

Rodbard, S., Compton, P., and Reimann, M.: Factors affecting the transvalvar pressure difference in a hydraulic model of aortic stenosis, Cardiology **61**:232, 1976.

Schmid-Schönbein, G. W., Skalak, R., Usami, S., and Chien, S.: Cell distribution in capillary networks, Microvasc. Res. **19**:18, 1980.

Schmid-Schönbein, H., and Wells, R. E., Jr.: Rheological properties of human erythrocytes and their influence upon the "anomalous" viscosity of blood, Ergeb. Physiol. **63**:146, 1971.

Talbot, L., and Berger, S. A.: Fluid-mechanical aspects of the human circulation, Am. Scientist **62**:671, 1974.

Varco, R. L., moderator: Mechanical surface and gas layer effects on moving blood (symposium), Fed. Proc. **30**:1477-1712, 1971.

Yen, R. T., and Fung, Y. C.: Inversion of Fåhraeus effect and effect of mainstream flow on capillary hematocrit, J. Appl. Physiol. **42**:578, 1977.

Zamir, M.: The role of shear forces in arterial branching, J. Gen. Physiol. **67**:213, 1976.

Books and monographs

Caro, C. G., Pedley, T. J., and Seed, W. A.: Mechanics of the circulation. In Guyton, A. C., editor: MTP International Review of Science; Physiology Series One, London, 1974, Butterworths, vol. I, pp. 1-48.

Charm, S. E., and Kurland, G. S.: Blood flow and microcirculation, New York, 1974, John Wiley & Sons, Inc.

Cokelet, G. R., Meiselman, H. J., and Brooks, D. E., editors: Erythrocyte mechanics and blood flow, New York, 1980, Alan R. Liss, Inc.

Dintenfass, L.: Rheology of blood in diagnostic and preventive medicine, London, 1976, Butterworths.

Hwang, N. H. C., and Normann, N. A.: Cardiovascular flow dynamics and measurements, Baltimore, 1977, University Park Press.

Noordergraaf, A.: Circulatory system dynamics, New York, 1979, Academic Press, Inc.

CHAPTER 4

THE CARDIAC PUMP

It is nearly impossible to contemplate the pumping action of the heart without being struck by its simplicity of design, its wide range of activity and functional capacity, and the staggering amount of work it performs relentlessly over the lifetime of an individual. To understand how the heart accomplishes its important task, it is first necessary to consider the relationships between the structure and function of its various components.

STRUCTURE OF THE HEART IN RELATION TO FUNCTION
Myocardial cell

A number of important morphological and functional differences exist between myocardial and skeletal muscle cells. However, the contractile elements within the two types of cells are quite similar; each skeletal and cardiac muscle cell is made up of sarcomeres (from Z line to Z line) containing thick filaments composed of myosin (in the A band) and thin filaments containing actin. The thin filaments extend from the Z line (through the I band) to interdigitate with the thick filaments. The sliding filament hypothesis appears to explain contraction satisfactorily in both. Furthermore, skeletal and cardiac muscle show similar length-tension relationships. The sarcomere length has been determined with electron microscopy in papillary muscles and intact ventricles rapidly fixed during systole or diastole. Maximal developed tension is observed at resting sarcomere lengths of 2.0 to 2.3 μ for car-

diac muscle. At such lengths, there is overlap of thick and thin filaments, and a maximal number of crossbridge attachments. Developed tension of cardiac muscle is less than the maximum value when the sarcomeres are stretched beyond the optimum length, probably because of less overlap of the filaments, and hence less interaction between crossbridges on the thick and thin filaments. At resting sarcomere lengths shorter than the optimum values, the thin filaments are either compressed or they overlap each other; this may interfere in some manner with the development of contractile force. At short sarcomere lengths there may be failure of excitation to release Ca^{++} from the sarcoplasmic reticulum, as is the case in skeletal muscle.

In general, the fiber length-tension relationship for the papillary muscle also holds true for fibers in the intact heart. This relationship may be expressed graphically, as in Fig. 4-1, by substituting ventricular systolic pressure for tension and end-diastolic ventricular volume for myocardial resting fiber (and hence sarcomere) length. The lower curve in Fig. 4-1 represents the increment in pressure produced by each increment in volume when the heart is in diastole. The upper curve represents the peak pressure developed by the ventricle during systole at each degree of filling and illustrates the *Frank-Starling relationship* of initial myocardial fiber length (or initial volume) to tension (or pressure) development by the ventricle. Note that the pressure-volume curve in

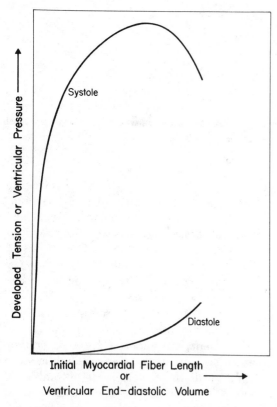

Fig. 4-1. Relationship of myocardial resting fiber length (sarcomere length) or end-diastolic volume to developed tension or peak systolic ventricular pressure during ventricular contraction in the intact dog heart. (Redrawn from Patterson, S. W., Piper, H., and Starling, E. H.: J. Physiol. **48:**465, 1914.)

diastole is initially quite flat, indicating that large increases in volume can be accommodated with only small increases in pressure, yet systolic pressure development is considerable at the lower filling pressures. However, the ventricle becomes much less distensible with greater filling, as evidenced by the sharp rise of the diastolic curve at large intraventricular volumes. In the normal intact heart, peak tension may be attained at a filling pressure of 12 mm. Hg. At this intraventricular diastolic pressure, which is about the upper limit observed in the normal heart, the sarcomere length is 2.2 μ. However, developed tension has been shown to peak at filling pressures as high as 30 mm. Hg in the isolated heart. Even at higher diastolic pressures (>50 mm. Hg) the sarcomere length is not greater than 2.6 μ in cardiac muscle, whereas in many skeletal muscles sarcomere lengths as great as 3.65 μ can be obtained with stretch. This resistance to stretch of the myocardium at high filling pressures probably resides in the noncontractile constituents of the tissue and may serve as a safety factor against overloading of the heart in diastole. Usually, ventricular diastolic pressure is about 0 to 7 mm. Hg and the average diastolic sarcomere length about 2.0 μ. Thus the normal heart operates on the ascending portion of the Frank-Starling curve depicted in Fig. 4-1.

Fig. 4-2. Low-power electron micrograph of right ventricular wall of mouse heart. Tissue fixed in a phosphate-buffered glutaraldehyde solution and postfixed in ferrocyanide-reduced osmium tetroxide. This procedure results in the deposition of electron-opaque precipitate in the extracellular space, thus outlining the sarcolemmal borders (*SL*) of the muscle cells and tracing out the intercalated discs (*ID*) and transverse tubules (*TT*). Mitochondria (*Mit*) are visible between the myofibrils. Several blood vessels of various sizes (*BV*) are present in this field. *Nu*, nucleus of myocardial cell. ×6000. (Electron micrograph courtesy Dr. Michael S. Forbes.)

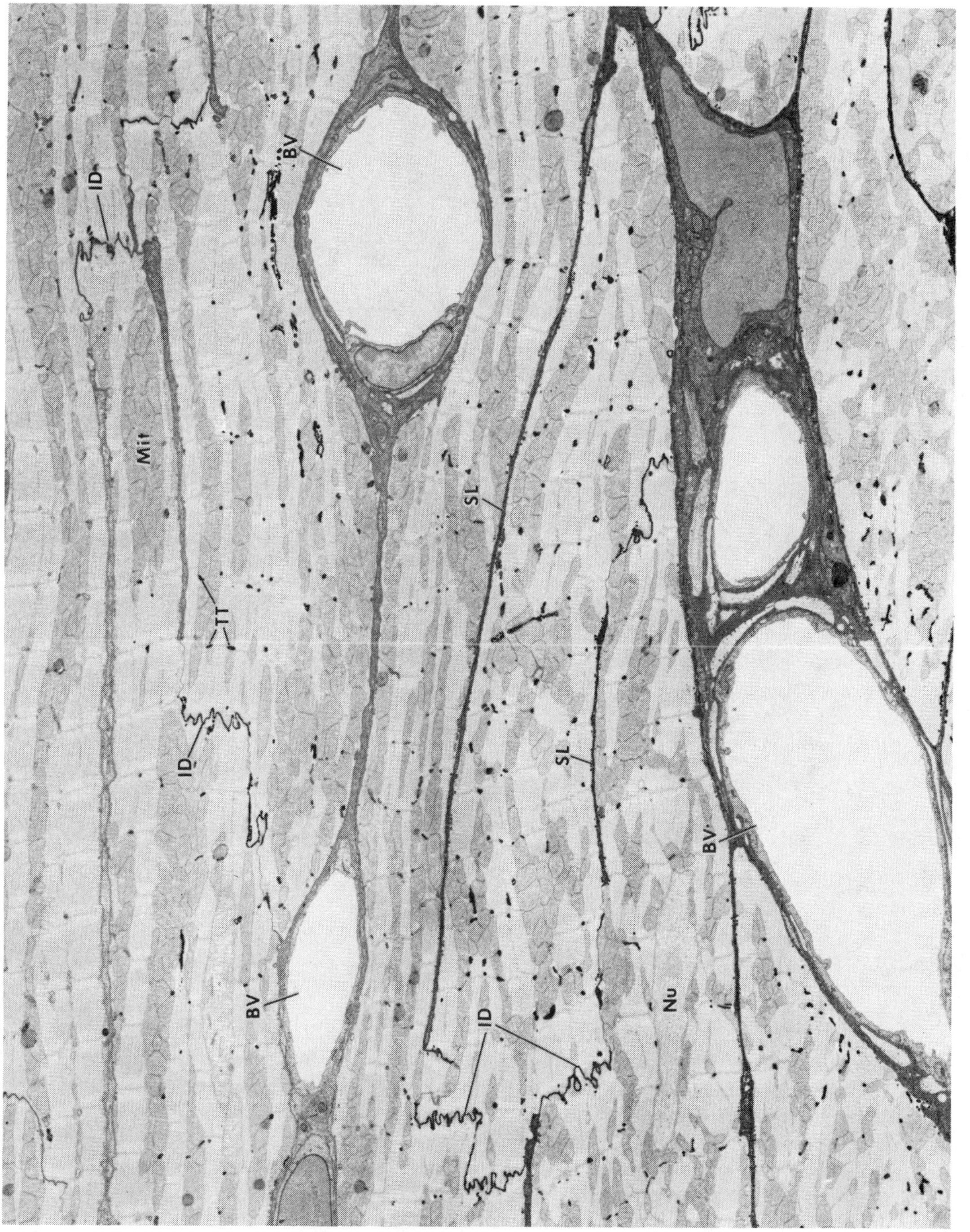

Fig. 4-2. For legend see opposite page.

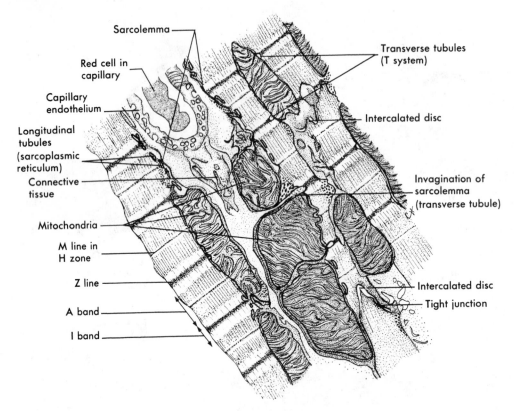

Sarcolemma

Red cell in capillary

Capillary endothelium

Longitudinal tubules (sarcoplasmic reticulum)

Connective tissue

Mitochondria

M line in H zone

Z line

A band

I band

Transverse tubules (T system)

Intercalated disc

Invagination of sarcolemma (transverse tubule)

Intercalated disc

Tight junction

Fig. 4-3. Diagram of an electron micrograph of cardiac muscle showing large numbers of mitochondria, the intercalated discs with tight junctions (nexi), the transverse tubules, and the longitudinal tubules. (Approximately ×30,000.)

Fig. 4-4. Electron micrograph of cardiac muscle in longitudinal section. Tissue treated as described in Fig. 4-2. Surface structures shown include the sarcolemma (*SL*), intercalated disc (*ID*) with its associated gap junctions (*GJ*), and sarcolemmal invaginations which may be oriented either transverse to the cell axis (transverse tubules, *TT*) or parallel to it (axial tubules, *AxT*). the sarcoplasmic reticulum is seen in the form of tubules (*SR*) or saccules which form couplings (*C*) with the sarcolemma or its invaginations. Mitochondria (*Mit*) are large and elongated, lying between the myofibrils. Z, Z line of sarcomere; *A*, A-band; *I*, I-band; *M*, M line in H zone of sarcomere. ×30,000. (Electron micrograph courtesy Dr. Michael S. Forbes.)

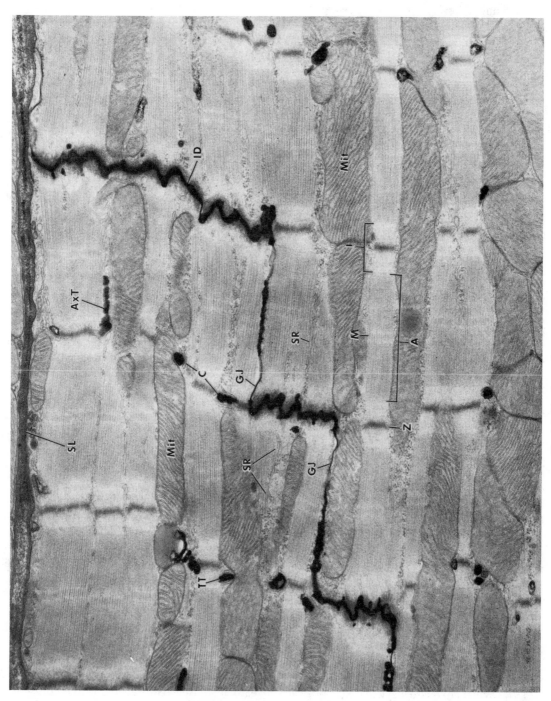

Fig. 4-4. For legend see opposite page.

A striking difference in the appearance of cardiac and skeletal muscle is the semblance of a syncytium in cardiac muscle with branching interconnecting fibers. However, the myocardium is not a true anatomical syncytium, since, laterally, the myocardial fibers are separated from adjacent fibers by their respective sarcolemmas, and the end of each fiber is separated from its neighbor by dense structures, *intercalated discs*, that are continuous with the sarcolemma (Fig. 4-2). Nevertheless, cardiac muscle functions as a syncytium, since a wave of depolarization followed by contraction of the entire myocardium (an all-or-none response) occurs when a suprathreshold stimulus is applied to any one focus. Graded contraction, as seen in skeletal muscle by activation of different numbers of fibers, does not occur in heart muscle. Whether the intercalated discs represent a high-resistance or a low-resistance conduction pathway is controversial. The *tight junctions*, where there appears to be fusion of the intercalated discs (Fig. 4-3), are thought by most investigators to represent low-resistance pathways between myocardial cells.

Another difference between cardiac and fast skeletal muscle fibers is in the abundance of mitochondria *(sarcosomes)* in the two tissues. Fast skeletal muscle, which is called on for relatively short periods of repetitive or sustained contraction and which can metabolize anaerobically and build up a substantial oxygen debt, has relatively few mitochondria in the muscle fibers. In contrast, cardiac muscle, which is required to contract repetitively for a lifetime and which is incapable of developing a significant oxygen debt, is very rich in sarcosomes (Figs. 4-3 and 4-4). Rapid oxidation of substrates with the synthesis of ATP can keep pace with the myocardial energy requirements by virtue of the large numbers of sarcosomes containing the respiratory enzymes necessary for oxidative phosphorylation.

To provide adequate oxygen and substrate for its metabolic machinery, the myocardium is also endowed with a rich capillary supply, about one capillary per fiber. Thus diffusion distances are short, and oxygen, carbon dioxide, substrates, and waste material can move rapidly between myocardial cell and capillary. With respect to exchange of substances between the capillary blood and the myocardial cells, electron micrographs of myocardium show deep invaginations of the sarcolemma into the fiber at the Z lines (Figs. 4-3 and 4-4). These sarcolemmal invaginations constitute the transverse or T tubular system. The lumina of these T tubules are continuous with the bulk interstitial fluid and they play a key role in excitation-contraction coupling. They are thought to provide a pathway for the rapid transmission of the electrical signal from the surface sarcolemma to the inside of the fiber, thus enabling nearly simultaneous activation of all myofibrils, including those deep within the interior of the fiber. In mammalian ventricular cells, adjacent transverse tubules are interconnected by longitudinally running or axial tubules, thus forming an extensively interconnected lattice of "intracellular" tubules. This T system is open to the interstitial fluid, is lined with a basement membrane continuous with that of the surface sarcolemma, and contains micropinocytotic-like vesicles (Fig. 4-5). A network of sarcoplasmic reticulum consisting of small diameter sarcotubules is also present surrounding the myofibrils; these sarcotubules are believed to be "closed," since colloidal tracer particles (20 to 100 Å in diameter) do not enter them. They do not contain basement membrane. Flattened elements of the sarcoplasmic reticulum are often found in close proximity to the T system as well as to the surface sarcolemma, forming so-called "diads" (Fig. 4-5). The sarcoplasmic reticulum sequesters calcium (during diastole) and releases some of the calcium involved in activation of the contractile proteins in myocardial contraction. The other source of calcium in cardiac muscle contraction is that present in the extracellular fluid. Thus in ventricular cells

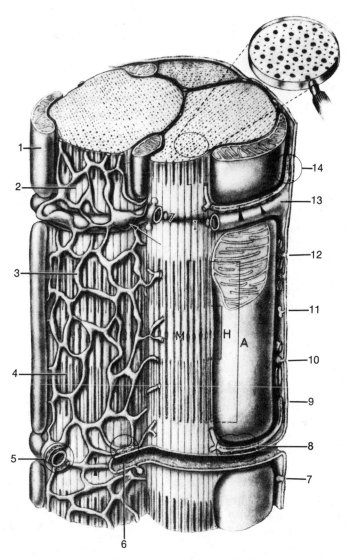

Fig. 4-5. Model of mammalian cardiac muscle showing a portion of a muscle fiber: *1*, mitochondrion; *2*, actin (thin) filament; *3*, sarcoplasmic reticulum; *4*, myosin (thick) filament; *5*, triad, made up of two couplings at a transverse tubule; *6*, diad, made up of one coupling at a transverse tubule. The couplings in *5*, *6*, and *8* are interior couplings. *7*, Pit (or coated vesicle) also commonly seen at transverse tubules; *8*, interior coupling; *9*, sarcolemma; *10*, pinocytotic vesicle or caveola; *11*, branched caveola; *12*, caveola containing dense granules; *13*, transverse tubule; *14*, peripheral coupling. Large arrowhead: junctional granules (or central membrane), only within junctional sarcoplasmic reticulum. Small arrowhead: junctional processes. Arrow: junctional sarcoplasmic reticulum with connections of the sarcoplasmic reticulum across the Z line. Also shown are the M line, A band, H band, and I band. Note the narrow junction of the sarcoplasmic reticulum of the couplings at *5*, *6*, *8*, *14*. (Reproduced by permission from Z. Zellforsch. **98:**437, 1969.)

the myofibrils and mitochondria have close access to a space that is continuous with the interstitial fluid. The T system is absent or poorly developed in atrial cells of many mammalian hearts, but these atrial cells do contain a small sarcotubular system.

Myocardial contractile machinery and contractility

Studies on the mechanisms of muscle contraction have, for the most part, been carried out on skeletal muscle. However, in recent years a great deal of attention has been focused on cardiac muscle in an attempt to determine to what extent the information gained about skeletal muscle applies to the myocardium. Despite the fact that none of the existing models for cardiac muscle contraction are compatible with all the data on muscle fiber length, force, and velocity, the model originally proposed for skeletal muscle by A. V. Hill has been found of some use in the consideration of cardiac muscle. It consists of a *contractile element,* a *series elastic element* (an elastic component in series with the contractile element), and a *parallel elastic element* (an elastic component in parallel with the contractile and series elastic elements). The elastic elements are defined for the present only in functional terms, since their anatomical counterparts have not been established.

The true meaning of the *active state* is not fully understood. In general it refers to the ability of the contractile system to bear a load after activation of the muscle. Its magnitude and duration depend on the concentration and persistence of free Ca^{++} in the myoplasm as well as that bound to active sites. However, it cannot be measured experimentally, and hence its value as a means of providing a better understanding of the contractile process is questionable. Velocity and force of contraction are a function of the intracellular concentration of free calcium ions as well as the mechanical load on the contractile system. Force and ve-

locity are inversely related, so that with no load, force is negligible and velocity is maximal, whereas with an isometric contraction, where no external shortening occurs, force is maximal and velocity is zero.

The sequence of events in an afterloaded isotonic contraction of a papillary muscle is illustrated in Fig. 4-6. Point A represents the resting state in which the preload is responsible for the existing initial stretch. With stimulation the contractile element begins to shorten, and at point B the series elastic element has been stretched but the load has not yet been lifted (the isometric phase of the contraction). This stretch of the series elastic element (an expression of the muscle extensibility) consumes a certain amount of energy. Therefore, the energy used for shortening of the muscle is actually less than the total energy expenditure in a single contraction. Stretch of the series elastic element is represented in the diagram at the lower right as a progressive rise in force with no external shortening. At point C the force developed by the contractile element has equaled the load and the load has been raised without further stretch of the series elastic element. This is represented in the diagram on the right as external shortening of the muscle without a further increase in force.

The initial slope (dashed tangent) of the shortening curve (Fig. 4-6, upper right section) depicts the initial rate of shortening (change in length with change in time—dl/dt). Since the initial velocity is dependent on the magnitude of the afterload, a series of initial velocities can be obtained from a papillary muscle by varying the afterload. This is portrayed in Fig. 4-7 where the slope of the initial velocity curve and the degree of shortening can be seen to decrease with increasing loads. Furthermore, the onset of shortening is delayed (longer isometric phase), but the time from stimulation to maximal shortening is unchanged.

If the initial velocity of shortening is plotted against the afterload, the force-velocity curves

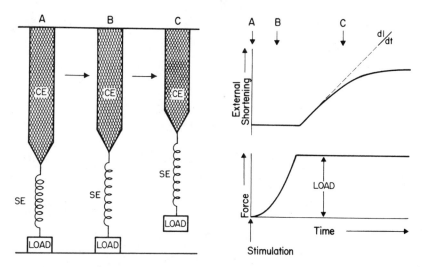

Fig. 4-6. Model for an afterloaded isotonic contraction of a papillary muscle. *A,* At rest; *B,* partial contraction of the contractile element *(CE)* with stretch of the series elastic element *(SE)* but without external shortening (the isometric phase of the contraction); *C,* further contraction of the *CE* with external shortening and lifting of the afterload. The tangent *(dl/dt)* to the initial slope of the shortening curve on the right is the velocity of initial shortening. (Redrawn from Sonnenblick, E. H.: The myocardial cell, Philadelphia, 1966, University of Pennsylvania Press.)

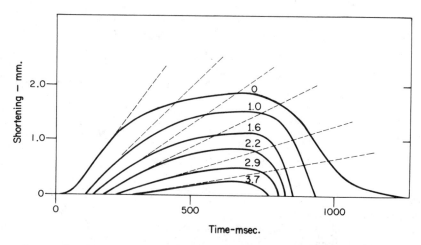

Fig. 4-7. Series of superimposed afterloaded contractions of a papillary muscle at a constant initial length. The numbers refer to the magnitude of the afterload in grams and the dashed lines represent the initial velocity of shortening. (Redrawn from Sonnenblick, E. H.: The myocardial cell, Philadelphia, 1966, University of Pennsylvania Press.)

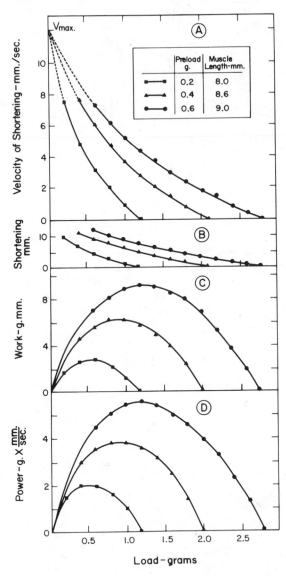

Fig. 4-8. The effect of increasing initial length of a cat papillary muscle on the force-velocity relationship, degree of shortening, muscle work, and muscle power. (Redrawn and reproduced by permission from Am. J. Physiol. **202**:931, 1962.)

shown in Fig. 4-8, *A*, are obtained. The maximal velocity (*Vmax*) may be estimated by extrapolation of the force-velocity curve back to zero load (as indicated by the dotted lines in Fig. 4-8, *A*) and represents the maximum rate of cycling of the crossbridges.

Contractility can be defined as a change in developed tension at a given resting fiber length (p. 162). However, it may also be defined in terms of a change in Vmax. Augmentation of contractility is observed with certain drugs, such as norepinephrine or digitalis, and with an increase in contraction frequency (*tachycardia*, when applied to the whole heart). The increase in contractility (*positive inotropic effect*) produced by any of the preceding interventions is reflected by increments in developed tension and Vmax.

An increase in initial fiber length produces a more forceful contraction, as shown in Fig. 4-1. However, this greater tension development is not associated with any change in contractility, as estimated by Vmax (Fig. 4-8, *A*). Fig. 4-8 also illustrates that at any given load, the degree of shortening, the work (shortening × load or force × distance), and the power (velocity × load or work/time) all increase with the initial length of the papillary muscle. It is apparent that with an increase in initial fiber length, greater force may be developed, but the estimated Vmax is the same for all three initial lengths. Hence changes in resting length may alter tension development but not contractility. This conclusion is, of course, based on the assumption that the displayed extrapolation of the force-velocity curves back to the vertical axis provide the true value for Vmax. This assumption has recently been challenged; some investigators have failed to obtain a hyperbolic force-velocity relationship and have observed changes in Vmax with changes in initial length (that is, Vmax is *length-dependent*). For this and several other reasons, estimates of Vmax may not be a reliable index of contractility.

Although the experiments from which the length-tension and force-velocity relationships have been derived were carried out on papillary muscles (essentially one-dimensional), the findings are to some extent applicable to the intact heart (three-dimensional); for the left ventricle the preload is the ventricular pressure at the end of diastole (end-diastolic pressure) and the afterload is the aortic pressure. However, the complex changes in ventricular shape that occur in systole and the fact that fibers branch at different angles and therefore do not all contract in their optimal longitudinal axis (Fig. 4-11) make accurate determinations of contractility in the whole heart very difficult.

A reasonable index of myocardial contractility can be obtained from the contour of ventricular pressure curves (Fig. 4-9). A hypodynamic

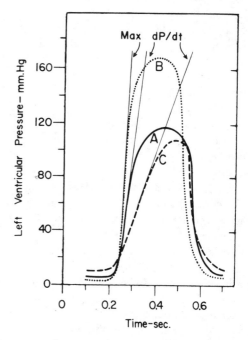

Fig. 4-9. Left ventricular pressure curves with tangents drawn to the steepest portions of the ascending limbs to indicate maximum dP/dt values. *A*, Control; *B*, hyperdynamic heart, as with norepinephrine administration; *C*, hypodynamic heart, as in cardiac failure.

heart is characterized by an elevated end-diastolic pressure, a slowly rising ventricular pressure, and a somewhat reduced ejection phase (curve C, Fig. 4-9), whereas a normal ventricle under adrenergic stimulation shows a reduced end-diastolic pressure, a fast rising ventricular pressure, and a brief ejection phase (curve B, Fig. 4-9). The slope of the ascending limb of the ventricular pressure curve indicates the maximum rate of force development by the ventricle (maximum rate of change in pressure with time; maximum dP/dt, as illustrated by the tangents to the steepest portion of the ascending limbs of the ventricular pressure curves in Fig. 4-9). The slope is maximal during the isovolumetric phase of systole (p. 89) and, at any given degree of ventricular filling, provides an index of the initial contraction velocity and hence of contractility. Similarly, one can obtain an indication of the contractile state of the myocardium from the initial velocity of blood flow in the ascending aorta (the initial slope of the aortic flow curve, Fig. 4-14). The *ejection fraction*, which is the ratio of the volume of blood ejected from the left ventricle per beat (stroke volume) to the volume of blood in the left ventricle at the end of diastole, is widely used clinically as an index of contractility. Other measurements or combinations of measurements that in general are concerned with the magnitude or velocity of the ventricular contraction have also been used to assess the contractile state of the cardiac muscle. There is no index that is entirely satisfactory at present, and this undoubtedly accounts for the large number of indices that are currently in use.

Excitation-contraction coupling. In addition to influences by neural, neurohumoral (for example, norepinephrine), and exogenous agents (for example, cardiac glycosides), myocardial contractility is also affected by the concentration of cations in the plasma. The earliest studies on isolated hearts perfused with isotonic salt solutions indicated the need for optimal concentrations of sodium, potassium, and calcium. In the absence of sodium, the heart will not beat because the action potential is dependent on extracellular sodium ions. In contrast, the resting membrane potential is independent of the sodium ion gradient across the membrane (Fig. 2-6). Under normal conditions the extracellular potassium concentration is low (4 mM.) and further reduction in extracellular potassium has little effect on myocardial excitation and contraction. However, increases in extracellular potassium, if great enough, produce depolarization and loss of excitability of the myocardial cells and cardiac arrest in diastole. In addition to effects on membrane excitability, calcium has a direct effect on myocardial contractility. Increases in extracellular calcium concentrations enhance myocardial contractility and, at unphysiologically high levels, produce cardiac arrest in systole (rigor), whereas decreases in extracellular calcium concentrations reduce contractility. That free intracellular calcium is the agent that initiates and sustains the contractile state of the myocardium is well documented. As indicated in Chapter 2, the electrical excitation alters the membrane permeability to Ca^{++}, thereby allowing Ca^{++} to enter the cell from the interstitial fluid (especially the T tubules) and also releasing Ca^{++} from its storage sites in the sarcoplasmic reticulum within the cell. This calcium binds to troponin and brings about some conformational change in the thin filament, which permits interaction between the actin and myosin filaments through the crossbridges (myosin heads). ATP is dephosphorylated to ADP by the ATPase present in the crossbridges and provides the energy for movement of the crossbridges in a ratchet-like fashion. This action increases the overlap of the actin and myosin filaments, resulting in shortening of the sarcomeres and contraction of the myocardium. Relaxation occurs when free calcium is taken up by the sarcoplasmic reticulum or is bound to the cell membrane and extruded from the cell. This

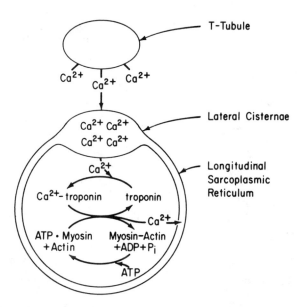

Fig. 4-10. Ca^{++} from the interstitial fluid (primarily from T tubules that penetrate the cell as invaginations of the sarcolemma but do not directly communicate with the intracellular compartment) enters the myoplasm of the cell. Ca^{++} is also released from the sarcoplasmic reticulum. The Ca^{++} reaches the affinity sites of troponin, which enables the actin and myosin filaments to interact and bring about contraction. Relaxation is initiated by the removal of Ca^{++} from the myoplasm by the sarcoplasmic reticulum. (Reproduced by permission from N. Engl. J. Med. **290**:445, 1974.)

oversimplification of excitation-contraction coupling is diagrammed in Fig. 4-10.

Cardiac chambers

The atria are thin-walled, low-pressure chambers that function more as large reservoir conduits of blood for their respective ventricles than as important pumps for the forward propulsion of blood. The ventricles were once thought to be made up of bands of muscle. However, it now appears that they are formed by a continuum of muscle fibers that take origin from the fibrous skeleton at the base of the heart (chiefly around the aortic orifice). These fibers sweep toward the apex at the epicardial surface and also pass toward the endocardium as they gradually undergo a 180-degree change

in direction to lie parallel to the epicardial fibers and form the endocardium and papillary muscles (Fig. 4-11). At the apex of the heart the fibers twist and turn inward to form papillary muscles, whereas at the base and around the valve orifices they form a thick powerful muscle that not only decreases ventricular circumference for ejection of blood but also narrows the A-V valve orifices as an aid to valve closure. In addition to a reduction in circumference, ventricular ejection is accomplished by a decrease in the longitudinal axis with descent of the base of the heart. The earlier contraction of the apical part of the ventricles coupled with approximation of the ventricular walls propels the blood toward the outflow tracts. The right ventricle, which develops a

ENDOCARDIUM

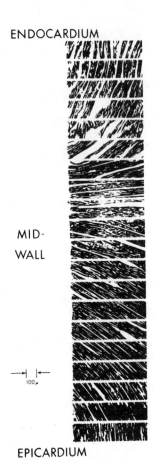

MID-
WALL

100μ

EPICARDIUM

Fig. 4-11. Sequence of photomicrographs showing fiber angles in successive sections taken from the middle of the free wall of the left ventricle from a heart in systole. The sections are parallel to the epicardial plane. Fiber angle is +90 degrees at the endocardium, running through 0 degrees at the midwall to −90 degrees at the epicardium. (From Streeter, D. D., Jr., Spotnitz, H. M., Patel, D. P., Ross, J., Jr., and Sonnenblick, E. H.: Circ. Res. **24:**339, 1969. By permission of The American Heart Association, Inc.)

mean pressure about one-seventh that developed by the left ventricle, is considerably thinner than the left.

Cardiac valves

The cardiac valves consist of thin flaps of flexible, tough endothelium-covered fibrous tissue firmly attached at the base to the fibrous valve rings. Movements of the valve leaflets are essentially passive, and the orientation of the cardiac valves is responsible for unidirectional flow of blood through the heart. There are two types of valves in the heart—the *atrioventricular valves*, or A-V valves, and the *semilunar valves* (Figs. 4-12 and 4-13).

Atrioventricular valves. The valve between the right atrium and right ventricle is made up of three cusps *(tricuspid valve)*, whereas that between the left atrium and left ventricle has two cusps *(mitral valve)*. The total area of the cusps of each A-V valve is approximately twice that of the respective A-V orifice so that there is considerable overlap of the leaflets in the closed position (Figs. 4-12 and 4-13). Attached to the free edges of these valves are fine, strong filaments *(chordae tendineae)*, which arise from the powerful papillary muscles of the respective ventricles and serve to prevent eversion of the valves during ventricular systole.

The mechanism of closure of the A-V valves has been the subject of considerable investigation, and a number of factors are thought to play a role in approximating the valve leaflets. In the normal heart the valve leaflets are relatively close during ventricular filling and provide a funnel for the transfer of blood from atrium to ventricle. This partial approximation of the valve surfaces during diastole is believed to be caused by eddy currents behind the leaflets and possibly also by some tension on the free edges of the valves, exerted by the chordae tendineae and papillary muscles that are stretched by the filling ventricle. The finding of muscle fibers of atrial origin in the mitral valve has suggested that contraction of these fi-

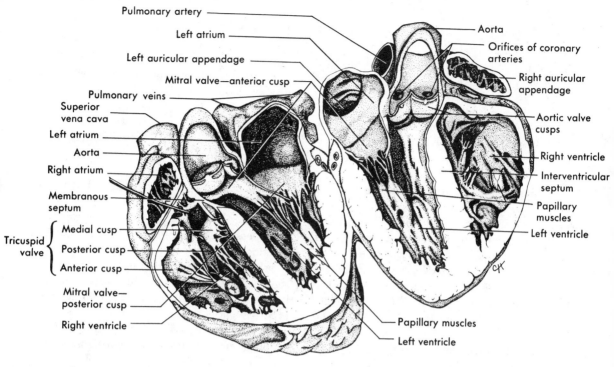

Pulmonary artery
Left atrium
Left auricular appendage
Mitral valve—anterior cusp
Pulmonary veins
Superior vena cava
Left atrium
Aorta
Right atrium
Membranous septum
Tricuspid valve { Medial cusp
Posterior cusp
Anterior cusp
Mitral valve—posterior cusp
Right ventricle

Aorta
Orifices of coronary arteries
Right auricular appendage
Aortic valve cusps
Right ventricle
Interventricular septum
Papillary muscles
Left ventricle
Papillary muscles
Left ventricle

Fig. 4-12. Drawing of a heart split perpendicular to the interventricular septum to illustrate the anatomical relationships of the leaflets of the A-V and aortic valves.

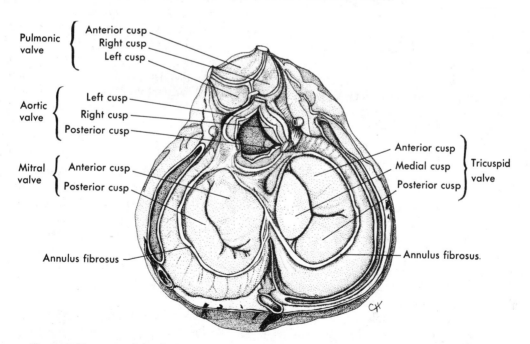

Pulmonic valve { Anterior cusp
Right cusp
Left cusp

Aortic valve { Left cusp
Right cusp
Posterior cusp

Mitral valve { Anterior cusp
Posterior cusp

Anterior cusp
Medial cusp } Tricuspid valve
Posterior cusp

Annulus fibrosus
Annulus fibrosus.

Fig. 4-13. Drawing of the four cardiac valves as viewed from the base of the heart. Note the manner in which the leaflets overlap in the closed valves.

bers during atrial systole may aid in this positioning of the valve leaflets. With atrial systole the abrupt ejection of additional blood into the distensible ventricles that are already filled with blood, coupled with the subsequent relaxation of the atria, results in a small reversal of the A-V pressure gradient sufficient to close the A-V valves. Thus the A-V valves may actually close before the onset of ventricular systole.

The rapid rise in ventricular pressure in early ventricular systole bulges the closed valves toward the atria and presses the leaflets tightly together. Simultaneous contraction of the papillary muscles restrains the edges of the valves from being everted into the atria. This concept that the A-V valves close prior to the onset of ventricular contraction has been challenged by studies employing *echocardiography* in humans. This technique consists of sending short pulses of high frequency sound waves (ultrasound) through the chest tissues and the heart and recording the echoes reflected from the various structures. The timing and the pattern of the reflected waves provide such information as the diameter of the heart, the ventricular wall thickness, and the magnitude and direction of the movements of various components of the heart. Echocardiography indicates that at normal P-R intervals the A-V valves are closed by the increase in ventricular pressures at the onset of ventricular contraction. Only when the P-R interval exceeds 0.18 seconds do the A-V valves close as a result of atrial relaxation (and pressure reversal) prior to the start of ventricular contraction.

The time interval between atrial and ventricular systole and the vigor and velocity of atrial contraction and relaxation determine the contribution of atrial systole to A-V valve closure. If ventricular systole immediately follows atrial systole, then A-V valve closure occurs by ventricular systole. However, if ventricular systole is delayed, the valve leaflets float apart as atrial and ventricular pressures equalize, and some regurgitation of blood may occur when the separated leaflets are closed by the pressure rise produced by ventricular systole. A strong, rapid atrial contraction induces a larger ventricular pressure elevation, and a quick relaxation reduces atrial pressure, thereby increasing the magnitude of the reversed A-V pressure gradient and facilitating A-V valve closure.

Studies with intravascular radiopaque dyes, in which no perceptible ventriculoatrial regurgitation occurred in the absence of atrial systole, have cast some doubt on the concept that closure of the A-V valves without regurgitation is dependent on a properly timed atrial contraction.

Semilunar valves. The valves between the right ventricle and the pulmonary artery and between the left ventricle and the aorta consist of three cuplike cusps attached to the valve rings (Figs. 4-12 and 4-13). At the end of the reduced ejection phase of ventricular systole there is a brief reversal of blood flow toward the ventricles (shown as a negative flow in the phasic aortic flow curve in Fig. 4-14) that snaps the cusps together and prevents regurgitation of blood into the ventricles. During ventricular systole the cusps do not lie back against the walls of the pulmonary artery and aorta but float in the bloodstream approximately midway between the vessel walls and their closed position. Behind the semilunar valves are small outpocketings of the pulmonary artery and aorta (*sinuses of Valsalva*), where eddy currents develop that tend to keep the valve cusps away from the vessel walls. The orifices of the right and left coronary arteries are located behind the right and the left cusps, respectively, of the aortic valve. Were it not for the presence of the sinuses of Valsalva and the eddy currents developed therein, the coronary ostia could be blocked by the valve cusps.

The pericardium

The pericardium is an epithelized fibrous sac. It closely invests the entire heart and the

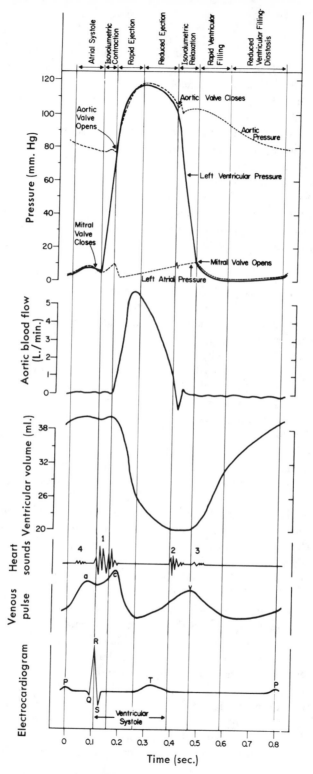

Fig. 4-14. Left atrial, aortic, and left ventricular pressure pulses correlated in time with aortic flow, ventricular volume, heart sounds, venous pulse, and electrocardiogram for a complete cardiac cycle in the dog.

cardiac portion of the great vessels and is reflected onto the cardiac surface as the epicardium. It normally contains a small amount of fluid, which provides lubrication for the continuous movement of the enclosed heart. The distensibility of the pericardium is small, so that it strongly resists a large, rapid increase in cardiac size. By virtue of this characteristic, the pericardium plays a role in preventing sudden overdistension of the heart. However, in congenital absence of the pericardium or after its surgical removal, cardiac function appears to be well within physiological limits.

HEART SOUNDS

There are usually four sounds produced by the heart, but only two are ordinarily audible through a stethoscope. With the aid of electronic amplification the less intense sounds can be detected and recorded graphically as a *phonocardiogram*. This means of registering heart sounds that may be inaudible to the human ear aids in delineating the precise timing of the heart sounds relative to other events in the cardiac cycle.

The first heart sound is initiated at the onset of ventricular systole (Fig. 4-14) and consists of a series of vibrations of mixed, unrelated, low frequencies (a noise). It is the loudest and longest of the heart sounds, has a crescendo-decrescendo quality, and is heart best over the apical region of the heart. The tricuspid valve sounds are heard best in the fifth intercostal space just to the left of the sternum, and the mitral sounds are heard best in the fifth intercostal space at the cardiac apex. The first heart sound is chiefly caused by the oscillation of blood in the ventricular chambers and vibration of the chamber walls. The vibrations are engendered in part by the abrupt rise of ventricular pressure with acceleration of blood back toward the atria, but primarily by sudden tension and recoil of the A-V valves and adjacent structures with deceleration of the blood by closure of the A-V valves. The vibrations of the ventricles and the contained blood are transmitted through surrounding tissues and reach the chest wall where they may be heard or recorded. The intensity of the first sound is primarily a function of the force of ventricular contraction but also of the interval between atrial and ventricular systoles (p. 86).

The second heart sound, which occurs on closure of the semilunar valves (Fig. 4-14), is composed of higher frequency vibrations (higher pitch), is of shorter duration and lower intensity, and has a more snapping quality than the first heart sound. The second sound is caused by abrupt closure of the semilunar valves, which initiates oscillations of the columns of blood and the tensed vessel walls by the stretch and recoil of the closed valve. The second sound caused by closure of the pulmonic valve is heard best in the second thoracic interspace just to the left of the sternum, whereas that caused by closure of the aortic valve is heard best in the same intercostal space but to the right of the sternum. Conditions that bring about a more rapid closure of the semilunar valves, such as increases in pulmonary artery or aortic pressure (for example, pulmonary or systemic hypertension), will increase the intensity of the second heart sound. In the adult the aortic valve sound is usually louder than the pulmonic, but in cases of pulmonary hypertension the reverse is often true.

A normal phonocardiogram taken simultaneously with an electrocardiogram is illustrated in Fig. 4-15. Note that the first sound, which starts just beyond the peak of the R wave, is composed of irregular waves and is of greater intensity and duration than the second sound, which appears at the end of the T wave. A third and fourth heart sound do not appear on this record.

The third heart sound, which is more frequently heard in children with thin chest walls or in patients with left ventricular failure, consists of a few low-intensity, low-frequency vibrations heard best in the region of the apex.

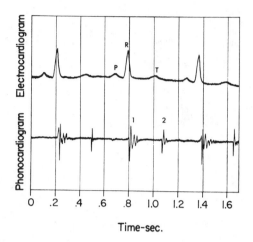

Fig. 4-15. Phonocardiogram illustrating the first and second heart sounds and their relationship to the P, R, and T waves of the electrocardiogram. (Time lines = 0.04 second.)

It occurs in early diastole and is believed to be the result of vibrations of the ventricular walls caused by abrupt acceleration and deceleration of blood entering the ventricles on opening of the atrioventricular valves (Fig. 4-14).

A fourth or atrial sound, consisting of a few low-frequency oscillations, is occasionally heard in normal individuals. It is caused by oscillation of blood and cardiac chambers created by atrial contraction (Fig. 4-14).

Since the onset and termination of right and left ventricular systoles are not precisely synchronous, differences in time of vibration of the two atrioventricular valves or of the two semilunar valves can sometimes be detected with the stethoscope. Such asynchrony of valve vibrations, which may sometimes indicate abnormal cardiac function, is manifest as a *split sound* over the apex of the heart for the atrioventricular valves and over the base for the semilunar valves. The heart sounds may also be altered by deformities of the valves: *murmurs* may be produced, and the character of the murmur serves as an important guide in the diagnosis of valvular disease. When the third

and fourth (atrial) sounds are accentuated, as occurs in certain abnormal conditions, triplets of sounds may occur, resembling the sound of a galloping horse. These *gallop rhythms* are essentially of two types—*presystolic gallop* caused by accentuation of the atrial sound, and *protodiastolic gallop* caused by accentuation of the third heart sound.

CARDIAC CYCLE
Ventricular systole

Isovolumetric contraction. The onset of ventricular contraction coincides with the peak of the R wave of the electrocardiogram and the initial vibration of the first heart sound. It is indicated on the ventricular pressure curve as the earliest rise in ventricular pressure after atrial contraction. The interval of time between the start of ventricular systole and the opening of the semilunar valves (when ventricular pressure rises abruptly) is termed *isovolumetric contraction*, since ventricular volume is constant during this brief period (Fig. 4-14). The curve depicting the ventricular volume is obtained from a *cardiometer* (Fig. 4-16), a rigid chamber (usually glass) that encloses the two ventricles and is sealed at the atrioventricular groove by a rubber membrane that is tight enough to prevent an air leak yet loose enough not to interfere with ventricular inflow. The side arm of the cardiometer is connected to a volume recorder that, in this instance (Fig. 4-14), is mounted in a manner to inscribe an upward deflection as an increase in ventricular volume.

The increment in ventricular pressure during isovolumetric contraction is transmitted across the closed valves and is evident in Fig. 4-14 as a small oscillation on the aortic pressure curve. Isovolumetric contraction has also been referred to as isometric contraction. However, some fibers shorten and others lengthen, as evidenced by changes in ventricular shape; it is therefore not a true isometric contraction.

Ejection. Opening of the semilunar valves

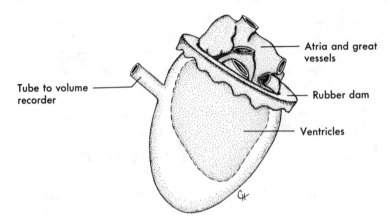

Tube to volume recorder

Atria and great vessels

Rubber dam

Ventricles

Fig. 4-16. Cardiometer, a device for measuring changes in ventricular volume.

marks the onset of the ejection phase, which may be subdivided into an earlier, shorter phase *(rapid ejection)* and a later, longer phase *(reduced ejection)*. The rapid ejection phase is distinguished from the reduced ejection phase by a sharper rise in ventricular and aortic pressures, a more abrupt decrease in ventricular volume, and a greater aortic blood flow (Fig. 4-14). The sharp decrease in the left atrial pressure curve at the onset of ejection results from the descent of the base of the heart and stretch of the atria. During the reduced ejection period, runoff of blood from the aorta to the periphery begins to exceed ventricular output and aortic pressure declines, whereas throughout ventricular systole the blood returning to the atria produces a progressive increase in atrial pressure. Note that during approximately the first third of the ejection period left ventricular pressure slightly exceeds aortic pressure and flow accelerates (continues to increase), whereas during the last two-thirds of ventricular ejection the reverse holds true. This reversal of the ventricular-aortic pressure gradient in the presence of continued flow of blood from left ventricle to aorta (caused by the momentum of the forward blood flow) is the result of the storage of potential energy in the stretched arterial walls and produces a deceleration of blood flow into the aorta. The peak of the flow curve coincides in time with the point at which the left ventricular pressure curve intersects the aortic pressure curve during ejection. Thereafter, flow decelerates (continues to decrease) because the pressure gradient has been reversed. With right ventricular ejection there is shortening of the free wall of the right ventricle (descent of the tricuspid valve ring) in addition to lateral compression of the chamber. However, with left ventricular ejection there is very little shortening of the base-to-apex axis, and ejection is accomplished chiefly by compression of the left ventricular chamber.

The venous pulse curve shown in Fig. 4-14 has been taken from a jugular vein and the *c* wave is caused by impact of the adjacent common carotid artery. Note that except for the *c* wave, the venous pulse closely follows the atrial pressure curve.

At the end of ejection, a volume of blood approximately equal to that ejected during systole remains in the ventricular cavities. This *residual volume* is fairly constant in normal hearts but is smaller with increased heart rate or reduced outflow resistance and larger when the opposite conditions prevail. An increase in myocardial contractility may decrease residual volume (or increase stroke volume), especially in the depressed heart. With severely hypodynamic and dilated hearts, as in *heart failure,* the residual volume can become many times greater than the stroke volume. In addition to serving as a small adjustable blood reservoir, the residual volume can, to a limited degree, permit transient disparities between the outputs of the two ventricles.

Ventricular diastole

Isovolumetric relaxation. Closure of the aortic valve produces the incisura on the descending limb of the aortic pressure curve and the second heart sound (with some vibrations evident on the atrial pressure curve) and marks the end of ventricular systole. The period between closure of the semilunar valves and opening of the A-V valves is termed *isovolumetric* (or *isometric*) relaxation and is characterized by a precipitous fall in ventricular pressure without a change in ventricular volume.

Rapid filling phase. The major part of ventricular filling occurs immediately on opening of the atrioventricular valves when blood that had returned to the atria during the previous ventricular systole is abruptly released into the relaxing ventricles. This period of ventricular filling is called the *rapid filling phase*. In Fig. 4-14 the onset of the rapid filling phase is indicated by the decrease in left ventricular pressure below left atrial pressure, resulting in the opening of the mitral valve. The rapid flow of blood from atria to relaxing ventricles produces a decrease in atrial and ventricular pressures and a sharp increase in ventricular volume.

The decrease in pressure from the peak of the *v* wave of the venous pulse is caused by transmission of the pressure decrease incident to the abrupt transfer of blood from the right atrium to the right ventricle with opening of the tricuspid valve. Rapid ventricular filling is responsible for the few small vibrations that

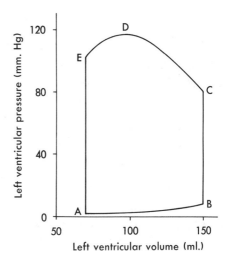

Fig. 4-17. Pressure-volume loop of the left ventricle for a single cardiac cycle *(ABCDE)*.

constitute the third heart sound. Elastic recoil of the previous ventricular contraction may aid in drawing blood into the relaxing ventricle when residual volume is small, but it probably does not play a significant role in ventricular filling under most normal conditions.

Diastasis. The rapid filling phase is followed by a phase of slow filling, called *diastasis*. During diastasis, blood returning from the periphery flows into the right ventricle and blood from the lungs into the left ventricle. This small, slow addition to ventricular filling is indicated by a gradual rise in atrial, ventricular, and venous pressures and in ventricular volume.

Pressure-volume relationship. The changes in left ventricular pressure and volume throughout the cardiac cycle are summarized diagrammatically in Fig. 4-17. The element of time is not considered in this pressure-volume loop. Diastolic filling starts at *A* and terminates at *B*, when the mitral valve closes. Note that there is only a small increase in pressure with the increase in ventricular volume during diastole.

With isovolumetric contraction (*B-C*) there is a steep rise in pressure with no change in ventricular volume. At *C* the aortic valve opens and during the first phase of ejection (rapid ejection (*C-D*), there is a large reduction in volume associated with a continued but lesser increase in ventricular pressure than that which occurred during isovolumetric contraction. This is followed by reduced ejection (*D-E*) and a small decrease in ventricular pressure. The aortic valve closes at *E*, and this event is followed by isovolumetric relaxation (*E-A*), which is characterized by a sharp drop in pressure with no change in volume. The mitral valve opens at *A* to complete one cardiac cycle. At constant aortic pressure (afterload) an increase in filling, which is associated with some increase in end-diastolic pressure (point *B* shifts to the right), results in a shift of point *C* to the right and a greater stroke volume (new segment *C-E*). An increase in contractility at constant aortic pressure results in greater emptying of the heart during systole (point *E* shifts to the left).

Atrial systole. The onset of atrial systole occurs at the peak of the P wave of the electrocardiogram (curve of atrial depolarization), and the transfer of blood from atrium to ventricle made by the peristalsis-like wave of atrial contraction completes the period of ventricular filling. Atrial systole is responsible for the small increases in atrial, ventricular, and venous (*a* wave) pressures as well as in ventricular volume shown in Fig. 4-14. Throughout ventricular diastole, atrial pressure barely exceeds ventricular pressure, indicating a low-resistance pathway across the open A-V valves during ventricular filling. A few small vibrations produced by atrial systole constitute the fourth or atrial heart sound.

Since there are no valves at the junctions of the venae cavae and right atrium or of the pulmonary veins and left atrium, atrial contraction can force blood in both directions. Actually, little blood is pumped back into the venous

tributaries during the brief atrial contraction, in part because of the inertia of the inflowing blood and possibly because of some constriction of the veins at their points of entrance into the atria.

Atrial contraction is certainly not essential for ventricular filling, as can be observed in atrial fibrillation or complete heart block. However, its contribution is governed to a great extent by the heart rate and the structure of the atrioventricular valves. At slow heart rates, filling practically ceases toward the end of diastasis, and atrial contraction contributes little additional filling. This is so because there is a smaller gradient of pressure from atrium to ventricle and less inertia to overcome in forcing blood in a retrograde direction toward the venae cavae and pulmonary veins. With tachycardia and an abbreviated diastasis the atrial contribution can become substantial, especially if it follows immediately after the rapid filling phase when the atrioventricular pressure gradient is maximal. Should tachycardia become so great that the rapid filling phase is encroached on, atrial contraction assumes great importance in rapidly propelling blood into the ventricle during this brief period of the cardiac cycle. Of course, if the period of ventricular relaxation is so brief that filling is seriously impaired by severe abbreviation of the rapid filling phase, even atrial contraction cannot prevent inadequate ventricular filling and a consequent reduction in cardiac output. Obviously, if atrial contraction occurs simultaneously with ventricular contraction, no atrial contribution to ventricular filling can occur. In certain disease states the atrioventricular valves may be markedly narrowed (*stenotic*). Under such conditions atrial contraction may play a much more important role in ventricular filling than it does in the normal heart.

Ventricular contraction has been shown to aid indirectly in right ventricular filling by its effect on the right atrium. Descent of the base of the heart stretches the right atrium downward, and pressure measurements indicate a sharp reduction in right atrial pressure associated with acceleration of blood flow in the venae cavae toward the heart. Enhancement of venous return by ventricular systole provides an additional supply of atrial blood for ventricular filling during the subsequent rapid filling phase of diastole. However, this mechanism is probably of little physiological importance, except possibly at rapid heart rates.

BIBLIOGRAPHY
Journal articles

Abbott, B. C., and Mommaerts, W. F. H. M.: A study of inotropic mechanisms in the papillary muscle preparation, J. Gen. Physiol. 42:533, 1959.

Alpert, N. R., Hamrell, B. B., and Mulieri, L. A.: Heart muscle mechanics, Ann. Rev. Physiol. 41:521, 1979

Armour, J. A., and Randall, W. C.: Structural basis for cardiac function, Am. J. Physiol. 218:1517, 1970.

Brutsaert, D. L., Claes, V. A., and Sonnenblick, E. H.: The velocity of shortening of unloaded heart muscle relative to the length-tension relation, Circ. Res. 29:63, 1971.

Burggraf, G. W., and Craige, E.: The first heart sound in complete heart block; phono-echocardiographic correlations, Circulation 50:17, 1974.

Fabiato, A., and Fabiato, F.: Calcium and cardiac excitation—contraction coupling, Ann. Rev. Physiol. 41:473, 1979.

Jewell, B. R.: A reexamination of the influence of muscle length on myocardial performance, Circ. Res. 40:221, 1977.

Lau, V. K., and Sagawa, K.: Model analysis of the contribution of atrial contraction to ventricular filling, Ann. Biomed. Eng. 7:167, 1979.

Little, R. C.: The mechanism of closure of the mitral valve: a continuing controversy, Circulation 59:615, 1979.

Nayler, W. G., and Seabra-Gomes, R.: Excitation-contraction coupling in cardiac muscle, Prog. Cardiovasc. Dis. 18:75, 1976.

Noble, M. I. M.: The contribution of blood momentum to left ventricular ejection in the dog. Circ. Res. 23:663, 1968.

Ross, J. Jr., and Sobel, B. E.: Regulation of cardiac contraction, Ann. Rev. Physiol. 34:47, 1972.

Sagawa, K.: The ventricular pressure-volume diagram revisited, Circ. Res. 43:677, 1978.

Sonnenblick, E.: Force-velocity relations in mammalian heart muscle, Am. J. Physiol. 202:931, 1962

Streeter, D. D., Jr., Spotnitz, H. M., Patel, D. P., Ross, J., Jr., and Sonnenblick, E. H.: Fiber orientation in the canine left ventricle during diastole and systole, Circ. Res. 24:339, 1969.

Weber, K. T., and Janicki, J. S., editors: Cardiac mechanics (symposium), Fed. Proc. 39:131, 1980.

Zaky, A., Steinmentz, E., and Feigenbaum, H.: Role of atrium in closure of mitral valve in man, Am. J. Physiol. 217:1652, 1969.

Books and monographs

Brady, A. J.: Mechanical Properties of cardiac fibers. In Handbook of physiology; Section 2: The cardiovascular system—the heart, vol. I, Bethesda, MD, 1979, American Physiological Society, pp. 461-474.

Braunwald, E., Ross, J., Jr., and Sonnenblick, E. H.: Mechanisms of contraction of the normal and failing heart, ed. 2, Boston, 1976, Little, Brown and Co.

Katz, A. M.: Physiology of the heart, New York, 1977, Raven Press.

Langer, G. A., and Brady, A. J.: The mammalian myocardium, New York, 1974, John Wiley & Sons, Inc.

McKusick, V. A.: Cardiovascular sound in health and disease, Baltimore, 1958, The Williams & Wilkins Co.

Mirsky, I., Ghista, D. N., Sandler, H.: Cardiac mechanics: Physiological, clinical, and mathematical considerations, New York, 1974, John Wiley & Sons, Inc.

Parmley W. W., and Talbot, L.: Heart as a pump. In Handbook of physiology; Section 2: The cardiovascular system—the heart, vol. I, Bethesda, MD, 1979, American Physiological Society, pp. 429-460.

Sommer, J. R., and Johnson, E. A.: Ultrastructure of cardiac muscle. In Handbook of physiology; Section 2: The cardiovascular system—the heart, vol. I, Bethesda, MD, 1979, American Physiological Society, pp. 113-186.

CHAPTER 5

THE ARTERIAL SYSTEM

HYDRAULIC FILTER

The principal function of the systemic and pulmonary arterial systems is to distribute blood to the capillary beds throughout the body. The arterioles, the terminal components of this system, serve to regulate the fractional distribution among the various capillary beds. Between the arterioles and the heart, the aorta and pulmonary artery and their major branches constitute a system of conduits of considerable volume and distensibility. An arterial system composed of elastic conduits and high-resistance terminals constitutes a *hydraulic filter* analogous to the resistance-capacitance (R-C) filters of electrical circuits.

Hydraulic filtering enables the intermittent output of the heart to be converted to a steady flow through the capillaries. This important function of the large elastic arteries was first recognized in 1834 by Ernst Weber of Leipzig. He explained the similarity between the distensible arteries and the *Windkessels* of the fire engines of that day. In those devices there was a large volume of air trapped in a container between the inflow and outflow ends of the manually operated engine. The Windkessel converted the intermittent inflow to a steady outflow of water at the nozzle of the fire hose.

The analogous function of the large elastic arteries is illustrated in Fig. 5-1. The entire stroke volume is discharged into the arterial system during systole, which occupies approximately one-third of the duration of the cardiac cycle at normal heart rates. In fact, as described on p. 90, most of the stroke volume is pumped during the rapid ejection phase, which constitutes only a fraction of total systole. Part of the energy of cardiac contraction is dissipated as forward capillary flow during systole; the remainder is stored as potential energy, in that much of the stroke volume is retained by the distensible arteries. During diastole the elastic recoil of the arterial walls converts this potential energy into capillary blood flow. If the arterial walls were rigid, then capillary flow would cease during diastole.

Hydraulic filtering also plays an important role in minimizing the work load of the heart. More work is required to pump a given flow intermittently than steadily, and the more effective the filtering, the less the magnitude of this excess work. This principle was well recognized by Weber over a century ago. A simple example will illustrate this point.

Consider first the steady flow of a fluid at a rate of 100 ml./sec. through a hydraulic system with a resistance of 1 mm. Hg/ml./sec. This would result in a constant pressure of 100 mm. Hg, as shown in Fig. 5-2, A. Neglecting any inertial effect, hydraulic work may be defined as $\int_{t_1}^{t_2} P\,dV$; that is, each small increment of volume pumped is multiplied by the pressure existing at the time, and the products, dW, are integrated over the time interval of interest, $t_2 - t_1$, to give the total work, W. For steady flow, W = PV. In the example in Fig. 5-2, A, the work done in pumping the fluid for 1 sec-

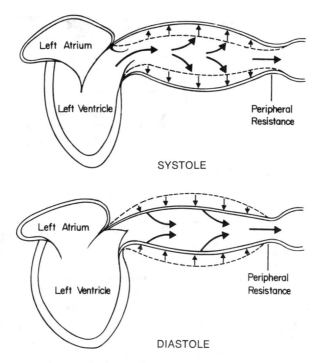

Fig. 5-1. During ventricular systole the stroke volume ejected by the ventricle results in some forward capillary flow, but most of the ejected volume is stored in the elastic arteries. During ventricular diastole the elastic recoil of the arterial walls maintains capillary flow throughout the remainder of the cardiac cycle.

ond would be 10,000 mm. Hg-ml. (or 1.33 × 10⁷ dyne-cm.).

Next, consider the example of an intermittent pump that puts out the same volume per second, but pumps the entire volume at a steady rate over 0.5 second and then pumps nothing during the next 0.5 second. Hence, it pumps at the rate of 200 ml./sec. for 0.5 second, as shown in Fig. 5-2, *B* and *C*. In *B* the conduit is rigid and the fluid is incompressible, but the system has the same resistance as in *A*. During the pumping phase of the cycle (systole) the flow of 200 ml./sec. through a resistance of 1 mm. Hg/ml./sec. would produce a pressure of 200 mm. Hg. During the filling phase of the pump (diastole) the pressure would be 0 mm. Hg in this rigid system. The

work done during systole would be 20,000 mm. Hg/ml., or twice that required in the example in *A*.

If the system were very distensible, hydraulic filtering would be very effective, and the pressure would remain virtually constant throughout the entire cycle (Fig. 5-2, *C*). Of the 100 ml. of fluid pumped during the 0.5 second of systole, only 50 ml. would be emitted through the high-resistance outflow end of the system during systole. The remaining 50 ml. would be stored by the distensible conduit during systole and would flow out during diastole. Hence the pressure would be virtually constant at 100 mm. Hg throughout the cycle. The fluid pumped during systole would be ejected at only half the pressure that prevailed in Fig.

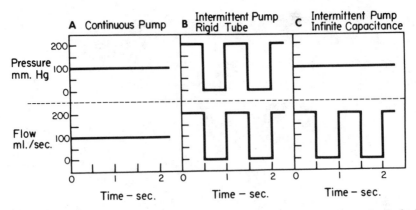

Fig. 5-2. The relationships between pressure and flow for three systems, in each of which the flow is 100 ml./sec. and the resistance is 1 mm. Hg/ml./sec. In system A the flow is steady and the distensibility of the conduit is immaterial. In systems B and C the flow is intermittent; it is steady for half the cycle and ceases for the remainder of the cycle. In system B the conduit is rigid, whereas in system C the conduit is infinitely distensible, resulting in perfect filtering of the pressure. In systems A and C the work per second is 10,000 mm. Hg-ml. (or 1.33×10^7 dyne-cm.); in system B the work per second is twice as great.

5-2, B, and, therefore, the work would be only half as great. With nearly perfect filtering, as in Fig. 5-2, C, the work would be identical to that for steady flow (Fig. 5-2, A).

Naturally, the filtering accomplished by the systemic and pulmonic arterial systems is at some level intermediate between the examples in Fig. 5-2, B and C. It has been estimated that under average normal conditions, the additional work imposed by intermittency of pumping, in excess of that for the steady flow case, is about 35% for the right ventricle and about 10% for the left ventricle. These fractions change, however, with variations in heart rate, peripheral resistance, and arterial distensibility.

ARTERIAL ELASTICITY

The elastic properties of the arterial wall may be appreciated by considering first the *static pressure—volume relationship* for the aorta. To obtain the curves shown in Fig. 5-3, aortas were obtained at autopsy from individuals in different age groups. All branches of the aorta were ligated and successive volumes of liquid were injected into this closed elastic system in the same manner that successive increments of water might be introduced into a balloon. After each increment of volume, the internal pressure was measured. In Fig. 5-3 it is apparent that the curve relating pressure to volume for the youngest age group (curve *a*) is sigmoidal. Although it is quite linear over most of its extent, the slope decreases at the upper and lower ends. At any given point the slope (dV/dP) represents the aortic *capacitance* (or *compliance*). Thus in normal individuals the aortic capacitance is least at extremely high and low pressures and greatest over the usual range of pressure variations. This resembles the familiar capacitance changes encountered in inflating a balloon, where the greatest difficulty in introducing air is experienced at the beginning of inflation and again at near-maximum volume, just prior to rupture of the balloon.

It is also apparent from Fig. 5-3 that the curves become displaced downward and the slopes diminish as a function of advancing age.

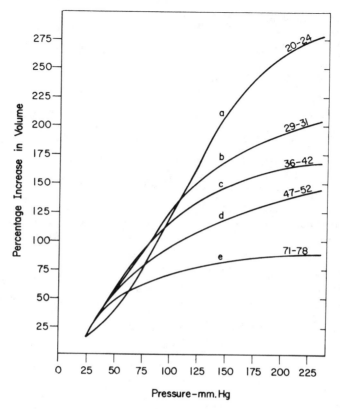

Fig. 5-3. Pressure-volume relationships for aortas obtained at autopsy from humans in different age groups (denoted by the numbers at the right end of each of the curves). (Redrawn from Hallock, P., and Benson, I. C.: J. Clin. Invest. **16:**595, 1937.)

Thus, for any given pressure above about 80 mm. Hg, the capacitance decreases with age, a manifestation of increased rigidity caused by progressive atherosclerosis. The heart is unable to eject its stroke volume into a rigid arterial system as rapidly as into a more compliant system. As capacitance diminishes, peak arterial pressure occurs progressively later in systole. Hence, there is a significant prolongation of the rapid ejection phase of systole.

A complete representation of arterial elasticity cannot be derived from such static pressure-volume curves. When *dynamic curves* are obtained, that is, when pressure is recorded during continuous injection or withdrawal of fluid,

it is found that the pressure is a function not only of volume but also of the rate of change of volume. Therefore the arterial wall has *visco-elastic* rather than purely elastic properties (where pressure would be a function of volume alone).

DETERMINANTS OF THE ARTERIAL BLOOD PRESSURE

The determinants of the pressure that may exist at any moment within the arterial system cannot be evaluated with great precision at present. Yet the arterial blood pressure is a quantitative measurement routinely obtained for diagnosis in most patients, and it provides a useful clue to their cardiovascular status. A simplified approach will therefore be undertaken in an attempt to gain a general understanding of the principal determinants of the arterial blood pressure. To accomplish this, the determinants of the *mean arterial pressure* (defined in the following section) will first be analyzed. The *systolic* and *diastolic arterial pressures* will then be considered as the upper and lower limits of periodic oscillations about this mean pressure. Finally, the more complex aspect of the arterial impedance will be considered to explain the changes in arterial pressure as the pulse wave progresses from the origin of the aorta toward the capillaries.

The determinants of the arterial blood pressure will be arbitrarily subdivided into "physical" and "physiological" factors. For the sake of simplicity, the arterial system will be considered as a static, elastic system, and the only two "physical" factors to be considered will be the blood volume within the arterial system and the elastic characteristics (capacitance) of the system. Several "physiological" factors will be considered, such as heart rate, stroke volume, cardiac output, and peripheral resistance. Such physiological factors will be shown to operate through one or both of the physical factors, however.

Mean arterial pressure

The *mean arterial pressure* is the average pressure during a given cardiac cycle that exists in the aorta and its major branches. It may be obtained from an arterial pressure tracing by measuring the area under the curve and dividing this area by the time interval involved, as shown in Fig. 5-4. During the registration of

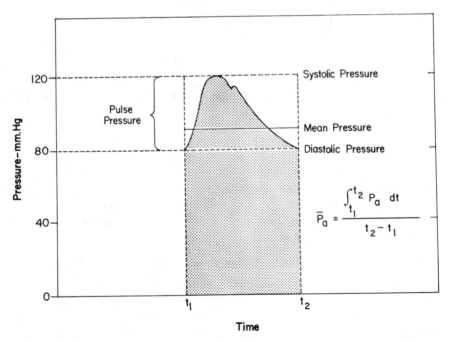

Fig. 5-4. Arterial systolic, diastolic, pulse, and mean pressures. The mean arterial pressure ($\bar{P}_a$) represents the area under the arterial pressure curve (shaded area) divided by the cardiac cycle duration ($t_2 - t_1$).

an arterial pressure curve, the tracing may be damped electronically to record the mean pressure. With a mercury manometer connected to a peripheral artery of an experimental animal, the mean pressure may be obtained by hydraulically damping the oscillations by means of a screw clamp on the tubing between the artery and manometer. The mean arterial pressure ($\overline{P}_a$) can usually be approximated satisfactorily from the measured values of the systolic (P_s) and diastolic (P_d) pressures by means of the following formula:

$$\overline{P}_a \cong P_d + \frac{1}{3}(P_s - P_d) \tag{1}$$

For the purposes of the present discussion, the mean pressure will be considered to be dependent only on the mean volume of blood in the arterial system and the elastic properties of the arterial walls. The arterial volume, V_a, in turn, is dependent on the rate of inflow, Q_i, from the heart into the arteries (*cardiac output*) and the rate of outflow, Q_o, from the arteries through the capillaries (*peripheral runoff*); expressed mathematically,

$$dV_a /dt = Q_i - Q_o \tag{2}$$

If arterial inflow exceeds outflow, then arterial volume increases, the arterial walls are stretched more, and pressure rises. The converse happens when arterial outflow exceeds inflow. When inflow equals outflow, then arterial pressure remains constant.

The change in pressure in response to an alteration of cardiac output can be better appreciated by considering some simple examples. Under control conditions, let cardiac output be 5 L./min. and mean arterial pressure ($\overline{P}_a$) be 100 mm. Hg. From the definition of total peripheral resistance

$$R = (\overline{P}_a - \overline{P}_{ra})/Q_o \tag{3}$$

If $\overline{P}_{ra}$ (right atrial pressure) is close to zero,

$$R \cong \overline{P}_a /Q_o \tag{4}$$

Therefore, in the example, R is 100/5, or 20 mm. Hg/L./min.

Now let cardiac output (Q_i) suddenly increase to 10 L./min. Instantaneously, $\overline{P}_a$ will be unchanged. Since the outflow (Q_o) from the arteries depends on $\overline{P}_a$ and R, instantaneously, Q_o will also remain unchanged. Therefore Q_i, now 10 L./min., will exceed Q_o, still only 5 L./min. This will result in an increase in the mean arterial blood volume ($\overline{V}_a$). From equation (2), when $Q_i > Q_o$, then $d\overline{V}_a /dt > 0$; that is, volume is increasing. When $Q_i < Q_o$, the converse is true. When $Q_i = Q_o$, then $d\overline{V}_a /dt = 0$ and $\overline{V}_a$ remains constant.

Since $\overline{P}_a$ is essentially dependent only on the mean arterial blood volume, $\overline{V}_a$, and the arterial capacitance, C_a, an increase in $\overline{V}_a$ will be accompanied by a rise in $\overline{P}_a$. By definition

$$C_a = d\overline{V}_a /d\overline{P}_a \tag{5}$$

Therefore,

$$d\overline{V}_a = C_a d\overline{P}_a \tag{6}$$

and

$$\frac{d\overline{V}_a}{dt} = C_a \frac{d\overline{P}_a}{dt} \tag{7}$$

From equation (2),

$$\frac{d\overline{P}_a}{dt} = \frac{Q_i - Q_o}{C_a} \tag{8}$$

Hence P_a will rise when $Q_i > Q_o$, will fall when $Q_i < Q_o$, and will remain constant when $Q_i = Q_o$.

In the previous example, where Q_i is suddenly doubled, $\overline{P}_a$ will continue to rise as long as Q_i exceeds Q_o. It is evident from equation (4) that Q_o will not attain a value of 10 L./min. until $\overline{P}_a$ reaches a level of 200 mm. Hg, as long as R remains constant at 20 mm. Hg/L./min. It is apparent then that as $\overline{P}_a$ approaches 200, Q_o will almost equal Q_i and $\overline{P}_a$ will rise very slowly. When Q_i is first raised, however, Q_i is greatly in excess of Q_o, and therefore $\overline{P}_a$ will

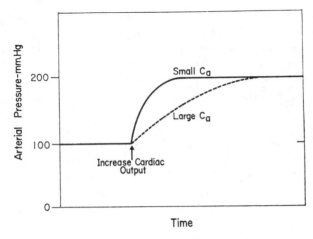

Fig. 5-5. When cardiac output is suddenly increased, the arterial capacitance *(C$_a$)* determines the *rate* at which the mean arterial pressure will attain its new, elevated value, but will not determine the *magnitude* of the pressure.

rise sharply. The pressure-time tracing in Fig. 5-5 shows that, regardless of the value of C$_a$, the initial slope is relatively steep; the slope diminishes as pressure rises, to approach a final value asymptotically.

Furthermore, the height to which $\overline{P}_a$ will rise is entirely independent of the elastic characteristics of the arterial walls. $\overline{P}_a$ must rise to a level such that Q$_o$ = Q$_i$. It is apparent from equation (4) that Q$_o$ depends only on pressure gradient and resistance to flow. Hence C$_a$ determines only the rate at which the new equilibrium value of $\overline{P}_a$ will be approached, as illustrated in Fig. 5-5. When C$_a$ is small (rigid vessels), a relatively slight increase in $\overline{V}_a$ (associated with the transient excess of Q$_i$ over Q$_o$) produces a relatively large increase in $\overline{P}_a$. Hence $\overline{P}_a$ attains its new equilibrium level quickly. Conversely, when C$_a$ is large, then considerable volumes can be accommodated with relatively small pressure changes, and the new equilibrium value of $\overline{P}_a$ is reached at a slower rate.

Similar reasoning may now be applied to explain the changes in $\overline{P}_a$ that accompany alterations in peripheral resistance. Let the control

conditions be identical with those of the preceding example, that is, Q$_i$ = 5, $\overline{P}_a$ = 100, and R = 20. Then let R suddenly be increased to 40. Instantaneously, $\overline{P}_a$ will be unchanged. With $\overline{P}_a$ = 100 and R = 40, Q$_o$ = $\overline{P}_a$/R = 2.5 L./min. If Q$_i$ remains constant at 5 L./min., Q$_i$ > Q$_o$, and $\overline{V}_a$ will increase; hence $\overline{P}_a$ will rise. $\overline{P}_a$ will continue to rise until it reaches 200 mm. Hg. At this level, Q$_o$ = 200/40 = 5 L./ min., which equals Q$_i$. $\overline{P}_a$ will then remain at this new elevated equilibrium level as long as Q$_i$ and R do not change.

It is clear, therefore that *the level of the mean arterial pressure is dependent only on cardiac output and peripheral resistance.* It is immaterial whether any change in cardiac output is accomplished by an alteration of heart rate, of stroke volume, or of both. Any change in heart rate that is balanced by a concomitant, oppositely directed change in stroke volume will not alter Q$_i$; hence $\overline{P}_a$ will not be affected.

Pulse pressure

If we assume that the arterial pressure, P$_a$, at any moment in time is dependent primarily on arterial blood volume, V$_a$, and arterial capacitance, C_a, then it can be shown that the arterial *pulse pressure* (difference between systolic and diastolic pressures) is principally a function of stroke volume and arterial capacitance.

Stroke volume. The effect of a change in stroke volume on pulse pressure may be analyzed under conditions in which C_a remains virtually constant over the range of pressures under consideration. In this situation the curve relating P_a to V_a is linear, as in Fig. 5-6. This curve would correspond fairly closely with the curve for the 20- to 24-year age group in Fig. 5-3, especially over the pressure range between 75 and 150 mm. Hg.

In an individual with such a $P_a:V_a$ curve, the arterial pressure would oscillate about some mean value ($\overline{P}_A$ in Fig. 5-6) that depends entirely on cardiac output and peripheral resistance, as explained previously. This mean pressure corresponds to some mean arterial blood volume, $\overline{V}_A$, and the coordinates $\overline{P}_A$, $\overline{V}_A$ define point $\overline{A}$ on the graph. During diastole, peripheral runoff occurs in the absence of ventricular ejection of blood, and P_a and V_a diminish to minimum values, P_1 and V_1, just prior to the next ventricular ejection. P_1 is then, by definition, the *diastolic pressure*.

During ventricular .ejection, there is rapidly introduced into the arterial system a volume of blood that greatly exceeds the peripheral runoff during this same portion of the cardiac cycle. Arterial pressure and volume therefore rise from point A_1 toward point A_2 in Fig. 5-6. As described on p. 90 the normal heart discharges most of its stroke volume during the early part of systole, the so-called rapid ejection phase. The maximum arterial volume, V_2, is reached at the end of the rapid ejection phase, and this volume corresponds to a peak pressure, P_2, which is the *systolic pressure*. The mean arterial pressure is ordinarily somewhat less than the arithmetic average of the systolic and diastolic pressures, as illustrated in Fig. 5-4.

The *pulse pressure* is the difference be-

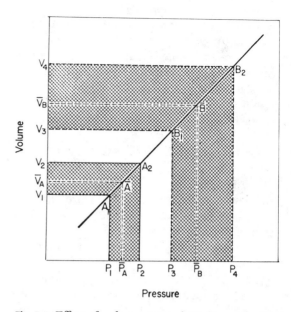

Fig. 5-6. Effect of a change in stroke volume on pulse pressure in a system in which arterial capacitance is constant over the range of pressures and volumes involved. A larger volume increment ($V_4 - V_3$ as compared to $V_2 - V_1$) results in a greater mean pressure ($\overline{P}_B$ as compared to $\overline{P}_A$) and a greater pulse pressure ($P_4 - P_3$ as compared to $P_2 - P_1$).

tween systolic and diastolic pressures ($P_2 - P_1$ in Fig. 5-6), and it corresponds to some *volume increment*, $V_2 - V_1$. This increment equals the volume of blood discharged by the left ventricle during the rapid ejection phase minus the volume that has run off to the periphery during this same phase of the cardiac cycle. When a normal heart beats at a normal frequency, this volume increment is a large fraction of the stroke volume (about 80%). It is this increment that will raise arterial volume rapidly from V_1 to V_2 and hence will cause the arterial pressure to rise from the diastolic to the systolic level (P_1 to P_2 in Fig. 5-6). During the remainder of the cardiac cycle, peripheral runoff will greatly exceed cardiac ejection. The resultant arterial blood volume decrement will cause volumes and pressures to fall from point A_2 back to point A_1.

If stroke volume is now doubled, while heart rate and peripheral resistance remain constant, the mean arterial pressure will be doubled, to $\overline{P}_B$ in Fig. 5-6. Thus the arterial pressure will now oscillate about this new value of the mean arterial pressure. A normal, vigorous heart will eject this greater stroke volume during a fraction of the cardiac cycle approximately equal to the fraction observed at the lower stroke volume. Therefore the volume increment, $V_4 - V_3$, will be a large fraction of the new stroke volume and hence will be approximately twice as great as the previous volume increment ($V_2 - V_1$). With a linear $P_a : V_a$ curve, the greater volume increment will be reflected by a pulse pressure ($P_4 - P_3$) that will be aproximately twice as great as the original pulse pressure ($P_2 - P_1$). With a rise in both mean and pulse pressures, it is evident from inspection of Fig. 5-6 that the rise in systolic pressure (from P_2 to P_4) exceeds the rise in diastolic pressure (from P_1 to P_3).

Arterial capacitance. To assess arterial capacitance as a determinant of pulse pressure, the relative effects of the same volume increment ($V_2 - V_1$ in Fig. 5-7) in a young person (curve A) and in an elderly person (curve B) will be compared. Let cardiac output and total peripheral resistance be the same in both cases; therefore $\overline{P}_a$ will be the same. It is apparent from Fig. 5-7 that the same volume increment will result in a greater pulse pressure in the less distensible arteries of the elderly individual ($P_4 - P_1$) than in the more compliant arteries of the young one ($P_3 - P_2$). For the reasons enunciated on p. 94, this will impose a greater work load on the left ventricle of the elderly than of the young person, even if the stroke volumes, total peripheral resistances, and mean arterial pressures are equivalent.

Fig. 5-8 displays the effects of a change in arterial capacitance on the arterial pressure in an isolated cat heart preparation. As the capac-

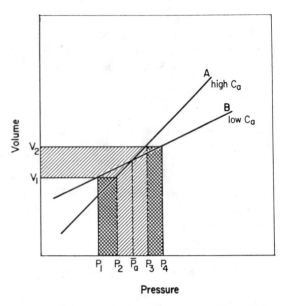

Fig. 5-7. For a given volume increment ($V_2 - V_1$) a reduced arterial capacitance (curve B as compared to curve A) results in an increased pulse pressure ($P_4 - P_1$ as compared to $P_3 - P_2$).

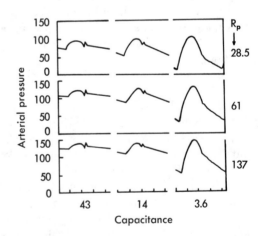

Fig. 5-8. The changes in aortic pressure with changes in arterial capacitance (C) and peripheral resistance (R_p) in an isolated cat heart preparation. (Modified from Elizinga, G., and Westerhof, N.: Pressure and flow generated by the left ventricle against different impedances, Circ. Res. **32**:178, 1973. By permission of the American Heart Association, Inc.)

itance *(C)* was reduced from 43 to 14 to 3.6 units (left to middle to right column, respectively), the pulse pressure increased significantly. In this preparation, the stroke volume decreased as the capacitance was diminished. This accounts for the failure of the mean arterial pressure to remain constant at the different levels of arterial capacitance.

Total peripheral resistance and arterial diastolic pressure. It is often stated that increased total peripheral resistance (TPR) affects primarily the level of the diastolic arterial pressure. The validity of such an assertion deserves close scrutiny. First, let TPR be increased in an individual with a linear $P_a : V_a$ curve, as depicted in Fig. 5-9, *A*. If heart rate and stroke volume remain constant, then an increase in TPR will evoke a proportionate increase in $\overline{P}_a$ (from P_2 to P_5). If the volume increments ($V_2 - V_1$ and $V_4 - V_3$) are equal at both levels of TPR, then the pulse pressures ($P_3 - P_1$ and $P_6 - P_4$) will also be equal. Hence systolic (P_6) and diastolic

(P_4) pressures will have been elevated by exactly the same amounts from their respective control levels (P_3 and P_1).

It is certainly conceivable that with a higher TPR a given stroke volume might not be ejected as rapidly as the same stroke volume against a lower TPR. In such a case, the volume increment might not be as great for the same stroke volume, since a greater fraction of the peripheral runoff might occur during ejection. A somewhat smaller volume increment would result in a proportionately smaller pulse pressure; in Fig. 5-9, *A*, $V_4 - V_3$ would be less, and so also would $P_6 - P_4$ Under such circumstances, an augmentation of TPR would be associated with a somewhat greater rise in diastolic than in systolic pressure.

In the experiment illustrated in Fig. 5-8, as the peripheral resistance (R_p) was raised from 28.5 to 61 to 137 units (top to middle to bottom row, respectively), the mean arterial pressure increased. The heart responded to the aug-

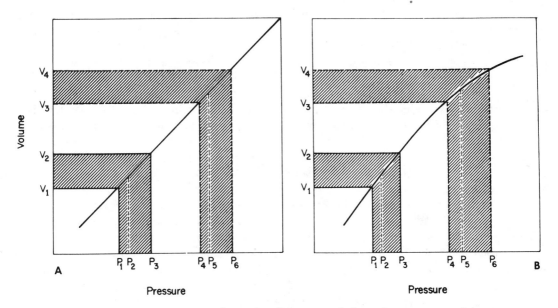

Fig. 5-9. Effect of a change in total peripheral resistance (volume increment remaining constant) on pulse pressure when the pressure-volume curve for the arterial system is rectilinear, **A,** or curvilinear, **B.**

mentation of resistance by pumping a smaller stroke volume. Hence, the volume increment undoubtedly decreased as the resistance was augmented. This was reflected by a progressive reduction in the pulse pressure, for any given level of arterial capacitance. Note that in the left and middle columns of Fig. 5-8, the pulse pressure decreased as R_p was raised.

Chronic hypertension, a condition characterized by a persistent elevation of TPR, occurs more commonly in middle-aged and elderly individuals than in younger persons. The $P_a : V_a$ curve for a hypertensive patient would, therefore, possess the configuration shown in Fig. 5-9, *B* (which resembles curves *b* to *e* in Fig. 5-3), rather than that displayed in Fig. 5-9, *A*. The type of curve in Fig. 5-9, *B*, reveals that C_a is less at higher than at lower pressures. As before, if cardiac output remains constant, an increase in TPR would cause a proportionate rise in $\overline{P}_a$ (from P_2 to P_5). For equivalent increases in TPR, the elevation of pressure from P_2 to P_5 will be the same in Fig. 5-9, *A* and *B*, for reasons discussed on p. 100. Assuming the volume increment ($V_4 - V_3$ in Fig. 5-9, *B*) at elevated TPR to be equal to the control increment ($V_2 - V_1$), it is evident that the pulse pressure ($P_6 - P_4$) in the hypertensive range will greatly exceed that ($P_3 - P_1$) at normal pressure levels. In other words a given volume increment will produce a greater pressure increment when the tube is more rigid than when it is more compliant. Hence the rise in systolic pressure ($P_6 - P_3$) will greatly exceed the increase in diastolic pressure ($P_4 - P_1$). These changes in arterial pressure closely resemble those seen in patients with hypertension. Diastolic pressure is indeed elevated in such individuals, but ordinarily not more than 10 to 40 mm. Hg above the average normal level of 80 mm. Hg, whereas it is not uncommon for systolic pressures to be elevated by 50 to 150 mm. Hg above the average normal level of 120 mm. Hg. The combination of an increased resistance and diminished ar-

terial capacitance would be represented in Fig. 5-8 by a shift in direction from the top left panel to the bottom right panel; that is, both the mean pressure and the pulse pressure would be increased significantly.

Fallacious logic is often used to argue that increased peripheral resistance affects diastolic pressure selectively. It is reasoned that with greater TPR, peripheral runoff occurs more slowly. Therefore P_a declines more slowly from the peak systolic level and does not fall to the usual diastolic level by the time of the next ventricular .ejection. Such reasoning would certainly be applicable to the first few beats after a sudden increase in TPR. However, once P_a became elevated to its equilibrium level, the arteriovenous pressure gradient would be increased proportionately and the rate of runoff at the higher TPR would equal the rate evidenced during control P_a and TPR.

It is apparent, therefore, that the *pulse pressure is principally dependent on volume increment* (a function primarily of stroke volume) and *arterial capacitance*. Stimuli that alter pulse pressure usually do so by changing one or both of these factors. If a change in heart rate, for example, is not accompanied by an alteration of stroke volume, then pulse pressure may or may not be modified, depending on the value of C_a at the new level of P_a. If C_a remains constant, then an increased heart rate will cause a rise in $\overline{P}_a$ but no change in pulse pressure. The situation is represented by Fig. 5-9, *A*, where $V_2 - V_1$ represents the volume increment at the normal heart rate and $V_4 - V_3$ the volume increment during tachycardia. If stroke volumes are equal, the volume increments will be approximately equal. Hence the pulse pressure ($P_6 - P_4$) during tachycardia will be similar to the pulse pressure ($P_3 - P_1$) at the control heart rate. If the increased P_a during tachycardia is associated with a reduction in arterial compliance, as in Fig. 5-9, *B*, the pulse pressure will be augmented ($P_6 - P_4$, as compared with the control pulse pressure, $P_3 - P_1$) as long as stroke volume remains constant. However, changes in heart rate are usually accompanied by changes in stroke volume, and this in turn will affect pulse pressure in accordance with the principles illustrated in Fig. 5-6.

PERIPHERAL ARTERIAL PRESSURE CURVES

The radial stretch of the ascending aorta brought about by left ventricular ejection initiates a pressure wave that is propagated down the aorta and its branches with a finite velocity that is considerably faster than the actual forward movement of the blood itself. It is this propagated pressure wave that one perceives in counting the pulse rate by palpating the radial artery.

The velocity of transmission of the pressure wave varies inversely with the vascular capacitance. With accurate measurement of the transmission velocity, valuable information has been derived concerning the elastic characteristics of the arterial tree. In general, transmission velocity increases with age, confirming the observation that the arteries become less compliant with advancing age. Also, velocity in-

creases progressively as the pulse wave travels from the ascending aorta toward the periphery. This indicates that vascular capacitance diminishes in the more distal portions of the arterial system, a fact that has also been confirmed by direct measurement.

The arterial pressure contour becomes progressively more distorted as the wave is transmitted down the arterial system; the changes in configuration of the pulse with distance are shown in Fig. 5-10. Aside from the increasing delay in the time of onset of the initial pressure rise associated with the transmission delay, three major changes occur in the arterial pulse contour as the pressure wave travels away from the aortic arch. First, the high-frequency components of the pulse, such as the incisura, are damped out and soon disappear. Second, the systolic portions of the pressure wave become narrowed and attain greater peak values. In the curves shown in Fig. 5-10, the systolic pressure at the level of the knee was 39 mm. Hg greater than that recorded in the aortic arch. Third, a hump may become prominent on the diastolic portion of the pressure wave. These changes in contour of the pulse wave are pronounced in

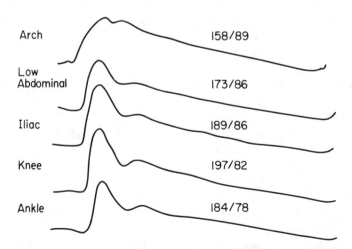

Arch   158/89

Low Abdominal   173/86

Iliac   189/86

Knee   197/82

Ankle   184/78

Fig. 5-10. Arterial pressure curves recorded from various sites in an anesthetized dog. (From Remington, J. W., and O'Brien, L. J.: Am. J. Physiol. **218**:437, 1970.)

young individuals, but the magnitude of the alterations diminishes with age. In elderly patients with severe atherosclerosis, the pulse wave may be transmitted virtually unchanged from the ascending aorta to the periphery.

The damping of the high-frequency components is largely ascribable to the viscoelastic properties of the arterial walls. The precise mechanism for the peaking of the pressure wave is controversial. Probably there are several contributory factors including (1) reflection, (2) tapering, (3) resonance, and (4) changes in transmission velocity with pressure level. Relative to the first of these mechanisms, whenever significant changes in configuration or in dimensions occur (such as at points of branching), pressure waves are reflected backward. Hence pressure at any point in time and space is determined by the algebraic summation of an antegrade incident wave and retrograde reflected waves. It has been calculated that about 80% of the incident wave is reflected back from the peripheral bed at normal levels of peripheral resistance. When the bed is dilated, the fraction decreases, whereas vasoconstriction has the opposite effect. The second factor, tapering, is involved in altering the pulse contour because in proceeding distally along the aorta or any large artery, the lumen progressively narrows beyond each successive branch. It has been shown that a pressure wave becomes amplified as it progresses down a tapered tube. With respect to the possible contribution of resonance, the complicated wave form of the arterial pulse may be considered to be composed of a series of sine waves (Fourier series) consisting of some fundamental frequency and its harmonics. The arterial tree resonates at certain of these frequencies, and other frequencies are effectively damped. Finally, as stated previously, transmission velocity varies inversely with arterial capacitance. Furthermore, capacitance varies inversely with pressure level (Fig. 5-3). Hence the points on the pressure curve at the higher levels of pressure tend to travel faster than those at lower pressure levels. Thus the peak of the arterial pressure curve tends to catch up with the beginning, or "foot," of the same curve. This contributes to the peaking and narrowing of the curve in more distal vessels such as the femoral artery. Reflection and resonance also account for the diastolic humps on these same peripheral curves.

BLOOD PRESSURE MEASUREMENT IN HUMANS

In certain instances, needles or catheters are introduced into peripheral arteries of patients and arterial blood pressure is measured *directly* by means of strain gauges. In the vast majority of cases, however, the blood pressure is estimated *indirectly* by means of a *sphygmomanometer*. This instrument consists of an inextensible cuff containing an inflatable bag. The cuff is wrapped around the extremity (usually the arm, occasionally the thigh) so that the inflatable bag lies between the cuff and the skin, directly over the artery to be compressed. The artery is occluded by inflating the bag, by means of a rubber squeeze-bulb, to a pressure in excess of arterial systolic pressure. The pressure in the bag is measured by means of a mercury manometer or an aneroid manometer. Pressure is released from the bag at a rate of 2 or 3 mm. Hg per heartbeat by means of a needle valve in the inflating bulb.

When blood pressure readings are taken from the arm, the systolic pressure may be estimated by palpating the radial artery at the wrist *(palpatory method)*. When pressure in the bag exceeds the systolic level, no pulse will be perceived. As the pressure falls just below the systolic level (Fig. 5-11, *A*), a spurt of blood will pass through the brachial artery under the cuff during the peak of systole and a slight pulse will be felt at the wrist.

The *auscultatory method* is a more sensitive and therefore a more precise method for measuring systolic pressure, and it also permits the

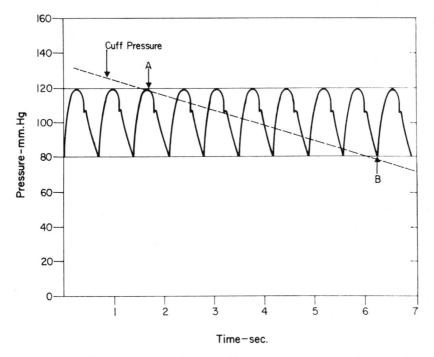

Fig. 5-11. Principle of measurement of arterial blood pressure with a sphygmomanometer. The oblique line represents pressure in the inflatable bag in the cuff. At cuff pressures greater than the systolic pressure (to the left of *A*), no blood progresses beyond the cuff and no sounds can be detected below the cuff. At cuff pressures between the systolic and diastolic levels (between *A* and *B*), spurts of blood traverse the arteries under the cuff and produce the Korotkoff sounds. At cuff pressures below the diastolic pressure (to the right of *B*), arterial flow past the region of the cuff is continuous, and no sounds are audible.

estimation of the diastolic level as well. The physician listens with a stethoscope applied to the skin of the antecubital space over the brachial artery. While the pressure in the bag exceeds the systolic pressure, the brachial artery is occluded and no sounds are heard. When the inflation pressure falls just below the systolic level (point *A* in Fig. 5-11), the small spurt of blood escapes through the cuff and a slight tapping sound is heard. This represents the systolic pressure. It usually corresponds closely with the systolic pressure when it is measured directly and exceeds by a few mm. Hg the pressure estimated by the palpatory method

(because the auscultatory method is more sensitive than the palpatory method). As inflation pressure continues to fall more blood escapes under the cuff per beat and the sounds (called *Korotkoff sounds*) are heard as louder thuds. As the inflation pressure approaches the diastolic level, the Korotkoff sounds become muffled. As they fall just below the diastolic level (point *B* in Fig. 5-11), the sounds disappear; this indicates the diastolic pressure. The origin of the Korotkoff sounds is related to the spurt of blood passing under the cuff and meeting a static column of blood; the impact and turbulence generate vibrations, some of which are in the audible range of frequencies. Once the inflation pressure is less than the diastolic pressure, flow is continuous in the brachial artery and sounds are no longer audible.

BIBLIOGRAPHY
Journal articles

Attinger, E. O., and Attinger, F. M.: Frequency dynamics of peripheral vascular blood flow, Ann. Rev. Biophys. Bioeng. 4:7, 1973.

Chungcharoen, D.: Genesis of Korotkoff sounds, Am. J. Physiol. 207:190, 1964.

Cox, R. H.: Determinants of systemic hydraulic power in unanesthetized dogs, Am. J. Physiol. 226:579, 1974.

Dobrin, P. B.: Mechanical properties of arteries, Physiol. Rev. 58:397, 1978.

Elzinga, G., and Westerhof, N.: Pressure and flow generated by the left ventricle against different impedances, Circ. Res. 32:178, 1973.

Farrar, D. J., Green, H. D., Bond, M. D., Wagner, W. D., and Gobbée, R. A.: Aortic pulse wave velocity, elasticity, and composition in a nonhuman primate model of atherosclerosis, Circ. Res. 43:52, 1978.

Hallock, P., and Benson, I. C.: Studies on the elastic properties of isolated human aorta, J. Clin. Invest. 16:595, 1937.

Milnor, W. R.: Arterial impedance as ventricular afterload, Circ. Res. 36:565, 1975.

Milnor, W. R.: Pulsatile blood flow, N. Engl. J. Med. 287:27, 1972.

Murgo, J. P., Westerhof, N., Giolmo, J. P., and Altobelli, S. A.: Aortic input impedance in normal man: relationship to pressure wave forms, Circulation 62:105, 1980.

Nichols, W. W., Conti, C. R., Walker, W. E., and Milnor, W. R.: Input impedance of the systemic circulation in man, Circ. Res. 40:451, 1977.

O'Rourke, M. F.: The arterial pulse in health and disease, Am. Heart J. 82:687, 1971.

Pepine, C. J., Nichols, W. W., and Conti, C. R.: Aortic input impedance in heart failure, Circulation 58:460, 1978.

Taylor, M. G.: Hemodynamics, Ann. Rev. Physiol. 34:87, 1973.

Van den Bos, G. C., Westerhof, N., Elzinga, G., and Sipkema, P.: Reflection in the systemic arterial system: effects of aortic and carotid occlusion, Cardiovasc. Res. 10:565, 1976.

Books and monographs

Attinger, E. O.: Pulsatile blood flow, New York, 1964, McGraw-Hill Book Co.

Bader, H.: The anatomy and physiology of the vascular wall. In Handbook of physiology; Section 2: Circulation, vol. II, Washington, D.C., 1963, American Physiological Society, pp. 865-889.

Bauer, R. D., and Busse, R., editors: Arterial system: dynamics, control theory and regulation, Heidelberg, 1978, Springer-Verlag.

McDonald, D. A.: Blood flow in arteries, London, 1960, Edward Arnold, Ltd.

Spencer, M. P., and Denison, A. B., Jr.: Pulsatile blood flow in the vascular system. In Handbook of physiology; Section 2: Circulation, vol. II, Washington, D.C., 1963, American Physiological Society, pp. 839-864.

CHAPTER 6

THE MICROCIRCULATION AND LYMPHATICS

The entire circulatory system is geared to supply the body tissues with blood in amounts commensurate with their requirements for oxygen and nutrients. The capillaries, consisting of a single layer of endothelial cells, permit rapid exchange of water and solutes with interstitial fluid. The muscular arterioles, which are the major *resistance vessels*, regulate regional blood flow to the capillary beds, and the venules and small veins serve primarily as collecting channels and storage or *capacitance vessels*.

The arterioles, which range in diameter from about 5 to 100 μ, have a thick smooth muscle layer, a thin adventitial layer, and an endothelial lining (Fig. 1-1). The arterioles give rise directly to the capillaries (5 to 10 μ diameter), or in some tissues to metarterioles (10 to 20 μ diameter), which then give rise to capillaries (Fig. 6-1). The metarterioles can serve either as thoroughfare channels to the venules, bypassing the capillary bed, or as conduits to supply the capillary bed. There are often cross connections between arterioles and between venules as well as in the capillary network. Arterioles that give rise directly to capillaries regulate flow through their cognate capillaries by constriction or dilation. At the points of origin of the capillaries in some tissues there is a small cuff of smooth muscle called the *precapillary sphincter* that controls the blood flow through the cognate capillaries (Fig. 6-1). The

capillaries form an interconnecting network of tubes of different lengths with an average length of 0.5 to 1 mm.

Capillary distribution varies from tissue to tissue. In metabolically active tissues, such as cardiac and skeletal muscle and glandular structures, capillaries are numerous, whereas in less active tissues, such as subcutaneous tissue or cartilage, *capillary density* is low. Also all capillaries are not of the same diameter, and since some capillaries have diameters less than that of the erythrocytes, it is necessary for the cells to become temporarily deformed in their passage through these capillaries. Fortunately, the normal red cells are quite flexible and readily change their shape to conform with that of the small capillaries.

Blood flow in the capillaries is not uniform and is chiefly dependent on the contractile state of the arterioles and, where present, precapillary sphincters. The average velocity of blood flow in the capillaries is approximately 1 mm./sec.; however, it can vary from zero to several millimeters per second in the same vessel within a brief period of time. These changes in capillary blood flow may be of random type or may show rhythmical oscillatory behavior of different frequencies that are caused by contraction and relaxation (*vasomotion*) of the precapillary vessels. This vasomotion is to some extent an intrinsic contractile behavior of the vascular smooth muscle and is independent of

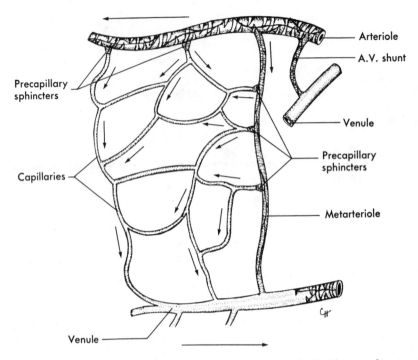

Fig. 6-1. Schematic drawing of the microcirculation (after Zweifach). The circular structures on the arteriole and venule represent smooth muscle fibers, and the branching solid lines represent sympathetic nerve fibers. The arrows indicate the direction of blood flow.

external input. Furthermore, changes in *transmural pressure* (intravascular minus extravascular pressure) at the precapillary vessels influences their contractile state; an increase in transmural pressure, whether produced by an increase in venous pressure or by dilation of arterioles, results in contraction of the terminal arterioles at the points of origin of the capillaries. In addition, humoral and possibly neural factors also affect vasomotion. For example, when the precapillary sphincters contract in response to increased transmural pressure, the contractile response can be overriden and vasomotion abolished. This effect is accomplished by metabolic (humoral) factors (p. 126) when the oxygen supply is reduced below the requirements of the parenchymal tissue, as occurs in muscle during exercise.

Reduction of transmural pressure will induce relaxation of the terminal arterioles, but blood flow through the capillaries obviously cannot increase if the reduction in intravascular pressure is caused by severe constriction of the parent arterioles or metarterioles. Large arterioles and metarterioles also exhibit vasomotion, but in the contraction phase they usually do not completely occlude the lumen of the vessel and arrest blood flow as is the case with contraction of the terminal arterioles and precapillary sphincters. Hence flow rate may be altered by arteriolar and metarteriolar vasomotion.

Since blood flow through the capillaries provides for exchange of gases and solutes between blood and tissue, it has been termed *nutritional flow*, whereas blood flow that bypasses the capillaries in traveling from the arterial to the venous side of the circulation has been termed *nonnutritional* or *shunt flow* (Fig. 6-1). In some areas of the body (such as skin) true arteriovenous (A-V) shunts exist (p. 223). However, in many tissues, such as muscle, evidence of anatomical shunts is lacking. Nevertheless, nonnutritional flow does occur and can be demonstrated in muscle by stimulation of the adrenergic sympathetic fibers to the muscle vessels during constant flow perfusion, or by stimulation of the sympathetic cholinergic fibers to muscle (p. 134). With adrenergic sympathetic fiber stimulation at constant flow, the capillary surface area available for solute exchange is reduced, as measured by decrease in intra-arterially administered ^{86}Rb uptake by the tissue. With sympathetic cholinergic fiber stimulation, blood flow increases but tissue clearance (washout) of radioactive isotope deposited in the tissue is unchanged. With this technique for the estimation of nutritional blood flow, a radioactive substance such as [22]Na[+] is injected into the tissue in a small volume of fluid to minimize tissue damage. The rapidity with which it is removed by the perfusing blood is indicative of the rate of nutritional blood flow. Increased flow through open capillaries in the absence of morphological A-V shunts has been termed a *physiological shunt*. It is the result of a greater flow of blood through open capillaries (shunts) with either no change or an increase in the number of closed capillaries. Adrenergic sympathetic fiber stimulation elicits arteriolar constriction, but when flow is set constant by a pump, all the perfusing blood passes through the decreased number of open channels during the sympathetic nerve stimulation. Cholinergic sympathetic fiber stimulation dilates arterioles, and the increased volume flow of blood caused by the decrease in vascular resistance occurs through the same capillaries that were open before nerve stimulation. In tissues that have metarterioles, shunt flow may be continuous from arteriole to venule during low metabolic activity when

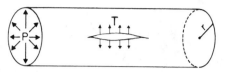

Fig. 6-2. Diagram of a small blood vessel to illustrate the law of Laplace—T = Pr, where P = intraluminal pressure, r = radius of the vessel, and T = wall tension as the force per unit length tangential to the vessel wall, tending to pull apart a theoretical longitudinal slit in the vessel.

many precapillary vessels are closed. When metabolic activity increases in such tissues and precapillary vessels open, blood passing through the metarterioles is readily available for capillary perfusion.

The true capillaries are devoid of smooth muscle and are therefore incapable of active changes in caliber. Changes in capillary diameter are passive and are caused by alterations in precapillary and postcapillary resistance. The thin-walled capillaries can withstand high internal pressures without bursting because of their narrow lumen. This can be explained in terms of the law of Laplace and is illustrated in the following comparison of wall tension of a capillary with that of the aorta. The Laplace equation is T = Pr, where T = tension in the vessel wall, P = transmural pressure, and r = the radius of the vessel. Wall tension is the force per unit length tangential to the vessel wall that opposes the distending force (Pr) that tends to pull apart a theoretical longitudinal slit in the vessel (Fig. 6-2). Transmural pressure is essentially equal to intraluminal pressure, since extravascular pressure is negligible. The Laplace equation applies to very thin wall vessels, such as capillaries. Wall thickness must be taken into consideration when the equation is applied to thick wall vessels, such as the aorta. This is done by dividing Pr (pressure × radius) by wall thickness (w). The equation now becomes σ (wall stress) = Pr/w.

	Aorta	Capillary
Radius (r)	1.5 cm.	5×10^{-4} cm.
Height of Hg column (h)	10 cm. Hg	2.5 cm. Hg
ρ	13.6 gm./cm.^{3}	13.6 gm./cm.^{3}
g	980 cm./sec.^{2}	980 cm./sec.^{2}
$\overline{P}$	$10 \times 13.6 \times 980 =$ 1.33×10^5 dynes/cm.^{2}	$2.5 \times 13.6 \times 980 =$ 3.33×10^4 dynes/cm.^{2}
w	0.2 cm.	1×10^{-4}cm.
$T = Pr$	$(1.33 \times 10^5)(1.5) =$ 2×10^5 dynes/cm.	$(3.33 \times 10^4)(5 \times 10^{-4}) =$ 16.7 dynes/cm.
$\sigma = \dfrac{Pr}{w}$	$\dfrac{2 \times 10^5}{0.2} =$ 1×10^6 dynes/cm.^{2}	$\dfrac{16.7}{1 \times 10^{-4}} =$ 1.67×10^5 dynes/cm.^{2}

To convert pressure in mm. Hg (height of mercury column) to dynes per square centimeter, $P = h\rho g$, where h = height of Hg column in cm., ρ = density of Hg in g/cm.^{3}, g = gravitational acceleration in cm./sec.^{2} (see p. 55), σ = force per unit area, w = wall thickness.

Thus at normal aortic and capillary pressures the wall tension of the aorta is about 12,000 times greater than that of the capillary. In a person standing quietly, capillary pressure in the feet may reach 100 mm. Hg. Under such conditions capillary wall tension increases to 66.5 dynes/cm., a value that is still only one three-thousandth that of the wall tension in the aorta at the same internal pressure.

In addition to providing an explanation for the ability of capillaries to withstand large internal pressures, the preceding calculations also point out that in dilated vessels, wall tension increases even when internal pressure remains constant and may, under certain circumstances (for example, syphilitic aneurysm of the aorta), be an important factor in rupture of the vessel. The above equation also indicates that as the wall of the vessel becomes thicker, the wall stress decreases. In *hypertension* (high blood pressure) the arterial vessel walls thicken (hypertrophy of the vascular smooth muscle),

thereby minimizing the arterial wall stress and hence the possibility of vessel rupture.

The diameter of the resistance vessels is determined by the balance between the contractile force of the vascular smooth muscle and the distending force produced by the intraluminal pressure. The greater the contractile activity of the vascular smooth muscle of an arteriole, the smaller its diameter, until a point is reached, in the case of small arterioles, when complete occlusion of the vessel occurs, in part because of infolding of the endothelium, and the cells trapped in the vessel. With progressive reduction in intravascular pressure, vessel diameter decreases as does tension in the vessel wall (Law of Laplace). When perfusion pressure is reduced, a point is reached where blood flow ceases even though there is still a positive pressure gradient. This phenomenon has been referred to as the "critical closing pressure" and its mechanism is still controversial. This critical closing pressure is low when vasomotor activity is reduced by inhibition of sympathetic nerve activity to the vessel, and it is increased when vasomotor tone is enhanced by activation of the vascular sympathetic nerve fibers. It has been suggested that flow stops because of vessel collapse when vascular smooth muscle contractile force exceeds intraluminal pressure. One difficulty with this hypothesis is that active tension in the vessel wall may decrease with reduction in vessel diameter. An equilibrium between distending force and wall tension is then established, and vessel diameter decreases progressively without abrupt collapse.

TRANSCAPILLARY EXCHANGE

Solvent and solute move across the capillary endothelial wall by three processes: diffusion, filtration, and via endothelial vessels (pinocytosis). By far the greatest number of molecules traverse the capillary endothelium by diffusion.

DIFFUSION

Under normal conditions only about 0.06 ml. of water per minute moves back and forth across the capillary wall per 100 gm. of tissue as a result of filtration and absorption, whereas 300 ml. of water per minute transfer across the endothelium per 100 gm. of tissue by diffusion, a 5,000-fold difference. Relating filtration and diffusion to blood flow, we find that about 2% of the plasma passing through the capillaries is filtered, whereas the diffusion of water is 40 times greater than the rate that it is brought to the capillaries by blood flow. The transcapillary exchange of solutes is also primarily governed by diffusion. Thus diffusion is the key factor in providing exchange of gases, substrates, and waste products between the capillaries and the tissue cells.

The process of diffusion is described by Fick's law:

$$\dot{N} = DA \frac{dc}{dx}$$

where

$\dot{N}$ = Quantity of a substance moved per unit time (t)
D = Free diffusion coefficient for a particular molecule (the value is inversely related to the square root of the molecular weight)
A = Cross-sectional area of the diffusion pathway
$\frac{dc}{dx}$ = Concentration gradient

Fick's law is also expressed as:

$$\dot{N} = PS\,(C_i - C_o)$$

where

P = Capillary permeability of the substance
S = Capillary surface area
C_i = Concentration of the substance inside the capillary
C_o = Concentration of the substance outside the capillary

Hence, the PS product provides a convenient expression of available capillary surface, since permeability is rarely altered under physiological conditions.

In the capillaries, diffusion of lipid-insoluble molecules is not free but is restricted to the pores whose mean size can be calculated by measurement of the diffusion rate of an uncharged molecule whose free diffusion coefficient is known. Movement of solutes across the endothelium is quite complex and involves corrections for attractions between solute and solvent molecules, interactions between solute molecules, pore configuration, and charge on the molecules relative to charge on the endothelial cells (as observed in the kidney). It is not simply a question of random thermal movements of molecules down a concentration gradient.

For small molecules, such as water, NaCl, urea, and glucose, the capillary pores offer little restriction to diffusion (low reflection coefficient) and diffusion is so rapid that the mean concentration gradient across the capillary endothelium is extremely small. However, with lipid-insoluble molecules of increasing size, diffusion through muscle capillaries becomes progressively more restricted, until diffusion becomes minimal with molecules of a molecular weight above about 60,000. With small molecules the only limitation to net movement across the capillary wall is the rate at which blood flow transports the molecules to the capillary surfaces (*flow limited*), whereas with larger molecules diffusion across the capillaries becomes the limiting factor (*diffusion limited*). The rate of diffusion of small lipid-insoluble molecules is so great that it is uninfluenced by filtration in the direction opposite to the concentration gradient of the diffusible substance. In fact, filtration *or* absorption accelerates the movement of tracer ions from interstitial fluid to blood (tissue clearance). The reason for this enhanced tissue clearance is not known, but it may be the result of a stirring effect on the interstitial fluid or to changes in its structure (for example, gel to sol transformation or "canals" in a gel matrix).

Movement of lipid-soluble molecules across the capillary wall is not limited to capillary pores (only about 0.02% of the capillary surface), since such molecules can pass directly through the lipid membranes of the entire capillary endothelium. Consequently, lipid-soluble molecules move with great rapidity between blood and tissue. The degree of lipid solubility (oil-to-water partition coefficient) provides a good index of the ease of transfer of lipid molecules through the capillary endothelium.

Oxygen and carbon dioxide are both lipid-soluble and readily pass through the endothelial cells. Calculations based on the diffusion coefficient for oxygen, capillary density and diffusion distances, blood flow, and tissue oxygen consumption indicate that the oxygen supply of normal tissue at rest and during activity is not limited by diffusion or the number of open capillaries. Recent measurements of PO_2 and saturation of blood in the microvessels indicate that in many tissues oxygen saturation at the entrance of the capillaries has already decreased to a saturation of about 80% as a result of diffusion of oxygen from arterioles. Such studies have also shown that CO_2 loading and the resultant intravascular shifts in the oxyhemoglobin dissociation curve occur in the precapillary vessels. These findings reflect not only the movement of gas to respiring tissue at the precapillary level, but also the direct flux of O_2 and CO_2 between adjacent arterioles, venules, and possibly arteries and veins (counter-current exchange). This exchange of gas represents a diffusional shunt of gas around the capillaries, and at low blood flow rates, it may limit the supply of oxygen to the tissue.

CAPILLARY FILTRATION

The direction and the magnitude of the movement of water across the capillary wall are determined by the algebraic sum of the hydrostatic and osmotic pressures existing across the membrane. An increase in intracapillary hydrostatic pressure favors movement of fluid from the vessel to the interstitial space, whereas an increase in the concentration of osmotically active particles within the vessels favors movement of fluid into the vessels from the interstitial space.

Hydrostatic forces

The hydrostatic pressure (blood pressure) within the capillaries is not constant and is dependent on the arterial pressure, the venous pressure, and the precapillary (arteriolar and precapillary sphincter) and postcapillary (venules and small veins) resistances. A gain in arterial or venous pressure increases capillary hydrostatic pressure, whereas a reduction in each has the opposite effect. Increase in arteriolar resistance or closure of precapillary sphincters reduces capillary pressure, whereas greater venous resistance (venules and veins) increases capillary pressure (Fig. 6-3).

The relationship of pre- and postcapillary pressure and resistance and their effect on capillary pressure can be expressed by the equation:

$$P_c = \frac{(R_v/R_a)\, P_a + P_v}{1 + (R_v/R_a)}$$

where

P_c = Capillary hydrostatic pressure
P_a = Arterial pressure
P_v = Venous pressure
R_a = Precapillary resistance (on the arterial side)
R_v = Postcapillary resistance (on the venous side)

A given increment in venous pressure produces a fivefold to tenfold greater effect on capillary hydrostatic pressure than the same increment in arterial pressure, and about 80% of the increase in venous pressure is transmitted back to the capillaries. The important variable in the preceding equation is the ratio of postcapillary resistance to precapillary resistance (R_v/R_a), where R_a is much greater than R_v (about 4 to 1). Hence, an increase in one resistance (R_a) may be offset by a proportional increase in the other (R_v).

Despite the fact that capillary hydrostatic pressure is variable from tissue to tissue, even within the same tissue, average values, obtained from many direct measurements in human skin, are about 32 mm. Hg at the arterial end of the capillaries and 15 mm. Hg at the venous end of the capillaries at the level of the heart (Fig. 6-4). The hydrostatic pressure in capillaries of the lower extremities will be higher and that of capillaries in the head will be lower in the standing position. Measurements of pressure at the arteriolar and venous

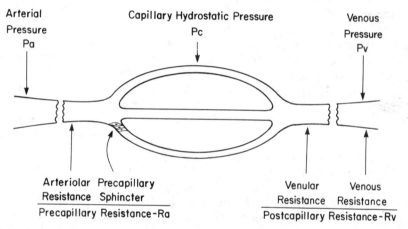

Fig. 6-3. Diagram of the terminal vascular bed, illustrating pre- and postcapillary pressures and resistances used in the calculation of capillary hydrostatic pressure.

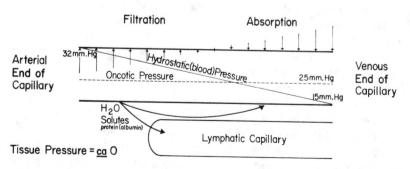

Fig. 6-4. Schematic representation of the factors responsible for filtration and absorption across the capillary wall and the formation of lymph.

ends of the capillaries in different tissues in several mammals give values that are reasonably close to the ones found in human skin capillaries. This hydrostatic pressure is the principal force in filtration across the capillary wall.

Tissue pressure, or more specifically interstitial fluid pressure (P_i) outside the capillaries, opposes capillary filtration and it is $P_c - P_i$ that constitutes the driving force for filtration. The true value of P_i is still controversial. For years it was assumed to be close to zero in the normal (nonedematous) state. However, recent studies, using perforated, plastic capsules implanted in the subcutaneous tissue or wicks inserted through the skin, indicate a negative P_i of about -7 mm. Hg. If the pressures recorded by these techniques are representative of interstitial fluid pressure in undisturbed tissue, then the hydrostatic driving force for capillary filtration is 7 mm. Hg greater than the value of P_c.

Osmotic forces

The key factor that restrains fluid loss from the capillaries is the osmotic pressure of the plasma proteins—usually termed the *colloid osmotic pressure* or *oncotic pressure* (π_p). The total osmotic pressure of plasma is about 6,000 mm. Hg, whereas the oncotic pressure is only about 25 mm. Hg. However, this small oncotic pressure plays an important role in fluid exchange across the capillary wall, since the plasma proteins are essentially confined to the intravascular space, whereas the electrolytes that are responsible for the major fraction of plasma osmotic pressure are practically equal in concentration on both sides of the capillary endothelium. The relative permeability of solute to water influences the actual magnitude of the osmotic pressure. The *reflection coefficient* (σ) is the index of the deviation from Van't Hoff's Law caused by permeability of the solute. Therefore, true osmotic pressure is defined by

$$\pi = \sigma RT (C_i - C_o)$$

where

σ = Reflection coefficient
R = Gas constant
T = Absolute temperature
C_i and C_o = Solute concentration inside and outside the capillary

For albumin, to which the endothelium is essentially impermeable, the reflection coefficient has a value of 1, whereas with small molecules the reflection coefficient is less than 1. Also, different tissues have different reflection coefficients for the same molecule, and therefore movement of a given solute across the endothelial wall varies with the tissue.

Of the plasma proteins, albumin is preponderate in determining oncotic pressure. The average albumin molecule (mol. wt. 69,000) is approximately half the size of the average globulin molecule (mol. wt. 150,000) and is present in almost twice the concentration as the globulins (4.5 gm. vs. 2.5 gm./100 ml. of plasma). Albumin also exerts a greater osmotic force than can be accounted for solely on the basis of the number of molecules dissolved in the plasma and for these reasons cannot be completely replaced by inert substances of appropriate molecular size such as dextran. This additional osmotic force becomes disproportionately greater at high concentrations of albumin (as in plasma) and is weak to absent in dilute solutions of albumin (as in interstitial fluid). One reason for this behavior of albumin is its negative charge at the normal blood pH and the attraction and retention of cations (principally Na^+) in the vascular compartment (the *Gibbs-Donnan effect*). Furthermore, albumin binds a small number of chloride ions, which increases its negative charge and, hence, its ability to retain more sodium ions inside the capillaries. The small increase in electrolyte concentration of the plasma over that of the interstitial fluid produced by the negatively charged albumin enhances its osmotic force to that of an ideal solution containing a solute of molecular weight of 37,000. If albumin did indeed have a molecular weight of 37,000, it would not be retained by the capillary endothelium because of its small size, and obviously could not function as a counter force to capillary hydrostatic pressure. If, however, albumin did not have an enhanced osmotic force, it would require a concentration of about 12 gm. of albumin/100 ml. of plasma to achieve a plasma oncotic pressure of 25 mm. Hg. Such a high albumin concentration would greatly increase blood viscosity and the resistance to blood flow through the vascular system. The other factors that contribute to the nonlinearity of the relationship of albumin concentration to osmotic force are not known. About 65% of plasma oncotic pressure is attributable to albumin, about 15% to the globulins, and the remainder to other ill-defined components of the plasma.

Small amounts of albumin escape from the capillaries and enter the interstitial fluid, where they exert a very small osmotic force (0.1 to 5 mm. Hg). This force (π_i) is small because of the low concentration of albumin in the interstitial fluid and because at low concentrations the osmotic force of albumin becomes simply a function of the number of albumin molecules per unit volume of interstitial fluid.

Balance of hydrostatic and osmotic forces—Starling hypothesis

The relationship between hydrostatic pressure and oncotic pressure and the role of these forces in regulating fluid passage across the capillary endothelium were expounded by Starling in 1896 and constitute the *Starling hypothesis*. It can be expressed by the equation:

$$\text{Fluid movement} = k[(P_c + \pi_i) - (P_i + \pi_p)]$$

where

P_c = Capillary hydrostatic pressure
P_i = Interstitial fluid hydrostatic pressure
π_p = Plasma protein oncotic pressure
π_i = Interstitial fluid oncotic pressure
k = Filtration constant for the capillary membrane

Filtration occurs when the algebraic sum is positive, and absorption when it is negative.

Classically, it has been thought that filtration occurs at the arterial end of the capillary and absorption at its venous end because of the gradient of hydrostatic pressure along the capillary. This is true for the idealized capillary as depicted in Fig. 6-4, but direct observations have revealed that many capillaries show filtration for their entire length whereas others show only absorption. In some vascular beds (for example, the renal glomerulus) hydrostatic pressure in the capillary is high enough to result in filtration along the entire length of the capillary. In other vascular beds, such as in the intestinal mucosa, the hydrostatic and oncotic forces are such that absorption occurs along the whole capillary. As discussed earlier in this chapter, capillary pressure is quite variable and depends on several factors, the principal one being the contractile state of the precapillary vessel. In the normal steady state, arterial pressure, venous pressure, postcapillary resistance, interstitial fluid hydrostatic and oncotic pressures, and plasma oncotic pressure are relatively constant, and change in precapillary resistance is the determining factor with respect to fluid movement across the wall for any given capillary. Since water moves so quickly across the capillary endothelium, the hydrostatic and osmotic forces are nearly in equilibrium along the entire capillary. Hence filtration and absorption in the normal state occur at very small degrees of imbalance of pressures across the capillary wall. Only a small percentage (2%) of the plasma flowing through the vascular system is filtered, and of this about 85% is absorbed in the capillaries and venules. The remainder returns to the vascular system as lymph fluid along with the albumin that escapes from the capillaries.

In the lungs the mean capillary hydrostatic pressure is only about 8 mm. Hg. Since the plasma oncotic pressure is 25 mm. Hg, the forces across the capillary membrane favor absorption. However, pulmonary lymph is formed and consists of fluid that is osmotically drawn out of the capillaries by the small amount of plasma protein that escapes through the capillary endothelium. Only in pathological conditions, such as left ventricular failure or stenosis of the mitral valve, does pulmonary capillary hydrostatic pressure exceed plasma oncotic pressure. When this occurs, it may lead to pulmonary edema, a condition that can seriously interfere with gas exchange in the lungs.

Capillary pores

The permeability of the capillary endothelial membrane is not the same in all body tissues. For example, the liver capillaries are quite permeable and albumin escapes at a rate severalfold greater than from the less permeable muscle capillaries. Also, there is not uniform permeability along the whole capillary; the venous ends are more permeable than the arterial ends, and permeability is greatest in the venules.

The sites where filtration occurs have been a controversial subject for a number of years. Some investigators believe that a fraction of the water flows through the capillary endothelial cell membranes. However, most investigators believe that water flows through apertures in the endothelial wall of the capillaries. Calculations based on the transcapillary movement of solutes of small molecular size led to the prediction of pore diameters of about 40 Å. However, electron microscopy failed to reveal pores, and the clefts at the junctions of endothelial cells appeared to be fused at the tight junctions (Fig. 6-5). However, studies on cardiac and skeletal muscle, with horseradish peroxidase, a protein with a molecular weight of 40,000, have demonstrated that many of the clefts between adjacent endothelial cells are open. Electron microscopy revealed filling of the clefts with peroxidase from the lumen side of the capillaries with a gap at the narrowest point of about 40 Å (*gap junctions*) providing morphological support of the physiological evidence for the existence of capillary pores. The clefts (pores) are sparse and represent only about 0.02% of the capillary surface area. In cerebral capillaries, where a blood-brain barrier to many small molecules exists, peroxidase studies do not reveal any pores. Some studies have failed to reveal the presence of interendothelial pores or clefts, even with microperoxidase (molecular weight of 1900), and transcapillary movement of solute (large and small molecules) is thought to occur through channels formed by the fusion of vesicles (vesicular channels) across the endothelial cells (Fig. 6-5). The basement membrane appears to restrict passage of large molecules (greater than 100 Å radius).

In addition to clefts, some of the more porous capillaries (for example, kidney, intestine) contain fenestrations (Fig. 6-5) 200 to 1,000 Å wide, whereas others (such as in the liver) have a discontinuous endothelium (Fig. 6-5). The fenestrations that appear to be sealed by a thin diaphragm are quite permeable to horseradish peroxidase as well as to a number of other tracers. Hence larger molecules can penetrate capillaries with fenestrations or gaps caused by discontinuous endothelium than can pass through the intercellular clefts of the endothelium.

Capillary filtration coefficient

The rate of movement of fluid across the capillary membrane (Q_f) is dependent not only on the algebraic sum of the hydrostatic and osmotic forces across the endothelium (ΔP), but also on the area of the capillary wall available for filtration (A_m), the distance across the capillary wall (Δx), the viscosity of the filtrate (η), and the filtration constant of the membrane (k). These factors may be expressed by the equation:

$$Q_f = \frac{kA_m\Delta P}{\eta\Delta x}$$

The dimensions are units of flow per unit of pressure gradient across the capillary wall per unit of capillary surface area. It should be apparent that this expression, which describes the flow of fluid through a membrane (pores), is essentially Poiseuille's law for flow through tubes (p. 58).

Since the thickness of the capillary wall and the viscosity of the filtrate are relatively constant, they can be included in the filtration constant, k, and if the area of the capillary membrane is not known, the rate of filtration can be expressed per unit weight of tissue. Hence the equation can be simplified to

$$Q_f = k_t \Delta P$$

where k_t is the capillary filtration coefficient (CFC) for a given tissue, and the units for Q_f are ml./min./100 gm. of tissue/mm. Hg.

The rate of filtration and absorption are determined by the rate of change in tissue weight or volume at different mean capillary hydrostatic pressures that are altered by adjustment of arterial and venous pressures. At the isogravimetric or isovolumetric point (constant weight or constant volume, respectively, as continuously measured with an appropriate scale or volume recorder), the hydrostatic and osmotic forces are balanced across the capillary wall and there is neither net filtration nor absorption. An abrupt increase in arterial pressure will increase capillary hydrostatic pressure and fluid will move from the capillaries to the interstitial fluid compartment. Since the pressure increment, the weight increase of the tissue per unit time, and the total weight of the tissue are known, the capillary filtration coefficient (CFC or k_t) in ml./min./100 gm. tissue/ mm. Hg can be calculated. With the isogravimetric and isovolumetric techniques, it is assumed that 80% of increments in venous pressure are transmitted back to the capillaries, that precapillary and postcapillary resistances are constant when venous pressure is changed, and that the weight or volume change that occurs immediately after raising venous pressure is the result of vascular distention and not filtration. These assumptions may not always be correct; nevertheless, the filtration coefficient constitutes a useful index of capillary permeability and surface area.

In any given tissue the filtration coefficient per unit area of capillary surface, and hence capillary permeability, is not changed by different physiological conditions, such as arteriolar dilation and capillary distension, or by such adverse conditions as hypoxia, hypercapnia, or reduced pH. With capillary injury (toxins, severe burns) capillary permeability increases greatly, as indicated by the filtration coefficient, and significant amounts of fluid and protein leak out of the capillaries into the interstitial space.

Since capillary permeability is constant under normal conditions, the filtration coefficient can be used to determine the relative number of open capillaries (total capillary surface area available for filtration in tissue). For example, increased metabolic activity of contracting skeletal muscle induces relaxation of precapillary resistance vessels with opening of more capillaries (*capillary recruitment*, resulting in an increased filtering surface).

Some protein is apparently required to maintain the integrity of the endothelial membrane. If the plasma proteins are replaced by nonprotein colloids so as to give the same oncotic pressure, the filtration coefficient is doubled and edema occurs. However, if as little as 0.2% protein is added, normal permeability is restored. Possibly the protein molecules are adsorbed on the endothelial membrane and alter pore dimensions.

Disturbances in hydrostatic-osmotic balance

Changes in arterial pressure per se may have little effect on filtration, since the change in pressure may be countered by adjustments of the precapillary resistance vessels (autoregulation, see p. 124), so that hydrostatic pressure in the open capillaries remains the same. However, with severe reduction in arterial pressure, as may occur in hemorrhage, there may be arteriolar constriction mediated by the sympathetic nervous system and a fall in venous pressure resulting from the blood loss. These changes will lead to a decrease in the capillary hydrostatic pressure. Furthermore, the low blood pressure causes a decrease in blood flow (and hence oxygen supply) to the tissue with the result that vasodilator metabolites accumulate and induce relaxation of arterioles. Precapillary vessel relaxation is also engendered by the reduced transmural pressure. As a consequence of these several factors, absorption predominates over filtration and occurs at a larger capillary surface area. This is one of the compensatory mechanisms employed by the body to restore blood volume (p. 264).

An increase in venous pressure alone, as occurs in the feet when one changes from the lying to the standing position, would elevate capillary pressure and enhance filtration. However, the increase in transmural pressure causes precapillary vessel closure, so that the capillary filtration coefficient actually decreases. This reduction in capillary surface available for filtration serves to protect against the extravasation of large amounts of fluid into the interstitial space (edema). With prolonged standing, particularly when associated with some elevation of venous pressure in the legs (such as that caused by tight garters or pregnancy) or with sustained increases in venous pressure as seen in congestive heart failure, filtration is greatly enhanced and exceeds the capacity of the lymphatic system to remove the capillary filtrate from the interstitial space.

A large amount of fluid can move across the capillary wall in a relatively short time. In a normal individual the filtration coefficient (k_t) for the whole body is about 0.0061 ml./min./100 gm. of tissue/mm. Hg. For a 70 kg. man, elevation of venous pressure of 10 mm. Hg for 10 minutes would increase filtration from cap-

illaries by 342 ml. This would not lead to edema formation, since the fluid is returned to the vascular compartment by the lymphatic vessels. When edema does develop, it usually appears in the dependent parts of the body, where the hydrostatic pressure is greatest, but its location and magnitude are also determined by the type of tissue. Loose tissues, such as the subcutaneous tissue around the eyes or in the scrotum, are more prone to collect larger quantities of interstitial fluid than are firm tissues, such as muscle, or encapsulated structures, such as the kidney.

The concentration of the plasma proteins may also change in different pathological states and, hence, alter the osmotic force and movement of fluid across the capillary membrane. The plasma protein concentration is increased in dehydration (for example, water deprivation, prolonged sweating, severe vomiting, and diarrhea), and water moves by osmotic forces from the tissues to the vascular compartment. In contrast, the plasma protein concentration is reduced in nephrosis (a renal disease in which there is loss of protein in the urine) and edema may occur. When there is extensive capillary injury, as in burns, leaks occur and plasma protein escapes into the interstitial space along with fluid and increases the oncotic pressure of the interstitial fluid. This greater osmotic force outside the capillaries leads to additional fluid loss and possibly to severe dehydration of the patient.

PINOCYTOSIS

Some transfer of substances across the capillary wall can occur in tiny pinocytotic vesicles (pinocytosis). These vesicles (Fig. 6-5), formed by a pinching off of the surface membrane, can take up substances on one side of the capillary wall, move across the cell, and deposit their contents at the other side. The amount of material that can be transported in this way is very small relative to that moved by diffusion. However, pinocytosis may be

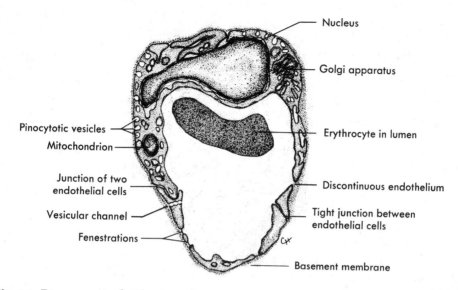

Fig. 6-5. Diagrammatic sketch of an electron micrograph of a composite capillary in cross section.

responsible for the movement of large lipid-insoluble molecules between blood and interstitial fluid.

LYMPHATICS

The terminal vessels of the lymphatic system consist of a widely distributed closed-end network of highly permeable lymph capillaries that are similar in appearance to blood capillaries. However, they are generally lacking in tight junctions between endothelial cells and possess fine filaments that anchor them to the surrounding connective tissue. With muscular contraction these fine strands may distort the lymphatic vessel to open spaces between the endothelial cells and permit the entrance of protein and large particles and cells present in the interstitial fluid. The lymph capillaries drain into larger vessels that finally enter the right and left subclavian veins at their junctions with the respective internal jugular veins. Only cartilage, bone, epithelium, and the tissues of the central nervous system are devoid

of lymphatic vessels. The plasma capillary filtrate is returned to the circulation by virtue of tissue pressure, facilitated by intermittent skeletal muscle activity, contractions of the lymphatic vessels, and an extensive system of one-way valves. In this respect they resemble the veins, although even the larger lymphatic vessels have thinner walls than the corresponding veins and contain only a small amount of elastic tissue and smooth muscle.

The volume of fluid transported through the lymphatics in 24 hours is about equal to the animal's total plasma volume and the protein returned by the lymphatics to the blood in a day is about one-fourth to one-half of the circulating plasma proteins. This is the only means whereby protein (albumin) that leaves the vascular compartment can be returned to the blood, since back diffusion into the capillaries cannot occur against the large albumin concentration gradient. Were the protein not removed by the lymph vessels, it would accumulate in the interstitial fluid and act as an oncotic force to draw fluid from the blood capillaries to produce increasingly severe edema. In addition to returning fluid and protein to the vascular bed, the lymphatic system filters the lymph at the lymph nodes and removes foreign particles, such as bacteria. The largest lymphatic vessel, the *thoracic duct,* in addition to draining the lower extremities, serves to return protein lost through the permeable liver capillaries and to carry substances absorbed from the gastrointestinal tract, principally fat in the form of chylomicrons, to the circulating blood.

Lymph flow varies considerably, being almost nil from resting skeletal muscle and increasing during exercise in proportion to the degree of muscular activity. It is increased by any mechanism that enhances the rate of blood capillary filtration, for example, increased capillary pressure or permeability or decreased plasma oncotic pressure. When either the volume of interstitial fluid exceeds the drainage capacity of the lymphatics or the lymphatic vessels become blocked, as may occur in certain disease states, interstitial fluid accumulates, chiefly in the more compliant tissues (for example, subcutaneous tissue) and gives rise to clinical edema.

BIBLIOGRAPHY
Journal articles

Bundgaard, M.: Transport pathways in capillaries—in search of pores, Ann. Rev. Physiol. **42**:325, 1980.

Duling, B. R., and Berne, R. M.: Longitudinal gradients in periarteriolar oxygen tension, Circ. Res. **27**:669, 1970.

Duling, B. R., and Klitzman, B.: Local control of microvascular function: role in tissue oxygen supply, Ann. Rev. Physiol **42**:373, 1980.

Garlick, D. G., and Renkin, E. M.: Transport of large molecules from plasma to interstitial fluid and lymph in dogs, Am. J. Physiol. **219**:1595, 1970.

Gauer, O. H., Crone, C., Guyton, A. C., Hammersen, F. Lauent, T. C., and Zweifach, B. W.: Proceedings of a symposium on capillary exchange and the interstitial space, Pfluegers. Arch. **336**:S1, 1972.

Gore, R. W., and McDonagh, P. F.: Fluid exchange across single capillaries, Ann. Rev. Physiol. **42**:337, 1980.

Karnovsky, M. J.: The ultrastructural basis of transcapillary exchanges, J. Gen. Physiol. **52**:645, 1968.

Krogh, A.: The number and distribution of capillaries in muscles with calculations of the oxygen pressure head necessary for supplying the tissue. J. Physiol. (London) **52**:409, 1919.

Leak, L. V.: Electron microscopic observations on lymphatic capillaries and the structural components of the connective tissue–lymph interface, Microvasc. Res. **2**:361, 1970.

Lewis, D. H., editor: Symposium on lymph circulation, Acta Physiol. Scand. Suppl. 463, 1979, pp. 9-127.

Palade, G. E.: Blood capillaries of the heart and other organs, Circulation **24**:368, 1961.

Rosell, S.: Neuronal control of microvessels, Ann. Rev. Physiol. **42**:359, 1980.

Starling, E. H.: On the absorption of fluids from the connective tissue spaces, J. Physiol. (London) **19**:312, 1896.

Books and monographs

Guyton, A. C., Taylor, A. E., and Granger, H. J.: Circulatory physiology II: dynamics and control of the body fluids, Philadelphia, 1975, W. B. Saunders Co.

Johnson, P. C.: The microcirculation, and local and humoral control of the circulation. In Guyton, A. C., and Jones, C. E., editors: Cardiovascular physiology, Series 1, London, 1974, Butterworths.

Johnson, P. C., editor: Peripheral circulation, New York, 1978, John Wiley & Sons, Inc.

Kaley, G., and Altura, A., editors: Microcirculation, vols. 1 and 2, Baltimore, 1977, 1978, University Park Press.

Krogh, A.: The anatomy and physiology of capillaries, New York, 1959, Hafner Co.

Yoffey, J. M., and Courtice, F. C.: Lymphatics, lymph and the lymphomyeloid complex, London, 1970, Academic Press, Inc.

CHAPTER 7

THE PERIPHERAL CIRCULATION AND ITS CONTROL

The peripheral circulation is essentially under dual control, centrally through the nervous system and locally in the tissues by the environmental conditions in the immediate vicinity of the blood vessels. The relative importance of these two control mechanisms is not the same in all tissues. In some areas of the body, such as the skin and the splanchnic regions, neural regulation of blood flow predominates, whereas in others, such as the heart and brain, this mechanism plays a minor role.

The vessels chiefly involved in regulating the rate of blood flow throughout the body are referred to as the *resistance vessels,* since these blood vessels offer the greatest resistance to the flow of blood pumped to the tissues by the heart and thereby are important in the maintenance of arterial blood pressure. Smooth muscle fibers constitute a large percentage of the composition of the walls of the resistance vessels (Fig. 1-1). Therefore, the vessel lumen can be varied from one that is completely obliterated by strong contraction of the smooth muscle, with infolding of the endothelial lining, to one that is maximally dilated as a result of full relaxation of the vascular smooth muscle. Some resistance vessels are closed at any given moment in time and partial contraction (or *tone*) of the vascular smooth muscle exists in the arterioles. Were all the resistance vessels in the body to dilate simultaneously, blood pressure would fall precipitously to very low levels.

VASCULAR SMOOTH MUSCLE

Vascular smooth muscle is the tissue responsible for the control of total peripheral resistance, arterial and venous tone, and the distribution of blood flow throughout the body. The smooth muscle cells are small, mononucleate, spindle shaped, and generally arranged in one or more helical or circular layers around the blood vessels. In general, the close association between action potentials and contraction observed in skeletal and cardiac muscle cells cannot be demonstrated in vascular smooth muscle. However, graded changes in membrane potential are often associated with increases or decreases in force. Contractile activity is generally elicited by neural or humoral stimuli. The behavior of smooth muscle in different vessels is quite variable. For example some vessels, particularly in the portal or mesenteric circulation, contain longitudinally oriented smooth muscle that is spontaneously active and that shows action potentials that are correlated with the contractions and the electrical coupling between cells.

The cells contain large numbers of thin, actin-containing filaments and comparatively small numbers of thick, myosin-containing filaments. These filaments are not arranged in transverse register to form visible sarcomeres, although the sliding filament mechanism is believed to operate in this tissue. Compared to skeletal muscle, the smooth muscle contracts very slowly, but it can develop high forces and operates over a considerable range of lengths under physiological conditions.

The interaction between myosin and actin, leading to contraction, is controlled by the myoplasmic Ca^{++} concentration as in other muscles but the molecular mechanism whereby Ca^{++} regulates contraction appears to differ. The increased myoplasmic Ca^{++} that initiates contraction can come from intracellular stores in the sarcoplasmic reticulum, be displaced from the plasma membrane, or pass into the cell following an increase in membrane Ca^{++} permeability. The relative importance of intracellular and extracelular Ca^{++} for activation varies with different vascular smooth muscles and different agonists.

Most of the arteries and veins of the body are supplied to different degrees solely by fibers of the sympathetic nervous system. These nerve fibers exert a tonic effect on the blood vessels, as evidenced by the fact that cutting or freezing the sympathetic nerves to a vascular bed (such as muscle) results in an increase in blood flow. Activation of the sympathetic nerves either directly or reflexly (see p. 129) enhances vascular resistance. In contrast to the sympathetic nerves the parasympathetic nerves tend to decrease vascular resistance, but they innervate only a small fraction of the blood vessels in the body, mainly in certain viscera and pelvic organs. Vascular smooth muscle also responds to humoral stimulation (hormones and drugs) without evidence of electrical excitation. This has been referred to as *pharmacomechanical coupling* and is mediated by Ca^{++} influx or release. In the category of pharmacological stimuli are such substances as catecholamines, histamine, acetylcholine, serotonin, angiotensin, adenosine, and the prostaglandins. Local environmental changes alter the contrac-

tile state of vascular smooth muscle and such alterations as increased temperature or increased CO_2 levels induce relaxation of this tissue.

In studies on this interesting and important type of muscle, great care should be taken in extrapolating results from one tissue to another or from the same tissue under different physiological conditions. For example, some agents elicit vasodilation in some vascular beds and vasoconstriction in others.

INTRINSIC OR LOCAL CONTROL OF PERIPHERAL BLOOD FLOW

In a number of different tissues the blood flow appears to be adjusted to the existing metabolic activity of the tissue. Furthermore, imposed changes in the perfusion pressure (arterial blood pressure) at constant levels of tissue metabolism, as measured by oxygen consumption, are met with vascular resistance changes that tend to maintain a constant blood flow. This mechanism is commonly referred to as *autoregulation* of blood flow and is illustrated graphically in Fig. 7-1. In the skeletal muscle preparation from which these data were gathered, the muscle was completely isolated from the rest of the animal and was in a resting state. From a control pressure of 100 mm. Hg, the pressure was abruptly increased or decreased and the blood flows observed immediately after changing the perfusion pressure are represented by the closed circles. Maintenance of the altered pressure at each new level was followed within 30 to 60 seconds by a return of flow to or toward the control levels; the open circles represent these steady state flows. Over the pressure range 20 to 120 mm. Hg, the steady state flow is relatively constant. Calculation of resistance across the vascular bed (pressure/flow) during steady state conditions indicates that with elevation of perfusion pressure the resistance vessels constricted, whereas with reduction of perfusion pressure, dilation occurred.

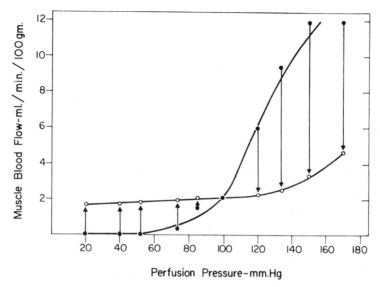

Fig. 7-1. Pressure-flow relationship in the skeletal muscle vascular bed of the dog. The closed circles represent the flows obtained immediately after abrupt changes in perfusion pressure from the control level (point where lines cross). The open circles represent the steady state flows obtained at the new perfusion pressure. (Redrawn from Jones, R. D., and Berne, R. M.: Circ. Res. **14:**126, 1964.)

The mechanism responsible for this constancy of blood flow in the presence of an altered perfusion pressure is not known. However, three explanations have been suggested—the *tissue pressure hypothesis,* the *myogenic hypothesis,* and the *metabolic hypothesis.* According to the tissue pressure concept, an increase in perfusion pressure produces an increase in blood volume of the tissue and a net transfer of fluid from the intravascular to the extravascular compartments. The resultant increase in tissue pressure (turgor) is believed to compress the very thin-walled vessels such as the capillaries, venules, and veins, and thereby to reduce the flow of blood into the tissue. A reduction in perfusion pressure would elicit the opposite response. Such a mechanism can operate only in an encapsulated structure where expansion of the tissue is restricted. However, even in the kidney,

which possesses a fairly rigid connective tissue capsule and shows a high degree of autoregulation of blood flow, conclusive evidence for a tissue pressure mechanism is lacking.

The myogenic hypothesis states that the vascular smooth muscle contracts in response to stretch and relaxes with a reduction in tension. Therefore, the initial flow increment produced by an abrupt increase in perfusion pressure that passively distends the blood vessels would be followed by a return of flow to the previous control level by contraction of the smooth muscles of the resistance vessels. One difficulty with the myogenic hypothesis is that for flow to remain constant following an increase in perfusion pressure, it is necessary for the caliber of the resistance vessels to be less than it was prior to the elevation of pressure. Should this occur, the stretch stimulus for the maintenance of increased resistance would be removed.

However, if an increase in vessel wall tension, and not stretch, triggers a contractile (constrictor) response, then it is possible, by consideration of the Laplace equation ($T = Pr$, see p. 111) to explain autoregulation in terms of the myogenic mechanism. Raising perfusion pressure increases the pressure across the vessel wall and by passive stretch increases the radius, resulting in a large increment in wall tension. Constriction of the vessel in response to the increase in tension results in a reduction of its diameter to less than the original value so that the product of pressure (increased) and radius (decreased) is restored to the control level. Since the resistance vessels also show intermittent contraction and relaxation, it is possible to avoid the paradox encountered when a simple maintained contractile response to stretch is postulated, by proposing an increase in the frequency of contractile responses (via stretch-induced increase in frequency of action potentials) with elevation of perfusion pressure. If such a mechanism were operative, the resistance vessels as a whole would spend more time in the contracted state when pressure was raised than they did prior to elevation of perfusion pressure.

The myogenic mechanism has been demonstrated in certain tissues and in isolated arterioles. Since blood pressure is reflexly maintained at a fairly constant level under normal conditions, operation of a myogenic mechanism would be expected to be minimized. However, when one changes position (from a lying to a standing position) a large change in transmural pressure occurs in the lower extremities. Were it not for the fact that the precapillary vessels constrict in response to this imposed stretch, the hydrostatic pressure in the lower parts of the legs would reach such high levels that large volumes of fluid would pass from the capillaries into the interstitial fluid compartment and produce edema.

According to the metabolic hypothesis, the blood flow is governed by the metabolic activity of the tissue, and any intervention that results in an oxygen supply that is inadequate for the requirements of the tissue gives rise to the formation of vasodilator metabolites. These metabolites are released from the tissue and act locally to dilate the resistance vessels. When the metabolic rate of the tissue increases or the oxygen delivery to the tissue decreases, more vasodilator substance is released and the metabolite concentration in the tissue increases. When metabolic activity at constant perfusion pressure decreases or perfusion pressure at constant metabolic activity increases, the tissue concentration of the vasodilator agent falls. A decrease in metabolite production or an increase in washout and/or inactivation of the metabolite elicits an increase in precapillary resistance. An attractive feature of the metabolic hypothesis is that, in most tissues, blood flow closely parallels metabolic activity. Hence, although blood pressure is kept fairly constant, metabolic activity and blood flow in the different body tissues vary together under physiological conditions (for example, exercise).

Many substances have been proposed as mediators of metabolic vasodilation. Some of the earliest ones suggested are lactic acid, CO_2, and hydrogen ion. However, the decrease in vascular resistance induced by supernormal concentrations of these dilator agents falls considerably short of the dilation observed under physiological conditions of increased metabolic activity.

Changes in oxygen tension can evoke changes in the contractile state of vascular smooth muscle; an increase in P_{O_2} elicits contraction, and a decrease in P_{O_2}, relaxation. If significant reductions in the intravascular P_{O_2} occur before the arterial blood reaches the resistance vessels (diffusion through the arterial and arteriolar walls, see p. 114), small changes in oxygen supply and/or consumption could elicit contraction or relaxation of the resistance vessels. However, direct measurements of P_{O_2} at the resistance vessels indicate that over a

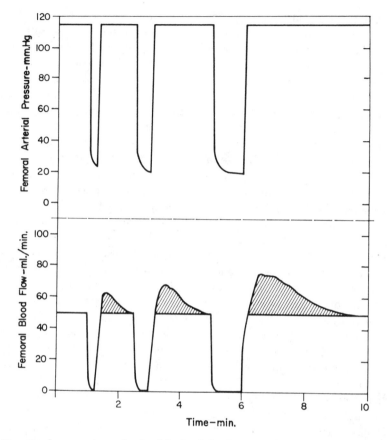

Fig. 7-2. Reactive hyperemia in the hind limb of the dog after 15-, 30-, and 60-second occlusions of the femoral artery.

wide range of P_{O_2} (11 to 343 mm. Hg) there is no correlation between oxygen tension and arteriolar diameter. Furthermore, if P_{O_2} were directly responsible for vascular smooth muscle tension, one would not expect to find a parallelism between the duration of arterial occlusion and the duration of the reactive hyperemia (Fig. 7-2) observed upon release of the occlusion. With either short occlusions (5 to 10 seconds) or long occlusions (1 to 3 minutes) the venous blood becomes bright red (well oxygenated) within 1 to 2 seconds after release of the arterial occlusion and hence the smooth muscle of the resistance vessels must be exposed to a

high P_{O_2} in each instance. Nevertheless, the longer occlusions result in longer periods of reactive hyperemia. These observations are more compatible with the release of a vasodilator metabolite from the tissue than with a direct effect of P_{O_2} on the vascular smooth muscle.

The potassium ion, inorganic phosphate, and interstitial fluid osmolarity can also induce vasodilation; since potassium and phosphate are released and osmolarity is increased during skeletal muscle contraction, it has been proposed that these factors contribute to *active hyperemia* (increased blood flow caused by enhanced tissue activity). However, significant increases of phosphate concentration and osmolarity are not consistently observed during muscle contraction and they may produce only transient increases in blood flow. Therefore, they are not likely candidates as mediators of the vasodilation observed with muscular activity. Potassium release occurs with the onset of skeletal muscle contraction or an increase in cardiac activity, and could be responsible for the initial decrease in vascular resistance observed with exercise or increased cardiac work. However, potassium release is not sustained, despite continued arteriolar dilation throughout the period of enhanced muscle activity. Therefore, some other agent must serve as mediator of the vasodilation associated with the greater metabolic activity of the tissue. Reoxygenated venous blood obtained from active cardiac and skeletal muscles under steady state conditions of exercise does not elicit vasodilation when infused into a test vascular bed. It is difficult to see how oxygenation of the venous blood could alter its potassium or phosphate content or its osmolarity and thereby destroy its vasodilator effect.

Recent evidence indicates that adenosine, which plays a role in the regulation of coronary blood flow, may also participate in the control of the resistance vessels in skeletal muscle; also, some of the prostaglandins have been proposed as important vasodilator mediators in certain vascular beds.

Thus there are a number of candidates for the mediator of metabolic vasodilation, and the relative contribution of each of the various factors remains the subject for future investigation. Several factors may be involved in any given vascular bed, and different factors preponderate in different tissues.

Metabolic control of vascular resistance via the release of a vasodilator substance is predicated on the existence of basal vessel tone. This tonic activity, or *basal tone,* of the vascular smooth muscle is readily demonstrable, but in contrast to tone in skeletal muscle it is independent of the nervous system. The factor re-

sponsible for basal tone in blood vessels is not known, but one or more of the following factors may be involved: (1) an expression of myogenic activity in response to the stretch imposed by the blood pressure, (2) the high oxygen tension of arterial blood, (3) the presence of calcium ions, or (4) some unknown factor in plasma, since addition of plasma to the bathing solution of isolated vessel segments evokes partial contraction of the smooth muscle.

A phenomenon that is mechanistically linked to autoregulation of blood flow is *reactive hyperemia*. If arterial inflow to a vascular bed is stopped for a few seconds to several minutes, the blood flow, on release of the occlusion, immediately exceeds the flow before occlusion and only gradually returns to the control level. This is illustrated in Fig. 7-2, where blood flow to the leg was stopped by clamping the femoral artery for 15, 30, and 60 seconds. Release of the 60-second occlusion resulted in a peak blood flow 70% greater than the control flow, with a return to control flow within about 110 seconds. When this same experiment is done in humans by inflating a blood pressure cuff on the upper arm, dilation of the resistance vessels of the hand and forearm, immediately after release of the cuff, is evident from the bright red color of the skin and the fullness of the veins. Within limits the peak flow and particularly the duration of the reactive hyperemia are proportional to the duration of the occlusion (Fig. 7-2). If the extremity is exercised during the occlusion period, reactive hyperemia is increased. These observations and the close relationship that exists between metabolic activity and blood flow in the unoccluded limb are consonant with a metabolic mechanism in the local regulation of tissue blood flow.

EXTRINSIC CONTROL OF PERIPHERAL BLOOD FLOW
Neural sympathetic vasoconstriction

There are a number of regions in the medulla that influence cardiovascular activity.

Some of the effects of stimulation of the dorsal lateral medulla are vasoconstriction, cardiac acceleration, and enhanced myocardial contractility. Caudal and ventromedial to the *pressor* region is a zone that upon stimulation produces a decrease in blood pressure. This *depressor area* exerts its effect by direct spinal inhibition and by inhibition of the medullary pressor region. However, the precise mechanism of its depressor actions is still unknown. These areas comprise a center not in an anatomical sense in that a discrete group of cells is discernible, but in a physiological sense in that stimulation of the pressor region produces the responses mentioned previously. From the vasoconstrictor regions, fibers descend in the spinal cord and synapse at different levels of the thoracolumbar region (T-1 to L-2 or L-3). Fibers from the intermediolateral gray matter of the cord emerge with the ventral roots, but leave the motor fibers to join the paravertebral sympathetic chains through the white rami communicans. These preganglionic white (myelinated) fibers may pass up or down the sympathetic chains to synapse in the various ganglia within the chains or in certain outlying ganglia. Postganglionic gray rami (unmyelinated) then join the corresponding segmental spinal nerves and accompany them to the periphery to innervate the arteries and veins. Postganglionic sympathetic fibers from the various ganglia join the large arteries and accompany them as an investing network of fibers to the resistance and capacitance vessels.

The vasoconstrictor regions are tonically active, and reflexes or humoral stimuli that enhance this activity result in an increase in frequency of impulses reaching the terminal branches to the vessels, where a constrictor neurohumor (norepinephrine) is released and elicits constriction (alpha adrenergic effect) of the resistance vessels. Inhibition of the vasoconstrictor areas reduces their tonic activity and hence diminishes the frequency of impul-

ses in the efferent nerve fibers, resulting in vasodilation. In this manner neural regulation of the peripheral circulation is accomplished primarily by alteration of the number of impulses passing down the vasoconstrictor fibers of the sympathetic nerves to the blood vessels. The vasomotor regions may show rhythmic changes in tonic activity manifested as oscillations of arterial pressure that occur at the frequency of respiration (*Traube-Hering waves*) or that are independent of and at a lower frequency than respiration (*Mayer waves*).

Sympathetic constrictor influence on resistance and capacitance vessels

The vasoconstrictor fibers of the sympathetic nervous system supply the arterioles and the veins. The arteries and larger veins also receive sympathetic innervation, but neural influence on the larger vessels is of far less functional importance than it is on the microcirculation. Capacitance vessels are apparently more responsive to sympathetic nerve stimulation than are resistance vessels, since they reach maximal constriction at a lower frequency of stimulation than do the resistance vessels. Norepinephrine is the neurotransmitter released at the sympathetic nerve terminals at the blood vessels, and many factors, such as circulating hormones and particularly locally released substances, modify the liberation of norepinephrine from the vesicles of the nerve terminals. The response of the resistance and capacitance vessels to stimulation of the sympathetic fibers is illustrated in Fig. 7-3. At constant arterial pressure, sympathetic fiber stimulation evoked a reduction of blood flow (constriction of the resistance vessels) and a decrease in blood volume of the tissue (constriction of the capacitance vessels). The abrupt decrease in tissue volume is caused by movement of blood out of the capacitance vessels and out of the hindquarters of the cat, whereas the late, slow, progressive decline in volume (to the right of the arrow) is caused by movement of extravascular fluid into the capil-

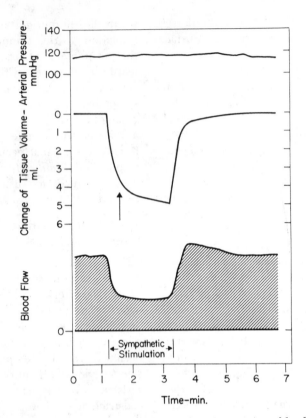

Fig. 7-3. Effect of sympathetic nerve stimulation (2 impulses/sec.) on blood flow and tissue volume in the hindquarters of the cat. The arrow denotes the change in slope of the tissue volume curve where the volume decrease due to emptying of capacitance vessels ceases and loss of extravascular fluid becomes evident. (Redrawn from Mellander, S.: Acta Physiol. Scand. **50**[Supp. 176]:1-86, 1960.)

laries and hence away from the tissue. The loss of tissue fluid is a consequence of the lowered capillary hydrostatic pressure brought about by constriction of the resistance vessels, with establishment of a new equilibrium of the forces responsible for filtration and absorption across the capillary wall.

In addition to active changes (contraction and relaxation of the vascular smooth muscle) in vessel caliber, there are also passive changes caused solely by alterations in intraluminal pressure; an increase in intraluminal pressure produces distension of the vessels and a decrease produces a reduction in caliber by recoil

of the elastic components of the vessel walls. The relative effects of the passive and active forces on the volume changes of the tissues are depicted in Fig. 7-4. With occlusion of the arterial blood supply to the tissue, perfusion pressure dropped almost to zero, blood flow became nil, and there was an abrupt 2 ml. decrease in tissue volume. This rapid reduction in tissue volume is a passive response to the reduction in arterial pressure, with expulsion of blood from the tissue. The slow decrease in tissue volume (between the first two arrows) is quite small, relative to that shown in Fig. 7-3, because of the brevity of the period between

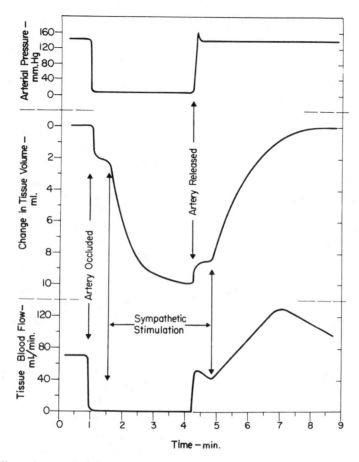

Fig. 7-4. Effect of arterial occlusion and sympathetic nerve stimulation (6 impulses/sec.) on the tissue volume in the hindquarters of the cat. Note the small changes in tissue volume with arterial occlusion and release compared to the large volume change obtained with sympathetic nerve stimulation. (Redrawn from Mellander, S.: Acta Physiol. Scand. 50[Supp. 176]:1-86, 1960.)

arterial occlusion and the start of sympathetic stimulation. This slight reduction in volume is caused by the decrease in capillary hydrostatic pressure incident to the reduction of the perfusion pressure. It represents loss of extravascular fluid via absorption from the capillaries and movement out through the venous channels. When the sympathetic nerve fibers were stimulated during the period of arterial occlusion, a large decrease in tissue volume was observed. Since blood flow had already been stopped by arterial occlusion, this change in tissue volume cannot be a passive response secondary to arteriolar constriction but must represent active constriction of the capacitance vessels. Note that once again a brief slow decrease in tissue volume followed the rapid change and indicates a continuation of the loss of extravascular fluid as a result of the reduction in capillary hydrostatic pressure. If the capillary pressure is held at low levels for 2 to 3 minutes, either by arterial occlusion or by constriction of the resistance vessels, a new pressure equilibrium is established across the capillary wall and further loss of tissue fluid becomes negligible. The protein concentration of the blood leaving the tissue during the period of slow decrease in tissue volume was found to be reduced, indicating dilution of the plasma proteins by the absorption of low-protein tissue fluid by the capillaries. When the arterial occlusion was released during continuous sympathetic nerve stimulation, a small abrupt increase in intravascular pressure occurred because of a passive stretch of the capacitance vessels by the increased intraluminal pressure. This rapid increment in tissue volume was followed by a small gradual increase (between pair of arrows to the right) that represents movement of fluid from the capillaries into the interstitial spaces in response to the elevation of capillary hydrostatic pressure. Cessation of sympathetic stimulation resulted in relaxation of the smooth muscle of the capacitance vessels and a restoration of their pre-existing blood volume. Note that blood flow increased above, and gradually returned toward, the control level, illustrating a typical reactive hyperemia response to the period of ischemia. From these observations it can be seen that the passive changes of the capacitance vessels are small, relative to those induced by active contraction and relaxation of vascular smooth muscle of the capacitance vessels. Under physiological conditions, in which such drastic reduction in perfusion pressure would not occur, the contribution of a passive factor to changes in volume of the capacitance vessels would be even less than shown in Fig. 7-4.

At basal tone approximately one-third of the blood volume of a tissue can be mobilized on stimulation of the sympathetic nerves at physiological frequencies. The basal tone is very low in capacitance vessels; with veins denervated, only small increases in volume are obtained with maximal doses of the vasodilator, acetylcholine. Therefore the blood volume at basal tone is close to the maximal blood volume of the tissue. The amount of blood that can be mobilized from skin and muscle at different frequencies of stimulation of the sympathetic nerves is depicted in Fig. 7-5. The greater mobilization of blood from the skin than from the muscle capacitance vessels may in part be caused by a greater sensitivity of these vessels to sympathetic stimulation, but it is also caused by the fact that basal tone is lower in skin than in muscle. Therefore, in the absence of neural influence, the skin capacitance vessels contain more blood than do the muscle capacitance vessels.

Blood is mobilized from capacitance vessels in response to physiological stimuli. In exercise, activation of the sympathetic nerve fibers produces constriction of veins and hence augments the cardiac filling pressure. Also, in arterial hypotension (as in hemorrhage), the capacitance vessels constrict to aid in overcoming the decreased central venous pressure associated with this condition. In addition the resistance vessels constrict in shock, thereby assisting in the maintenance or restoration of arterial pressure. With arterial hypotension the enhanced arteriolar constriction also leads to a small mobilization of blood from the tissue by virtue of recoil of the postarteriolar vessels when intraluminal pressure is reduced. Further, there is mobilization of extravascular fluid because of greater absorption in response to the lowered capillary hydrostatic pressure.

Clear dissociation between responses of the resistance and capacitance vessels can be demonstrated with nerve stimulation or with the use of epinephrine and acetylcholine. In Fig. 7-6, A, a small intravenous dose of epinephrine elicited an increase in blood flow (active dilation of resistance vessels) and a small abrupt increase in tissue volume (passive expansion of the capacitance vessels by the increased intraluminal pressure). Following the passive increase in tissue volume (to the left of the first

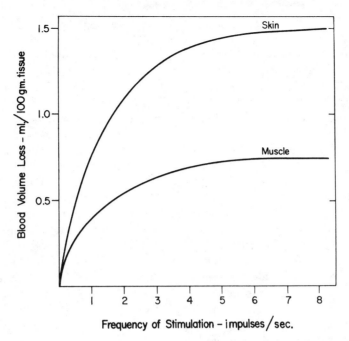

Fig. 7-5. Comparison of the blood volume loss from skin and muscle with sympathetic nerve stimulation. (Redrawn from Mellander, S.: Acta Physiol. Scand. **50**[Supp. 176]:1-86, 1960.)

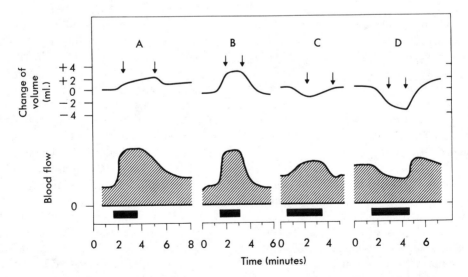

Fig. 7-6. Effect of acetylcholine and different doses of epinephrine on resistance and capacitance vessels in skeletal muscle. *A*, Epinephrine 0.3 μg/kg./min. *B*, Acetylcholine 1.7 μg/kg./min. (to give same degree of dilation of resistance vessels as in *A*). *C*, Epinephrine 0.7 μg/kg./min. *D*, Epinephrine 3.0 μg/kg./min. Pairs of arrows indicate periods of filtration and absorption. (Redrawn from Mellander, S.: Acta Physiol. Scand. **50**[Supp. 176]:1-86, 1960.)

arrow) a steady gradual tissue volume increase occurred (between arrows), attributable to movement of fluid from the capillaries to the interstitial space as a result of the elevated capillary hydrostatic pressure. Fig. 7-6, *B*, represents the effect of a dose of acetylcholine that elicited the same degree of resistance vessel dilation as that obtained with epinephrine in Fig. 7-6, *A*. Note that the acetylcholine produced a large increment in tissue volume (blood volume) and only a small increase in tissue volume attributable to increased capillary filtration (between arrows). In this case the concomitant dilation of the capacitance and resistance vessels resulted in a smaller increment in capillary pressure; hence a slower rate of capillary filtration. In Fig. 7-6, *C*, the dose of epinephrine was about twice that given in *A;* it elicited dilation of the resistance vessels but constriction of the capacitance vessels, resulting in a small decrease in tissue blood volume. These opposite effects on the resistance and capacitance vessels served to increase capillary hydrostatic pressure and enhance the rate of capillary filtration, as evidenced from the slowly rising portion of the tissue volume curve (between arrows). A still larger dose of epinephrine (Fig. 7-6, *D*) produced constriction of both resistance and capacitance vessels with a decrease in tissue blood volume (to left of first arrow) and movement of fluid from the extravascular to the intravascular compartment (between arrows) because of a decrease in net capillary hydrostatic pressure.

From these observations it becomes apparent that neural and humoral stimuli can exert similar or dissimilar effects on different segments of the vascular tree, and in so doing can alter blood flow, tissue blood volume, and extravascular volume to meet the physiological requirements of the organism.

Active sympathetic vasodilation

Although the major neural control of the peripheral vessels is provided by the adrenergic vasoconstrictor fibers of the sympathetic nervous system, there are also *sympathetic cholinergic fibers* innervating the resistance vessels in skeletal muscle and skin. Electrical stimulation of the sympathetic nerves to blood vessels produces vasoconstriction. However, if the adrenergic constrictor effect is blocked by a suitable adrenergic receptor blocking agent or if the neural stores of norepinephrine are depleted by prior treatment with reserpine, the stimulation results in *active dilation*, which can be blocked by the administration of atropine. There is no evidence that these cholinergic sympathetic dilator fibers innervate the capacitance vessels. However, a small fraction of the vasodilation observed with activation of the baroreceptors may occur through the sympathetic cholinergic neurons.

Fibers of the cholinergic sympathetic dilator system arise in the motor cortex of the cerebrum and pass through the hypothalamus and the ventral medulla before joining the other sympathetic outflow in the spinal cord. Activation of the cholinergic sympathetic dilator system produces a relatively large transient initial increase in blood flow followed by a smaller sustained increment in flow during the period of nerve stimulation. There is no evidence for any tonic activity of these fibers. Since excitement or apprehension seems to activate this system and induce vasodilation in skeletal muscle, it has been suggested that the role of the sympathetic vasodilators is to provide the muscles with an increased blood flow in anticipation of the use of the muscles (for example, flight or fight). These cholinergic sympathetic fibers have been demonstrated in dog and cat, but their existence in primates is questionable.

In addition to the cholinergic sympathetic vasodilator system, there are beta adrenergic receptors on the resistance vessels that can be demonstrated by sympathetic nerve stimulation after the administration of atropine and an alpha adrenergic receptor blocker. The vasodilation observed in muscle with intra-arterial injection of small doses of epinephrine or with isoproterenol is caused by stimulation of these beta adrenergic receptors. However, conclusive evidence of neural reflex activation of the vascular beta receptor is lacking. Finally, two other active vasodilator mechanisms mediated by the sympathetic nerve fibers have been suggested. One produces a transitory active vasodilation in response to baroreceptor or direct sympathetic nerve stimulation, and the other evokes a slow sustained vasodilation on sympathetic nerve stimulation that can be demonstrated only when the overriding constrictor response is blocked by adrenergic receptor blocking agents. The neurohumor mediating the transient response is reputed to be histamine, whereas that mediating the sustained vasodilation is unknown.

Parasympathetic neural influence

The efferent fibers of the cranial division of the parasympathetic nervous system supply blood vessels of the head and viscera, whereas fibers of the sacral division supply blood vessels of the genitalia, bladder, and large bowel. Skeletal muscle and skin do not receive parasympathetic innervation. Thus only a small proportion of the resistance vessels of the body receives parasympathetic fibers and therefore the effect of these cholinergic fibers on total vascular resistance is small. Stimulation of the parasympathetic fibers to the salivary glands induces marked vasodilation. However the increase in submaxillary gland blood flow seen with chorda tympani stimulation is believed by some investigators to be secondary to the increase in metabolic activity of the gland and by others to be a primary action on the arterioles of the gland. A vasodilator polypeptide, *bradykinin*, formed locally from the action of an enzyme on a plasma protein substrate present in the glandular lymphatics has been considered to be the metabolic mediator of the vasodilation produced by chorda tympani stimulation. Whether vasodilation in salivary glands results from the release of a cholinergic neurohumor from nerve endings, from the formation and release of bradykinin, or from both is unsettled. Bradykinin has also been reported to be formed in other exocrine glands, such as the lacrimal glands and the sweat glands. Its presence in sweat is thought to be partly responsible for the dilation of cutaneous blood vessels with sweating.

Dorsal root dilators

Vasomotor impulses were at one time thought to travel antidromically in the spinal sensory nerves. Other than the antidromic impulses seen with the axon reflex (p. 226), there is no good evidence that the spinal nerves transmit impulses from the spinal cord to the peripheral blood vessels.

Humoral factors

Epinephrine and norepinephrine exert a profound effect on the peripheral blood vessels. In skeletal muscle, epinephrine in low concentrations dilates resistance vessels (beta adrenergic effect). In skin, only vasoconstriction is obtained with epinephrine, whereas in all vascular beds the primary effect of norepinephrine is vasoconstriction. The adrenal gland, when stimulated, can release epinephrine and norepinephrine into the systemic circulation. However, the concentrations reached are relatively low and the effect of catecholamines released from the adrenal medulla is negligible compared to the effect produced by direct sympathetic innervation.

Vascular reflexes

In addition to the vasomotor areas that initiate sympathetic nerve activity, there are also areas located in the dorsal motor nucleus of the vagus and in the region of the nucleus ambiguus that initiate cardioinhibition. These areas are tonically active and serve essentially as brakes on the heart; interruption of fibers from these areas results in tachycardia, whereas

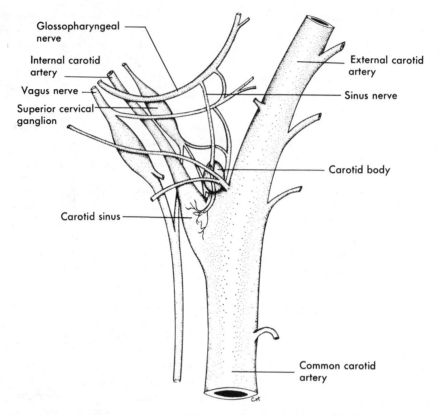

Fig. 7-7. Diagrammatic representation of the carotid sinus and carotid body and their inner-vation in the dog. (Redrawn from Adams, W. E.: The comparative morphology of the carotid body and carotid sinus, Springfield, Ill., 1958, Charles C Thomas, Publisher.)

stimulation of them produces bradycardia. Areas of the medulla that mediate sympathetic and vagal influences are under the influence of neural impulses arising in the baroreceptors, chemoreceptors, hypothalamus, cerebral cortex, and skin, and can also be altered by changes in the blood concentrations of CO_2 and O_2.

Baroreceptors. The *baroreceptors* (or *pressoreceptors*) are stretch receptors located in the carotid sinuses (slightly widened areas of the internal carotid arteries at their points of origin from the common carotid arteries) and in the aortic arch (Figs. 7-7 and 7-8). Impulses arising in the carotid sinus travel up the sinus nerve

(nerve of Hering) to the glossopharyngeal nerve and, via the latter, to the medulla, whereas impulses arising in the pressoreceptors of the aortic arch reach the medulla via afferent fibers in the vagus nerves. The pressoreceptor nerve terminals in the walls of the carotid sinus and aortic arch respond to the stretch and deformation of the vessel induced by the arterial pressure. With increase in blood pressure the frequency of firing is enhanced, whereas the converse is true with a reduction of blood pressure. An increase in frequency of impulses, as occurs with a rise in arterial pressure, acts to inhibit the vasoconstrictor regions, resulting in peripheral vasodilation

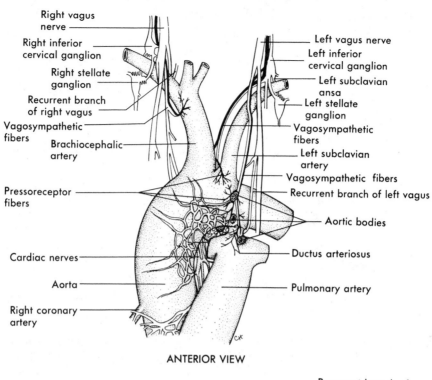

Right vagus nerve

Right inferior cervical ganglion

Right stellate ganglion

Recurrent branch of right vagus

Vagosympathetic fibers

Brachiocephalic artery

Pressoreceptor fibers

Cardiac nerves

Aorta

Right coronary artery

Left vagus nerve

Left inferior cervical ganglion

Left subclavian ansa

Left stellate ganglion

Vagosympathetic fibers

Left subclavian artery

Vagosympathetic fibers

Recurrent branch of left vagus

Aortic bodies

Ductus arteriosus

Pulmonary artery

ANTERIOR VIEW

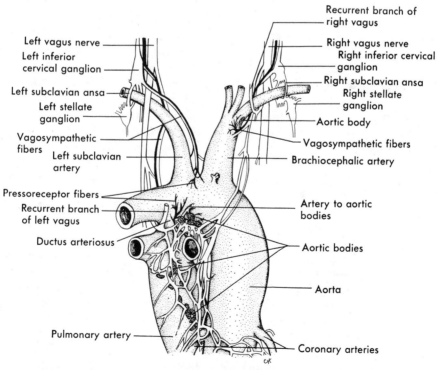

Left vagus nerve

Left inferior cervical ganglion

Left subclavian ansa

Left stellate ganglion

Vagosympathetic fibers

Left subclavian artery

Pressoreceptor fibers

Recurrent branch of left vagus

Ductus arteriosus

Pulmonary artery

Recurrent branch of right vagus

Right vagus nerve

Right inferior cervical ganglion

Right subclavian ansa

Right stellate ganglion

Aortic body

Vagosympathetic fibers

Brachiocephalic artery

Artery to aortic bodies

Aortic bodies

Aorta

Coronary arteries

POSTERIOR VIEW

Fig. 7-8. Anterior and posterior views of the aortic arch showing the innervation of the aortic bodies and pressoreceptors in the dog. (Modified from Nonidez, J. F.: Anat. Rec. **69**:299, 1937.)

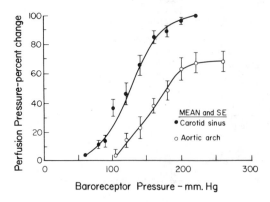

Fig. 7-9. The effects of nonpulsatile pressure changes in the isolated carotid sinus and aortic arch on the changes in perfusion pressure during constant flow in the hind limb of the dog. Note the greater sensitivity of the carotid pressor receptors. (Slightly modified from Donald, D. E., and Edis, A. J.: Comparison of aortic and carotid baroreflexes in the dog, J. Physiol. 215:521, 1971 [publisher: Cambridge University Press].)

and a lowering of blood pressure. Contributing to a lowering of the blood pressure is a bradycardia brought about by stimulation of the vagal regions. The carotid sinus and aortic baroreceptors are not equipotent in their effects on peripheral resistance in response to nonpulsatile alterations in blood pressure. In the experiments depicted in Fig. 7-9 the hind limb of the animal was perfused at a constant flow and the effect of increments or decrements of the pressure in the isolated carotid sinus or aortic arch on hind limb perfusion pressure was recorded. As can be gleaned from Fig. 7-9 the carotid sinus baroreceptors are more sensitive than are those in the aortic arch. Changes in pressure in the carotid sinus evoke greater alterations in perfusion pressure, and hence in resistance, than do equivalent changes in aortic arch pressure. However, with pulsatile changes in blood pressure the two sets of baroreceptors respond similarly.

The carotid sinus with the sinus nerve intact can be isolated from the rest of the circulation and perfused by either a donor dog or an artificial perfusion system. Under these conditions changes in the pressure within the carotid sinus are associated with reciprocal changes in the blood pressure of the experimental animal. The receptors in the walls of the carotid sinus show some adaptation and therefore are more responsive to constantly changing pressures than to sustained constant pressures. This is illustrated in Fig. 7-10, where at normal levels of blood pressure a barrage of impulses from a single fiber of the sinus nerve is initiated in early systole by the pressure rise and only a few spikes are observed during late systole and early diastole. At lower pressures these phasic changes are even more evident, but the overall frequency of discharge is reduced. The blood pressure threshold for eliciting sinus nerve impulses is about 50 mm. Hg, and a maximal sustained firing is reached at around 200 mm. Hg.

Since the pressoreceptors show some degree of adaptation, it is to be expected that their response at any level of mean arterial pressure would be greater with a large than with a small pulse pressure. This is indeed the case and is diagramatically illustrated in Fig. 7-11. At any level of mean arterial pressure, the lower the pulse pressure the greater the systemic vascular resistance. This modulation of the baroreceptor reflex by pulse pressure can play a significant role in compensatory adjustments of the circulatory system. For example, if mean blood pressure decreases (as in hemorrhage) from P_1 to P_2 (Fig. 7-11) and pulse pressure remains at 60 mm. Hg, systemic vascular resistance increases reflexly from point A to point B. However pulse pressure decreases in hypotension caused by hemorrhage, and if it decreased to 40 mm. Hg (point C) there would be a greater reflex constriction of the systemic resistance vessels. These changes in resistance would tend to restore blood pressure more closely toward normal levels and constitute an important protective mechanism in maintaining blood flow to vital tissues such as the brain and heart (p. 263). The resistance increases that occur in the peripheral vascular beds in response

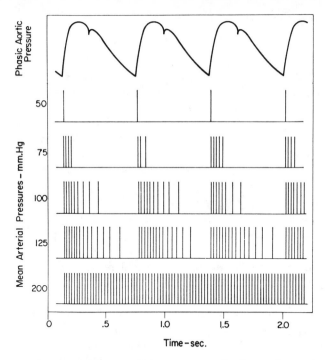

Fig. 7-10. Relationship of phasic aortic blood pressure to the firing of a single afferent nerve fiber from the carotid sinus at different levels of mean arterial pressure.

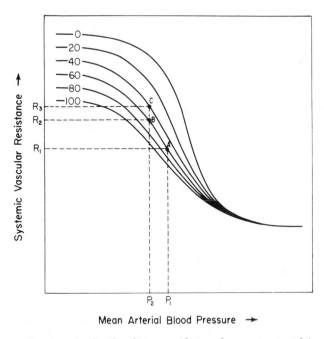

Fig. 7-11. Diagram showing the family of curves relating the mean arterial (carotid sinus and aortic arch) pressure and systemic vascular resistance at pulse pressures of 0, 20, 40, 60, 80, and 100 mm. Hg. (Redrawn from Angell James, J. E., and Daly, M. deB.: J. Physiol. [London] **209:**257, 1970.)

to a reduced pressure in the carotid sinus vary from one vascular bed to another and thereby produce a redistribution of blood flow. For example in some vessels studied, the resistance changes elicited by altering carotid sinus pressure around the normal operating sinus pressure were greatest in the femoral vessels, less in the renal, and least in the mesenteric and celiac vessels. The sensitivity of the carotid sinus reflex can be altered by changing the distensibility of the sinus. Local application of epinephrine causes contraction of the smooth muscle in the sinus wall and increases the response to a given rise in intravascular pressure.

In some individuals the carotid sinus is quite sensitive to pressure. Hence tight collars or other forms of external pressure over the region of the carotid sinus may elicit marked hypotension and fainting. In some patients with severe coronary artery disease and chest pain (*angina pectoris*), symptoms have been temporarily relieved by stimulation of the sinus nerve by means of a chronically implanted stimulator that can be activated externally. The reduction in blood pressure achieved by sinus nerve stimulation decreases the pressure work of the heart and hence the myocardial ischemia responsible for the pain. As would be expected, denervation of the carotid sinus can produce temporary, and in some instances prolonged, hypertension.

The baroreceptors play a key role in short term adjustments of blood pressure when relatively abrupt changes in blood volume, cardiac output, or peripheral resistance (as in exercise) occur. However, long term control of blood pressure—that is, over days, weeks, and longer—is determined by the fluid balance of the individual, namely the balance between fluid intake and fluid output. By far the single most important organ in the control of body fluid volume, and hence blood pressure, is the kidney. With overhydration, excessive fluid intake is excreted, whereas with dehydration there is a marked reduction in urine output.

Cardiopulmonary baroreceptors. In addition to the carotid sinus and aortic baroreceptors, there are also cardiopulmonary receptors that are tonically active and can reflexly alter peripheral resistance with changes in venous pressure or in pulmonary vascular pressure as might occur with alterations of blood volume. These receptors are located in the atria, ventricles, and pulmonary vessels, and when stimulated, inhibit vasoconstrictor tone of the resistance vessels. The atrial receptors are of two types: those that are stimulated by atrial contraction (A receptors), and those stimulated by atrial filling and distension (B receptors). The atrial receptors also play a role in the regulation of vasomotor tone and of urine output by means of stimulation or inhibition of angiotensin and of vasopressin (antidiuretic hormone). The cardiopulmonary receptors as well as the carotid and aortic baroreceptors are necessary for the full expression of blood pressure regulation. The major responses elicited by stimulation of these receptors differ in different vascular beds. For example, activation of the carotid pressoreceptors has a large effect on the splanchnic vascular bed but only a little influence on forearm vascular resistance. The reverse is true for the cardiopulmonary receptors.

Chemoreceptors. The chemoreceptors consist of small, highly vascular bodies in the region of the aortic arch and just medial to the carotid sinuses (Figs. 7-8 and 7-9). They are sensitive to changes in the P_{O_2}, P_{CO_2}, and pH of the blood. Although they are primarily concerned with the regulation of respiration, they reflexly influence the vasomotor regions to a minor degree. A reduction in arterial blood oxygen tension (Pa_{O_2}) stimulates the chemoreceptors, and the increase in the number of impulses in the afferent nerve fibers from the carotid and aortic bodies stimulates the vasoconstrictor regions, resulting in increased tone of the resistance and capacitance vessels. The chemoreceptors are also stimulated by increased arterial blood carbon dioxide tension (Pa_{CO_2}) and reduced pH, but the reflex effect induced is quite small compared to the direct effect of hypercapnia and hydrogen ions on the vasomotor regions in the medulla. When hypoxia and hypercapnia occur at the same time, the stimulation of the chemoreceptors is greater than the sum of the two stimuli when they act alone. When the chemoreceptors are stimulated simultaneously with a reduction in pressure in the baroreceptors, the chemoreceptors potentiate the vasoconstriction observed in the peripheral vessels. However, when the baroreceptors and chemoreceptors are both stimulated (for example, high carotid sinus pressure and low arterial PO_2), the effects of the baroreceptors predominate.

Hypothalamus. Optimal function of the cardiovascular reflexes requires the integrity of pontine and hypothalmic structures. Furthermore, these structures are responsible for behavioral and emotional control of the cardiovascular system. Stimulation of the anterior hypothalamus produces a fall in blood pressure and bradycardia, whereas stimulation of the posterolateral region of the hypothalamus produces a rise in blood pressure and tachycardia. The hypothalamus also contains a temperature-regulating center that affects the skin vessels. Stimulation by cold applications to skin or by cooling of the blood perfusing the hypothalamus results in constriction of the skin vessels and heat conservation, whereas warm stimuli result in cutaneous vasodilation and enhanced heat loss.

Cerebrum. The cerebral cortex can also exert a significant effect on blood flow distribution in the body. Stimulation of the motor and premotor areas can affect blood pressure—usually a pressor response is obtained. However, vasodilation and depressor responses may be evoked, as in blushing or fainting in response to an emotional stimulus.

Skin and viscera. Painful stimuli can elicit either pressor or depressor responses, depending on the magnitude and location of the stimulus. Distension of the viscera often evokes a depressor response, whereas painful stimuli on the body surface usually evoke a pressor response. In the anesthetized animal, strong electrical stimulation of a sensory nerve will produce a strong pressor response. However it is sometimes possible to obtain a depressor response with low-intensity and low-frequency stimulation. Furthermore, all vascular beds do not exhibit the same response; in some, resistance increases, whereas in others it decreases. In addition, muscle contractions can elicit reflex changes in the magnitude of vasoactivity in the muscle. For the most part, these reflexes are mediated through the vasomotor areas in the medulla, but there are also spinal areas that can aid in the regulation of peripheral resistance.

Pulmonary reflexes

Inflation of the lungs reflexly induces systemic vasodilation and a decrease in arterial blood pressure. Conversely, collapse of the lungs evokes systemic vasoconstriction. Afferent fibers mediating this reflex run in the vagus nerves and possibly to a limited extent in the sympathetic nerves. Their stimulation by stretch of the lungs inhibits the vasomotor areas. The magnitude of the depressor response to lung inflation is directly related to the degree of inflation and to the existing level of vasoconstrictor tone.

Chemosensitive regions of the medulla

Increases of Pa_{CO_2} stimulate the vasoconstrictor regions, thereby increasing peripheral resistance, whereas reduction in Pa_{CO_2} below normal levels (as with hyperventilation) decreases the degree of tonic activity of these areas thereby decreasing peripheral resistance. The chemosensitive regions are also affected by changes of pH. A lowering of blood pH stimulates and a rise in blood pH inhibits these areas. These effects of changes in Pa_{CO_2} and blood pH possibly operate through changes in cerebrospinal fluid pH, as appears to be the case for the respiratory center. Whether there are special hydrogen ion chemoreceptors mediating pH-induced vasomotor effects has not been established.

Oxygen tension has relatively little direct effect on the vasomotor region. The primary effect of hypoxia is reflexly mediated via the carotid and aortic chemoreceptors. Moderate reduction of Pa_{O_2} will stimulate the vasomotor region, but severe reduction will depress vasomotor activity in the same manner that other areas of the brain are depressed by very low oxygen tensions.

Cerebral ischemia, which may occur because of excessive pressure exerted by an expanding intracranial tumor, results in a marked increase in vasocontriction. The stimulation is probably caused by a local accumulation of CO_2 and reduction of O_2, and possibly by excitation of intracranial baroreceptors. With prolonged, severe ischemia, central depression eventually supervenes, and the blood pressure falls.

BALANCE BETWEEN EXTRINSIC AND INTRINSIC FACTORS IN REGULATION OF PERIPHERAL BLOOD FLOW

Dual control of the peripheral vessels by virtue of intrinsic and extrinsic mechanisms makes possible a number of vascular adjustments that enable the body to direct blood flow to areas where it is needed in greater supply and away from areas whose immediate requirements are less. In some tissues a more or less fixed relative potency of extrinsic and intrinsic mechanisms exists, and in other tissues, the ratio is changeable, depending on the state of activity of that tissue.

In the brain and the heart, both vital structures with very limited tolerance for a reduced blood supply, intrinsic flow-regulating mechanisms are dominant. For instance, massive discharge of the vasoconstrictor region over the sympathetic nerves, which might occur in se-

vere, acute hemorrhage, has negligible effects on the cerebral and cardiac resistance vessels, whereas skin, renal, and splanchnic blood vessels become greatly constricted.

In the skin the extrinsic vascular control is dominant. Not only do the cutaneous vessels participate strongly in a general vasoconstrictor discharge, but they also respond selectively through hypothalamic pathways to subserve the heat loss and heat conservation function required in body temperature regulation. However, intrinsic control can be demonstrated by local changes of temperature that can modify or override the central influence on resistance and capacitance vessels.

In skeletal muscle the interplay and changing balance between extrinsic and intrinsic mechanisms can be clearly seen. In resting skeletal muscle, neural control (vasoconstrictor tone) is dominant, as can be demonstrated by the large increment in blood flow that occurs immediately after section of the sympathetic nerves to the tissue. In anticipation of and at the start of exercise, such as running, there is an increase in blood flow in the leg muscles, possibly mediated by activation of the cholinergic sympathetic dilator system. After the onset of exercise the intrinsic flow-regulating mechanism assumes control, and because of the local increase in metabolites, vasodilation occurs in the active muscles. Vasoconstriction occurs in the inactive tissues as a manifestation of the general sympathetic discharge, but constrictor impulses reaching the resistance vessels of the active muscles are overridden by the local metabolic effect. Operation of this dual control mechanism thus provides increased blood where it is required and shunts it away from relatively inactive areas. Similar effects may be achieved with an increase in Pa_{CO_2}. Normally the hyperventilation associated with exercise keeps Pa_{CO_2} at normal levels. However, were Pa_{CO_2} to increase, a generalized vasoconstriction would occur by virtue of stimulation of the vasoconstrictor region by CO_2, but

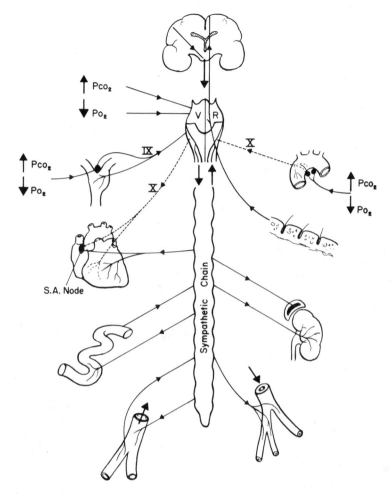

Fig. 7-12. Schematic diagram illustrating the neural input and output of the vasomotor region *(V R)*. *IX*, Glossopharyngeal nerve; *X*, vagus nerve.

in the active muscles, where the CO_2 concentration is highest, the smooth muscle of the arterioles would relax in response to the local P_{CO_2}. Factors affecting and affected by the vasomotor region are summarized in Fig. 7-12.

BIBLIOGRAPHY
Journal articles

Abboud, F. M., Eckberg, D. L., Johannsen, U. J., and Mark, A. L.: Carotid and cardiopulmonary baroreceptor control of splanchnic and forearm vascular resistance during venous pooling in man, J. Physiol. (London) **286:**173, 1979.

Abboud, F. M., Heistad, D. D., Mark, A. L., and Schmid, P. G.: Reflex control of the peripheral circulation, Prog. Cardiovasc. Dis. **18:**371, 1976.

Angell James, J. E., and Daly, M. de B.: Comparison of the reflex vasomotor responses to separate and combined stimulation of the carotid sinus and aortic arch baroreceptors by pulsatile and nonpulsatile pressures in the dog, J. Physiol. (London) **209:**257, 1970.

Belloni, F. L., and Sparks, H. V.: The peripheral circulation, Ann. Rev. Physiol. **40:**67, 1978.

Bohr, D. F.: Vascular smooth muscle updated, Circ. Res. **32:**665, 1973.

Brown, A. M.: Receptors under pressure—an update on baroreceptors, Circ. Res. **46:**1, 1980.

Browse, N. L., Lorenz, R. R., and Shepherd, J. T.: Response of capacity and resistance vessels of dog's limb to sympathetic nerve stimulation, Am. J. Physiol. **210:**95, 1966

Coleridge, H. M., and Coleridge, J. C. G.: Cardiovascular afferents involved in regulation of peripheral vessels, Ann. Rev. Physiol. **42:**413, 1980.

Donald, D. E., and Edis, A. J.: Comparison of aortic and carotid baroreflexes in the dog, J. Physiol. (London) **215:**521, 1971.

Donald, D. E., and Shepherd, J. T.: Autonomic regulation of the peripheral circulation, Ann. Rev. Physiol. **42:**429, 1980.

Donald, D. E., and Shepherd, J. T.: Reflexes from the heart and lungs: physiological curiosities or important regulatory mechanisms, Cardiovasc. Res. **12:**449, 1978.

Duling, B. R.: Oxygen sensitivity of vascular smooth muscle. II. In vivo studies, Am. J. Physiol. **227:**42, 1974.

Fox, R., and Hilton, S. M.: Bradykinin formation in human skin as a factor in heat vasodilatation, J. Physiol. (London) **142:**219, 1958.

Guyton, A. C., Coleman, T. G., Cowley, A. W., Jr., Manning, R. D., Jr., Norman, R. A., Jr., and Ferguson, J. D.: A systems analysis approach to understanding long-range arterial blood pressure control and hypertension, Circ. Res. **35:**159, 1974.

Hilton, S. M., and Spyer, K. M.: Central nervous regulation of vascular resistance, Ann. Rev. Physiol. **42:**399, 1980.

Kirchheim, H. R.: Systemic arterial baroreceptor reflexes, Physiol. Rev. **56:**100, 1976.

Mellander, S., and Johansson, B.: Control of resistance, exchange, and capacitance functions in the peripheral circulation, Pharmacol. Rev. **20:**117, 1968.

Somlyo, A. P., and Somlyo, A. V.: Vascular smooth muscle, Pharmacol. Rev. **20:**197, 1968.

Books and monographs

Bevan, J. A., Bevan, R. D., and Duckles, S. P.: Adrenergic regulation of vascular smooth muscle. In Handbook of physiology; Section 2: The cardiovascular system—smooth muscle, vol. II, Bethesda, MD, 1980, American Physiological Society, pp. 515-566.

Brown, A. M.: Cardiac reflexes. In Handbook of physiology; Section 2: The cardiovascular system—the heart, vol. I, Bethesda, MD, 1979, American Physiology Society, pp. 677-690.

Johnson, P. C.: The myogenic response. In Handbook of physiology; Section 2: The cardiovascular system—vascular smooth muscle, vol. II, Bethesda, MD, 1980, American Physiological Society, pp. 409-442.

Johnson, P. C., editor: Peripheral circulation, New York, 1978, John Wiley & Sons, Inc.

Korner, P. I.: Central nervous control of autonomic cardiovascular function. In Handbook of Physiology; Section 2: The cardiovascular system—the heart, vol. I, Bethesda, MD, 1979, American Physiological Society, pp. 691-739.

Rhodin, J. A. G.: Architecture of the vessel wall. In Handbook of physiology; Section 2: The cardiovascular system—vascular smooth muscle, vol. II, Bethesda MD, 1980, American Physiological Society, pp. 1-31.

Sparks, H. V., Jr.: Effect of local metabolic factors on vascular smooth muscle. In Handbook of physiology; Section 2: The cardiovascular system—Vascular smooth muscle, vol. II, Bethesda, MD, 1980, American Physiological Society, pp. 475-566.

Zelis, R., editor: The peripheral circulation, New York, 1975, Grune & Stratton, Inc.

CHAPTER 8

CONTROL OF THE HEART

The quantity of blood pumped by the heart may be varied by changing the frequency of its beats or the volume ejected per stroke. A discussion of the control of cardiac activity may therefore be subdivided into a consideration of the regulation of pacemaker activity and the regulation of myocardial performance. However, in the intact organism, a change in the behavior of one of these features of cardiac activity almost invariably produces an alteration in the other.

Experimentally it has been shown that certain local factors, such as temperature changes and tissue stretch, can affect the discharge frequency of the sinoatrial node. Under natural conditions, however, the principal control of heart rate is relegated to the autonomic nervous system, and the discussion will be restricted to this aspect of heart rate control. Relative to the performance of the myocardium, both intrinsic and extrinsic factors warrant extensive consideration.

NERVOUS CONTROL OF HEART RATE

In normal adults the average heart rate at rest is approximately 70 beats per minutes, but it is significantly greater in children. During sleep the heart rate diminishes by 10 to 20 beats per minute, but during emotional excitement or muscular activity it may accelerate to rates considerably above 100. In well-trained athletes at rest the rate is usually only 50 to 60 beats per minute. During diving the heart decelerates, and bradycardia persists even during the vigorous exercise of underwater swimming.

This bradycardia associated with diving is considerably more pronounced in certain aquatic mammals, such as the seal.

Under most conditions the sinoatrial node is under the tonic influence of both divisions of the autonomic nervous system. The sympathetic system exerts a facilitatory influence on the rhythmicity of the pacemaker, whereas the parasympathetic system imposes an inhibitory effect. It was first demonstrated by the Webers in 1845 that changes in heart rate usually involve a reciprocal action of the two divisions of the autonomic nervous system. Thus an acceleration of the heart rate is produced by a diminution of parasympathetic activity and a concomitant increase in sympathetic activity; deceleration is usually evoked by a reversal of these mechanisms. Under certain conditions, changes in heart rate may be achieved by selective action of just one division of the autonomic nervous system rather than by reciprocal changes in both divisions.

Ordinarily, in healthy, resting individuals the parasympathetic tone is predominant. Abolition of parasympathetic influences by transection of the vagus nerves or by the administration of atropine usually elicits a pronounced tachycardia, whereas abrogation of sympathetic activity usually results in only slight slowing of the heart. When both divisions of the autonomic nervous system are blocked, the heart rate of young adults has been found to average about 105 beats per minute. The rate that prevails after complete autonomic blockade is called the *intrinsic heart rate*.

Parasympathetic pathways

The cardiac parasympathetic fibers originate in the medulla, in a column of cells that lie in the dorsal motor nucleus or in the region of the nucleus ambiguus. The precise location varies from species to species. The medullary center on a given side projects to both the ipsilateral and contralateral vagus nerves. Centrifugal vagal fibers pass inferiorly through the neck in close proximity to the common carotid arteries and through the mediastinum to synapse with postganglionic cells located within the heart itself. Most of the cardiac ganglion cells are located near the S-A node and A-V conduction tissue. Stimulation of the cardiac ends of the severed vagus nerves of experimental animals has revealed that the right and left vagi are usually distributed differentially to the various cardiac structures. The right vagus nerve affects the S-A node predominantly, to produce sinus bradycardia or even complete cessation of S-A nodal activity for several seconds. The left vagus nerve exerts its greatest influence on the A-V conduction tissue, to produce various degrees of A-V block. However, there is a considerable overlap of distribution, so that left vagal stimulation also depresses the S-A node and right vagal stimulation evokes some impairment of A-V conduction.

The S-A and A-V nodes are rich in cholinesterase. Hence the effects of any given vagal impulse are ephemeral because the acetylcholine released at the nerve terminals is rapidly hydrolyzed. Also parasympathetic influences preponderate over sympathetic effects at the S-A node, as shown in Fig. 8-1. As the frequency of vagal stimulation in an anesthetized dog was increased from 0 to 8 pulses per second (left panel), the heart rate decreased by 75 beats per minute in the absence of concomitant cardiac sympathetic stimulation (S = 0). Sympathetic stimulation at 4 pulses per second (S = 4) caused an increase in heart rate of 80 beats per minute in the absence of vagal stimulation. However, when vagal stimulation at 8 pulses per second was combined with sympathetic stimulation at 4 pulses per second (S = 4), the heart rate declined by 155 beats per minute. From the right panel it is evident that increasing the sympathetic stimulation frequency from 0 to 4 pulses per second produced a substantial cardioacceleration in the absence of vagal stimulation (V = 0). However, when the vagi were stimulated at 8 pulses per second (V = 8), in-

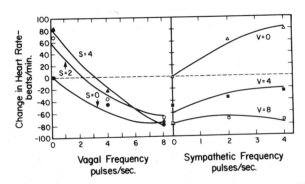

Fig. 8-1. The changes in heart rate in an anesthetized dog when the vagus and cardiac sympathetic nerves were stimulated simultaneously. The sympathetic nerves were stimulated at 0, 2, and 4 pulses/sec.; the vagus nerves at 0, 4, and 8 pulses/sec. The symbols represent the observed changes in heart rate; the curves represent a derived regression equation. (Modified from Levy, M. N., and Zieske, H.: J. Appl. Physiol. **27:**465, 1969.)

creasing the sympathetic stimulation frequency from 0 to 4 pulses per second had a negligible influence on heart rate.

Sympathetic pathways

The cardiac sympathetic fibers originate in the intermediolateral columns of the upper five or six thoracic and lower one or two cervical segments of the spinal cord. They emerge from the spinal column through the white rami communicantes and enter the paravertebral chain of ganglia. The anatomical details of the sympathetic innervation of the heart vary among mammalian species; the innervation has been elaborated in greatest detail in the dog (Fig. 8-2). In that species, virtually all of the preganglionic neurons ascend in the paravertebral chain and funnel through the stellate ganglia. In many species, including the cat, the synapse between pre- and postganglionic neurons takes place mainly in the stellate ganglia. In other species, such as the dog, the preganglionic neurons traverse the two limbs of the ansa subclavia and then synapse with the postganglionic neurons in the caudal cervical ganglia. These latter ganglia lie close to the vagus nerves in the superior portion of the mediastinum. Sympathetic and parasympathetic fibers then join to form a complex network of mixed efferent nerves to the heart (Fig. 8-2). Other sympathetic fibers ascend to the head and neck—as the cervical sympathetic chain in some species, and in a common bundle with the vagus nerve (as the vagosympathetic trunk) in other species.

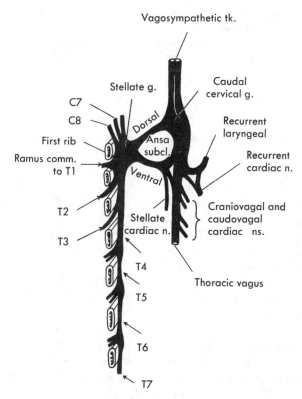

Fig. 8-2. Upper thoracic sympathetic chain and the cardiac autonomic nerves on the right side in the dog. (Modified from Mizeres, N. J.: Anat. Rec. **132:**261, 1958.)

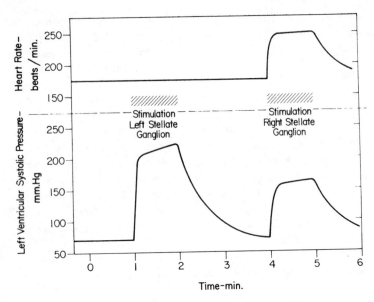

Fig. 8-3. In the dog, stimulation of the left stellate ganglion produces a greater effect on ventricular contractility than does right-sided stimulation, but it has a lesser effect on heart rate. In this example, traced from an original record, left stellate ganglion stimulation had no detectable effect at all on heart rate but had a considerable effect on ventricular performance in an isovolumetric left ventricle preparation.

The postganglionic cardiac sympathetic fibers approach the base of the heart along the adventitial surface of the great vessels. On reaching the base of the heart, these fibers are distributed to the various chambers as an extensive epicardial plexus. They then penetrate the myocardium, usually accompanying the branches of the coronary vessels. The adrenergic receptors in the nodal regions and in the myocardium are predominantly of the beta type; that is, they are responsive to beta adrenergic agonists, such as isoproterenol, and are inhibited by specific beta adrenergic blocking agents, such as propranolol.

As with the vagus nerves, there tends to be a differential distribution of the left and right sympathetic fibers. In the dog, for example, the fibers on the left side have more pronounced effects on myocardial contractility (so-called *augmentor fibers*) than on heart rate (so-

called *accelerator fibers*). In some dogs left cardiac sympathetic nerve stimulation may produce no detectable effects on heart rate, even though it may exert pronounced facilitatory effects on ventricular performance, as shown in Fig. 8-3. Conversely, right cardiac sympathetic nerve stimulation elicits considerably greater cardiac acceleration and less augmentation of contractile force than does stimulation of equivalent strength of sympathetic fibers on the left side. This bilateral asymmetry probably also exists in humans. It has recently been found that right stellate ganglion blockade caused a mean reduction in heart rate of 14 beats/min., whereas left-sided blockade elicited an average reduction of only 2 beats/min.

It is evident from Fig. 8-3 that the effects of sympathetic stimulation decay very gradually after the cessation of stimulation, in contrast to the abrupt termination of the response after vagal activity (not shown). Most of the norepinephrine released during sympathetic stimulation is taken up again by the nerve terminals, and much of the remainder is carried away by the bloodstream. Relatively little of the released norepinephrine is degraded in the tissues, in contrast to the fate of the acetylcholine released at the vagus nerve endings.

Control by higher centers

Dramatic alterations in cardiac rate, rhythm, and contractility have been induced experimentally by stimulation of various regions of the brain. In the cerebral cortex the centers regulating cardiac function are located mostly in the anterior half of the brain, principally in the frontal lobe, the orbital cortex, the motor and premotor cortex, the anterior part of the temporal lobe, the insula, and the cingulate gyrus. In the thalamus, tachycardia may be induced by stimulation of the midline, ventral, and medial groups of nuclei. Variations in heart rate may also be evoked by stimulating the posterior and posterolateral regions of the hypothalamus. Stimuli applied to the H_2 fields of Forel in the diencephalon elicit a variety of cardiovascular responses, including tachycardia; such changes simulate closely those observed during muscular exercise. Undoubtedly the cortical and diencephalic centers are responsible for initiating the cardiac reactions that occur during excitement, anxiety, and other emotional states. The hypothalamic centers are also involved in the cardiac response to alterations in environmental temperature. Recent studies have shown that localized temperature changes in the preoptic anterior hypothalamus evoke pronounced changes in heart rate and peripheral resistance.

Stimulation of the parahypoglossal area of the medulla produces a reciprocal activation of cardiac sympathetic and inhibition of cardiac parasympathetic pathways. In certain dorsal regions of the medulla, distint cardiac accelerator and augmentor points have been detected in animals with transected vagi. The accelerator regions were found to be more abundant on the right and the augmentor sites more prevalent on the left. A similar distribution has also been reported to exist in the hypothalamus. It appears, therefore, that for the most part the sympathetic fibers descend the brainstem ipsilaterally.

Bainbridge reflex and atrial receptors

In 1915 Bainbridge reported that infusions of blood or saline resulted in cardiac acceleration. This increase in heart rate occurred whether arterial pressure was unaffected or became somewhat elevated. Acceleration was observed whenever central venous pressure rose sufficiently to distend the right side of the heart, and the effect was abolished by bilateral transection of the vagi. Bainbridge postulated that increased cardiac filling elicited tachycardia reflexly and that the afferent impulses were conducted by the vagi.

Numerous investigators have confirmed the acceleration of the heart in response to the intravenous administration of fluid. However, the magnitude and direction of the response is dependent on the prevailing heart rate. At relatively slow rates, intravenous infusions usually result in cardiac acceleration. At more rapid initial rates, however, infusions will ordinarily slow the heart. Increases in blood volume not only evoke the Bainbridge reflex, but they also activate other reflexes, notably the baroreceptor reflex, which tend to elicit oppositely directed heart rate changes. The actual change in heart rate that occurs in response to an alteration of blood volume under a given set of conditions is therefore the resultant of these antagonistic reflex effects.

In a recent study on unanesthetized dogs, volume loading with blood resulted in increases in heart rate that were proportional to the augmentations of cardiac output (Fig. 8-4). Consequently, stroke volume remained virtually constant. Conversely, reductions in

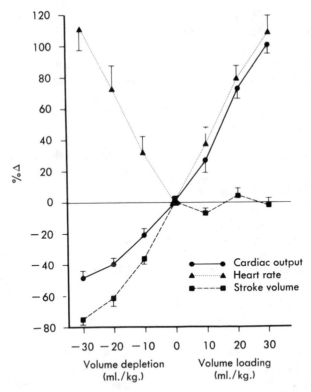

Fig. 8-4. Effects of blood transfusion and of bleeding on cardiac output, heart rate, and stroke volume in unanesthetized dogs. (From Vatner, S. F., and Boettcher, D. H.: Circ. Res. **42:**557, 1978. By permission of the American Heart Association, Inc.)

blood volume diminished the cardiac output but also evoked a tachycardia. Undoubtedly, the Bainbridge reflex was prepotent over the baroreceptor reflex when the blood volume was raised, but the baroreceptor reflex prevailed over the Bainbridge reflex under conditions of hypovolemia.

Receptors that influence heart rate have been detected in both atria. They are located principally in the venoatrial junctions—in the right atrium at its junctions with the venae cavae and in the left atrium at its junctions with the pulmonary veins. Distension of these receptors sends impulses centripetally in the vagi. The efferent impulses are carried by fibers from both autonomic divisions to the S-A node, and they produce an increase in heart rate. The cardiac response appears to be highly

selective. Even when the reflex increase in heart rate is quite large, changes in ventricular contractility have been negligible. Furthermore, the increase in heart rate is unattended by any augmentation of sympathetic activity to the peripheral arterioles. In fact, the only observed change in sympathetic activity to the peripheral vasculature is a significant reduction in sympathetic tone in the renal vessels.

Stimulation of the atrial receptors causes an increase in urine volume, in addition to the cardiac acceleration. The reduction in activity in the renal sympathetic nerve fibers might be partially responsible for this diuresis. However, the principal mechanism appears to be a neurally mediated reduction in the secretion of vasopressin (antidiuretic hormone) by the posterior pituitary gland.

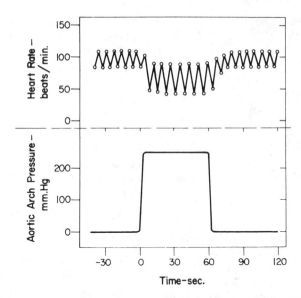

Fig. 8-5. Effect of a stepwise pressure change in the isolated aortic arch on heart rate. When pressure is raised, the mean heart rate decreases and there is an increase in the magnitude of the rhythmic fluctuations of heart rate at the frequency of respiratory movements. (Redrawn from Levy, M. N., Ng, M. L., and Zieske, H.: Circ. Res. **19**:930, 1966.)

Baroreceptor reflex

The inverse relationship between arterial blood pressure and heart rate was first described in 1859 by the French physician Étienne Marey, and is often referred to as *Marey's law of the heart*. Largely through the work of Hering, Koch, and Heymans it was demonstrated that the alterations in heart rate evoked by changes in blood pressure were dependent on the baroreceptors located in the aortic arch and carotid sinuses. These two sets of baroreceptors appear to be about equally potent in the regulation of heart rate. An example of the effects on heart rate elicited by a stepwise elevation of pressure in an isolated aortic arch preparation is illustrated in Fig. 8-5. An abrupt rise in pressure in the aortic arch from 0 to 250 mm. Hg resulted in bradycardia and an exaggeration of the respiratory cardiac arrhythmia (defined in the following section). The subsequent decline of pressure was followed by the return of heart rate to control levels.

It has generally been assumed that the changes in heart rate that occur in response to alterations in baroreceptor stimulation are mediated by means of reciprocal changes in activity in the two divisions of the autonomic nervous system. Recent investigations have challenged this concept. Currently the predominant idea is that reciprocal reflex changes in sympathetic and vagal activity do occur for small deviations in blood pressure within the normal range of pressures. However, when blood pressure is gradually increased to high levels, cardiac sympathetic tone is completely suppressed when only a fraction of the blood pressure elevation has been attained. Thereafter the additional reduction in heart rate with further increases in blood pressure is evoked entirely by an augmentation of vagal activity. The converse applies during the development of severe hypotension. Vagal tone virtually disappears after the initial, relatively small drop in blood pressure. As the pressure continues to decline, the further acceleration of the heart is

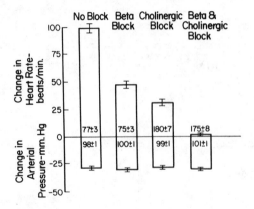

Fig. 8-6. The effects of beta adrenergic receptor blockade, cholinergic blockade, and combined blockade on the change in heart rate in a group of thirteen resting dogs when the mean arterial blood pressure was decreased by about 30 mm. Hg by means of nitroglycerin. The control values of heart rate and blood pressure prior to hypotension are shown in each column. (From Vatner, S. F., Higgins, C. B., and Braunwald, E.: Cardiovasc. Res. **8**:153, 1974.)

ascribable solely to a progressive increase in sympathetic neural activity.

The participation of both divisions of the autonomic nervous system in the heart rate response to a moderate hypotension is shown in Fig. 8-6. In a group of thirteen normal unanesthetized dogs, a 30 mm. Hg decrease in mean arterial blood pressure was produced by nitroglycerin, which dilates peripheral arterioles. With efferent vagal and sympathetic pathways both intact, this reduction in pressure was accompanied by a reflex increase in heart rate of about 100 beats per minute. After beta adrenergic receptor blockade with propranolol, the heart rate increased by only 50 beats per minute in response to the same blood pressure reduction. After cholinergic blockade with atropine, hypotension increased the heart rate by about 30 beats per minute. When the two drugs were given simultaneously to block both autonomic divisions, there was virtually no change in heart rate in response to the 30 mm.

Hg decline in mean arterial pressure. It is apparent that both autonomic divisions participated in the cardioacceleration evoked by a 30 mm. Hg blood pressure reduction in these experiments. Had more severe degrees of hypotension been induced, any additional cardiac acceleration would probably have been mediated solely by efferent sympathetic pathways. Once the mean arterial pressure has declined to about 20 mm. Hg below the normal level, vagal tone is usually negligible. Hence, no additional increase in heart rate can be achieved through a further reduction in vagal tone.

Respiratory cardiac arrhythmia

Rhythmic variations in heart rate, occurring at the frequency of respiration, are detectable in most individuals and tend to be especially pronounced in children. Typically the cardiac rate accelerates during inspiration and decelerates during expiration. The vagus nerves are principally responsible for mediating this respiratory cardiac arrhythmia, and the extent of the arrhythmia varies with the degree of vagal tone. Respiratory sinus arrhythmia becomes more pronounced when vagal tone is enhanced. An example is shown in Fig. 8-5, where vagal tone was increased by elevating pressure in the aortic arch. Under these conditions the amplitude of the cyclic variations in heart rate was twice as great as when the pressure in the aortic arch was low.

Both reflex and central factors contribute to the genesis of the respiratory cardiac arrhythmia. During inspiration the intrathoracic pressure decreases and venous return to the right side of the heart is accelerated. It has been postulated that this elicits the Bainbridge reflex. After the time delay required for the increased venous return to reach the left side of the heart, left ventricular output is increased and produces a rise in arterial blood pressure. This in turn will reduce heart rate through baroreceptor stimulation. Vasomotor tone also varies periodically at the frequency of respiration and results in rhythmic fluctuations in arterial blood pressure with consequent baroreceptor reflex effects on heart rate. Stretch receptors located in the lungs are capable of affecting heart rate. With moderate degrees of pulmonary inflation, cardioacceleration may be evoked reflexly. The afferent and efferent limbs of this reflex are located in the vagus nerves.

It was postulated by Traube in 1865 that respiratory cardiac arrhythmia was ascribable to an influence of the brainstem respiratory center on the cardiac autonomic centers. Abundant experimental evidence has subsequently been provided to demonstrate that central as well as reflex mechanisms play an important role. In the experiment displayed in Fig. 8-5, a total heart-lung bypass was employed in order to eliminate variations in respiration, venous return, or arterial blood pressure. Yet in that experiment a respiratory cardiac arrhythmia was still prominent.

Chemoreceptor reflex

It might be anticipated that the question of the directional change in heart rate that is evoked in response to peripheral chemoreceptor stimulation could be answered by a few simple experiments. Yet this problem has been controversial for several decades. It merits consideration because it illustrates the activation of several distinct mechanisms by the stimulation of one type of sensory receptor (in this case the peripheral chemoreceptors). These mechanisms in turn may exert disparate effects on a given effector organ (in this case the cardiac pacemaker).

In intact animals, stimulation of the carotid chemoreceptors elicits a consistent augmentation of ventilatory rate and depth but ordinarily evokes only slight increases or decreases in heart rate. The directional change in heart rate is related to the magnitude of the enhancement of pulmonary ventilation, as shown in Fig. 8-7. When respiratory stimulation is relatively mild, heart rate usually diminishes; when the in-

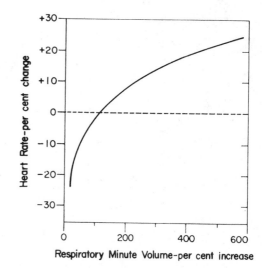

Fig. 8-7. Relationship between the change in heart rate and the change in respiratory minute volume during carotid chemoreceptor stimulation in spontaneously breathing cats and dogs. When respiratory stimulation was relatively slight, heart rate usually diminished; when respiratory stimulation was more pronounced, heart rate usually increased. (Modified from Daly, M. deB., and Scott, M. J.: J. Physiol. [London] **144:**148, 1958.)

crease in pulmonary ventilation is more pronounced, heart rate usually accelerates.

The cardiac response to peripheral chemoreceptor stimulation represents the resultant of primary and secondary reflex mechanisms. The primary reflex effect of carotid chemoreceptor excitation on the S-A node is inhibitory. Secondary effects are largely related to the concomitant stimulation of respiration.

An example of the primary inhibitory influence is displayed in Fig. 8-8. In this experiment on an anesthetized dog, the lungs were completely collapsed and blood oxygenation was accomplished by an artificial oxygenator. When the carotid chemoreceptors were stimulated, an intense bradycardia ensued that was usually accompanied by some degree of A-V block. Such effects are mediated primarily by efferent vagal fibers.

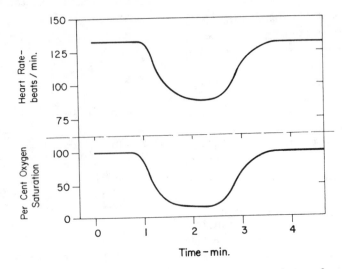

Fig. 8-8. Changes in heart rate during carotid chemoreceptor stimulation when the lungs are deflated and respiratory gas exchange is accomplished by an artificial oxygenator. The lower tracing represents the oxygen saturation of the blood perfusing the carotid chemoreceptors. The blood perfusing the remainder of the animal, including the myocardium, was fully saturated with oxygen throughout the experiment. (Modified from Levy, M. N., DeGeest, H., and Zieske, H.: Circ. Res. **18**:67, 1966.)

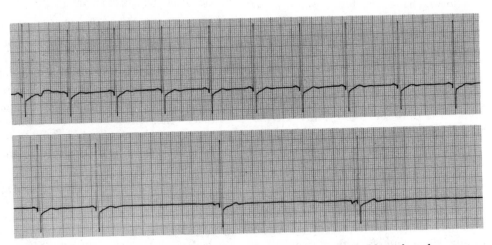

Fig. 8-9. Electrocardiogram of a 30-year-old quadriplegic man who could not breathe spontaneously and required tracheal intubation and artificial respiration. The two strips are continuous. The tracheal catheter was temporarily disconnected from the respirator at the beginning of the top strip, at which time his heart rate was 65 beats/min. In less than 10 sec., his heart rate decreased to about 20 beats/min. (Modified from Berk, J. L., and Levy, M. N.: Eur. Surg. Res. **9**:75, 1977.)

The identical primary inhibitory effect has recently been shown to operate in humans. The electrocardiogram in Fig. 8-9 was recorded from a quadriplegic patient who could not breathe spontaneously, but required tracheal intubation and artificial respiration. When the tracheal catheter was briefly disconnected to permit nursing care, the patient quickly developed a profound bradycardia. His heart rate was 65 beats/min. just before the tracheal catheter was disconnected. In less than 10 sec. after cessation of artificial respiration, his heart rate dropped to about 20 beats/min. This bradycardia could be prevented with atropine, and its onset could be delayed considerably by hyperventilating the patient prior to disconnecting the tracheal catheter.

The potent enhancement of pulmonary ventilation that is ordinarily evoked by carotid chemoreceptor stimulation influences heart rate secondarily, both by initiating more pronounced pulmonary inflation reflexes and by producing hypocapnia. Each of these influences has been found to accelerate the heart per se and to depress the primary cardiac response to chemoreceptor stimulation. Hence, when there is no experimental regulation of pulmonary ventilation, carotid chemoreceptor stimulation produces only a slight change (in either direction) in heart rate instead of the profound bradycardia that occurs when ventilation is held constant.

INTRINSIC REGULATION OF CARDIAC PERFORMANCE

Just as the heart has the inherent ability to initiate its own beat in the absence of any nervous or hormonal control, so also does it possess the capacity to adapt to changing hemodynamic conditions by virtue of mechanisms that are intrinsic to cardiac muscle itself. Experiments on completely denervated hearts reveal that this organ is able to adjust remarkably well to a variety of stressful circumstances. In racing greyhounds, for example, it was found

that animals with denervated hearts perform essentially as well as those with intact innervation. Their maximal running speed was reduced on the average by only 5% after complete cardiac denervation. In these dogs, exercise induced a threefold to fourfold increase in cardiac output, achieved principally by means of an increase in stroke volume. In normal dogs the increase of cardiac output with exercise is accomplished mainly by an acceleration of heart rate. It is unlikely that the cardiac adaptation in the denervated animals is achieved entirely by intrinsic mechanisms; circulating catecholamines undoubtedly play a role. If the beta adrenergic receptors are blocked in greyhounds with denervated hearts, their racing performance is severely impaired.

The heart is partially or completely denervated in a variety of clinical situations: (1) the surgically transplanted heart is totally denervated; (2) atropine is capable of blocking vagal effects on the heart, and propranolol and other beta adrenergic receptor blocking agents can abrogate the sympathetic influences; (3) certain drugs, such as reserpine, deplete cardiac norepinephrine stores, and thereby restrict or abolish sympathetic control; and (4) in severe, chronic congestive heart failure, cardiac norepinephrine stores are often severely diminished, thereby attenuating any sympathetic influences.

Historically the intrinsic cardiac adaptive mechanism that has received the greatest attention involves changes in the resting length of the myocardial fibers. This type of adaptation is frequently designated *Starling's law of the heart* or the *Frank-Starling mechanism*. The mechanical and structural bases for this mechanism have been explained in Chapter 4. The term *heterometric autoregulation* has been coined by Sarnoff and his collaborators to refer to those adaptive mechanisms that involve changes in myocardial fiber length. They also suggested the term *homeometric autoregulation* to be applied to those other intrinsic adjustments of cardiac performance that are independent of changes in myocardial fiber length.

Heterometric autoregulation

Studies on isolated hearts. In 1895 the German physiologist Otto Frank described the response of the isolated heart of the frog to alterations in the tension of the myocardial fibers just prior to contraction—the so-called *initial tension*. He recognized however that such changes in the initial tension were probably accompanied by changes in the resting fiber length. Representative isovolumetric pressure curves from his experiments are reproduced in Fig. 8-10. Curves *1* through *6* illustrate the response of the heart to increased filling. It is evident that the initial tension increases with greater de-

grees of filling and, at each succeeding level, contraction produces a progressively greater peak pressure. Frank recognized that such behavior of cardiac muscle was similar to that of skeletal muscle when it is stretched to progressively greater initial lengths prior to contraction.

In 1914 the noted English physiologist Ernest Starling and his collaborators described the intrinsic response of the heart to changes in venous return and arterial pressure in the canine heart-lung preparation, which is depicted in Fig. 8-11. In this preparation the right atrium is filled with blood from the reservoir that is connected to a cannula tied into the atrial end of the ligated superior vena cava. Venous return is varied either by altering the height of the reservoir or by adjusting a screw-

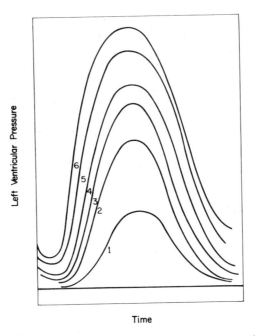

Fig. 8-10. Response of the frog ventricle to increased filling. The initial tension (intraventricular pressure at the onset of each contraction) increases as the filling volume is increased (denoted by the successively higher numbers in the figure). As initial tension is raised, the peak pressure developed during systole is also augmented. (Redrawn from Frank, O.: Z. Biol. **32:**370, 1895.)

clamp on the connecting tube. Right atrial pressure is recorded by means of a water manometer in the atrial end of the ligated inferior vena cava. From the right atrium, blood enters the right ventricle, which then pumps it though the pulmonary vessels. The trachea is cannulated and the lungs are artificially ventilated by means of an intermittent respiration pump. The pulmonary venous return enters the left atrium, in which the pressure is recorded by a water manometer. The aorta is ligated distal to the arch, and a cannula is inserted into the brachiocephalic artery. Blood is pumped by the left ventricle through this cannula and through artificial tubing that ultimately conducts the blood back through a heat-

ing coil to the right atrial reservoir. Arterial pressure is recorded by means of a mercury manometer. The volume of the right and left ventricles is recorded by means of a cardiometer (Fig. 4-16). Peripheral resistance is adjusted by means of a pressure-limiting device, which has become known as a Starling resistance. This device consists of a piece of collapsible tubing in a rigid chamber. Any desired air pressure is applied to the collapsible tubing though an inlet and is measured with a manometer. The cardiac output is measured by temporarily diverting the flow, which is returning to the venous return reservoir, into a graduated cylinder for a measured interval of time.

The response of the isolated heart to a sud-

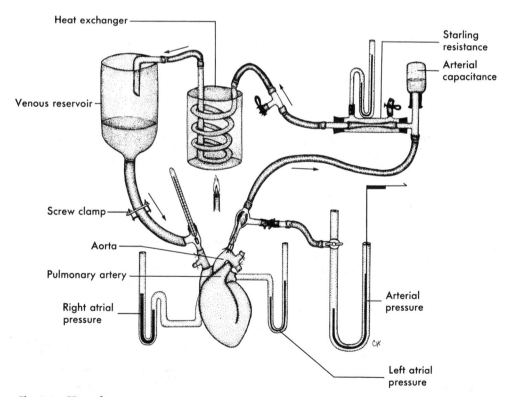

Fig. 8-11. Heart-lung preparation. (Redrawn from Patterson, S. W., and Starling, E. H.: J. Physiol. [London] **48**:357, 1914.)

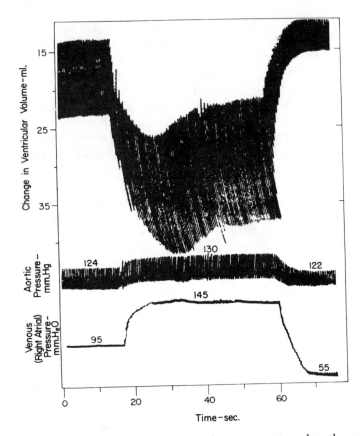

Fig. 8-12. Changes in ventricular volume in a heart-lung preparation when the venous reservoir was suddenly raised (right atrial pressure increased from 95 to 145 mm. H_2O) and subsequently lowered (right atrial pressure decreased from 145 to 55 mm. H_2O). Note that an increase in ventricular volume is registered as a downward shift in the tracing. (Redrawn from Patterson, S. W., Piper, H., and Starling, E. H.: J. Physiol. [London] **48**:465, 1914.)

den augmentation of venous return is shown in Fig. 8-12. Right atrial pressure increased rapidly and appreciably, whereas aortic pressure was permitted to increase only slightly. The top tracing in Fig. 8-12 reveals the changes in ventricular volume that resulted from the abrupt acceleration of venous return. Increased ventricular volume is registered as a downward deflection. Hence the upper border of the tracing represents the systolic volume, the lower border indicates the diastolic volume, and the amplitude of the deflections reflects the stroke

volume. For several beats after augmentation of venous return, ventricular volume progressively increased. This indicates that a disparity must have existed between ventricular inflow during diastole and ventricular output during systole; that is, during systole the ventricles did not expel an amount of blood equal to that which entered during diastole, until a new equilibrium was attained. It is this progressive accumulation of blood that produced the dilation of the ventricles and the lengthening of the individual myocardial fibers that constitute

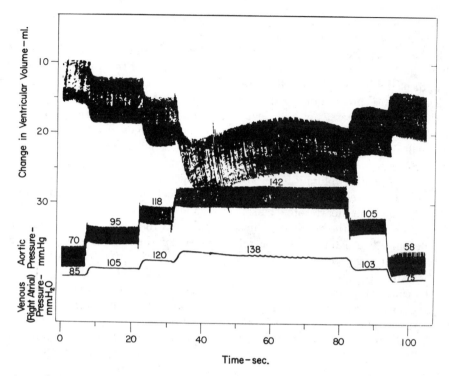

Fig. 8-13. Changes in ventricular volume, aortic pressure, and right atrial pressure in a heart-lung preparation when peripheral resistance was raised and subsequently lowered in several steps. Note that an increase in ventricular volume is registered as a downward shift in the tracing. (Redrawn from Patterson, S. W., Piper, H., and Starling, E. H.: J. Physiol. [London] **48**:465, 1914.)

the walls of the ventricles. The increased diastolic fiber length somehow facilitates ventricular contraction and enables the ventricles to pump a greater stroke volume, so that, at equilibrium, cardiac output exactly matches the augmented venous return. An optimal fiber length apparently exists, beyond which contraction is actually impaired. Therefore excessively high levels of venous return may depress rather than enhance the pumping capacity of the ventricles by overstretching the myocardial fibers.

Changes in diastolic fiber length also permit the isolated heart to compensate for an increase of peripheral resistance. In the experiment depicted in Fig. 8-13 the arterial resistance was abruptly raised in three steps, whereas venous inflow was held constant. Each step increase in resistance was associated with an increase in arterial pressure and ventricular volume. With each abrupt elevation of arterial pressure, the left ventricle was at first unable to pump a normal stroke volume. Since venous return was held constant, diminution of stroke volume was attended by a rise in ventricular diastolic volume and therefore in the length of the myocardial fibers. This change in end-diastolic fiber length finally enabled the ventricle to pump a given stroke volume against a greater peripheral resistance.

The external work performed per stroke by the left ventricle is approximately equal to the product of the mean arterial pressure and stroke volume (p. 94). Therefore the increased diastolic length of the cardiac muscle fiber results in a greater work production by the left ventricle. However, at excessively high levels of peripheral resistance, further augmentation of resistance will result in reductions of stroke volume and stroke work.

Changes in ventricular volume have also been shown to be involved in the cardiac adaptation to alterations in heart rate. During bradycardia, for example, the increased duration of diastole permits greater ventricular filling. The consequent augmentation of myocardial fiber length results in an increased stroke volume. With constant venous return, therefore, the reduction in heart rate is fully compensated by virtue of an increase in stroke volume, such that cardiac output remains constant.

When cardiac compensation involves ventricular dilation, the tension required by each myocardial fiber to generate a given intraventricular systolic pressure must be appreciably greater than that developed by the fibers in a ventricle of normal size. The relationship between wall tension and cavity pressure resembles that for cylindrical tubes (p. 111) in that for a constant internal pressure, wall tension varies directly with the radius. As a consequence the dilated heart has been found to have a considerably greater oxygen requirement to perform a given amount of external work than does the normal heart.

In many experiments Starling and his coworkers recorded left ventricular pressure as well as ventricular volume. In certain experiments, augmented venous return or peripheral resistance produced definite ventricular diastolic distension, yet left ventricular pressure at the end of diastole (*left ventricular end-diastolic pressure*) was not detectably different from that observed during the control conditions. Such findings led Starling and his associates to emphasize that greater fiber length rather than increased tension was the critical determinant of the strength of myocardial contraction. Subsequent investigations have shown that, under most conditions, changes in resting fiber length are accompanied by concordant changes in resting fiber tension. However at smaller ventricular volumes the myocardium is considerably more distensible, and changes in fiber length may be accompanied by scarcely detectable changes in tension. This is illustrated by Fig. 4-1, in which the slope (which is the reciprocal of ventricular distensibility) of the "diastole" curve is seen to increase with progressively greater myocardial fiber length. Thus the ventricle is quite distensible in the normal range (<12 mm. Hg) of end-diastolic pressures, but becomes relatively rigid at abnormally high levels of pressure. The precise relationship between pressure and volume of the ventricles during diastole is complex because the elastic nature of the ventricular walls is not the sole consideration. Account must also be taken of the viscous and inertial properties of the ventricular walls, so that diastolic ventricular pressure is not only a function of volume per se but also of the rate of change of volume.

In the intact animal, of course, the heart is enclosed in the pericardial sac. Thus, the relatively rigid pericardium determines the pressure-volume relationship at the higher levels of pressure and volume. It is likely that this limitation by the pericardium is exerted even under normal conditions, when an individual is at rest and his heart rate is slow. In the cardiac dilation and hypertrophy that usually accompanies chronic heart failure, the pericardium is stretched considerably, and its limitation is exerted at pressures and volumes that are entirely different from those in normal individuals.

Studies on more intact preparations. The major problem involved in assessing the role of the

Frank-Starling mechanism in intact animals and humans resides in the difficulty of obtaining a representative measure of end-diastolic myocardial fiber length. The Frank-Starling mechanism has been represented graphically with some index of ventricular performance usually plotted along the ordinate and some index of fiber length along the abscissa. The most common indices of ventricular performance that have been assessed are cardiac output, stroke volume, and stroke work. The indices of fiber length include ventricular end-diastolic volume, ventricular end-diastolic pressure, ventricular circumference, and mean atrial pressure.

The Frank-Starling mechanism is usually represented by a family of so-called ventricular function curves, rather than by a single curve. To construct a given ventricular function curve, blood volume is altered over a wide

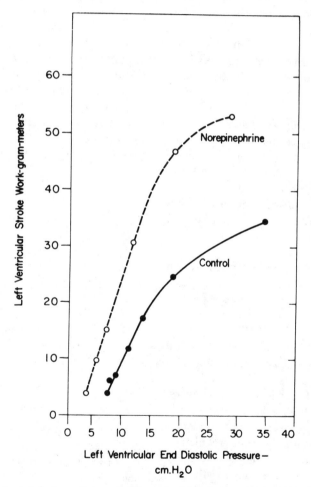

Fig. 8-14. A constant infusion of norepinephrine in the dog causes the ventricular function curve to shift to the left, signifying an enhancement of ventricular contractility. (Redrawn from Sarnoff, S. J., Brockman, S. K., Gilmore, J. P., Linden, R. J., and Mitchell, J. H.: Circ Res. **8**:1108, 1960.)

range of values, and stroke work and end-diastolic pressure are measured at each step. A similar series of observations are then made during the desired experimental intervention. For example, the ventricular function curve obtained during a norepinephrine infusion lies above and to the left of a control ventricular function curve (Fig. 8-14). It is evident that, for a given level of left ventricular end-diastolic pressure, the left ventricle performs more work during a norepinephrine infusion than during control conditions. Hence a shift of the ventricular function curve to the left usually signifies an improvement of ventricular contractility (p. 81); a shift to the right usually indicates an impairment of contractility and a consequent tendency toward *cardiac failure*. A shift in a ventricular function curve cannot be interpreted uniformly as an indication of a change in contractility, however. Contractility is a measure of cardiac performance at a given level of preload and afterload. The end-diastolic pressure is ordinarily a good index of the preload. In assessing changes in myocardial contractility, the cardiac afterload must be held constant as the end-diastolic pressure is varied over a range of values.

It is very difficult at present to determine the precise position on a Frank-Starling curve at which the heart of an intact, conscious person or animal may be operating. Recent studies on instrumented, conscious dogs indicate that when the dog is reclining and relaxed, the left ventricle seems to be at nearly its maximal size at the end of diastole; it is probably limited by the pericardium. When blood volume is reduced, the end-diastolic volume diminishes, and the Frank-Starling mechanism is undoubtedly involved in cardiac output regulation. Conversely, when blood volume is acutely expanded, end-diastolic volume apparently cannot increase appreciably. Hence, extrinsic mechanisms probably play a more crucial role in cardiac output regulation. In the experiment illustrated in Fig. 8-4, for example, the trans-

fusion of blood caused a parallel increase in heart rate and cardiac output; stroke volume did not change appreciably. The Bainbridge reflex seemed to play the dominant role in this adaptation to an increased blood volume.

The Frank-Starling mechanism is the one that is certainly most ideally suited for matching the cardiac output to the venous return. Any sudden, excessive output by one ventricle soon results in a greater venous return to the other ventricle. The consequent increase in diastolic fiber length serves as the stimulus to increase the output of the second ventricle to correspond with that of its mate. For this reason it is the Frank-Starling mechanism that maintains a precise balance between the outputs of the right and left ventricles. Since the two ventricles are arranged in series in a closed circuit, it is apparent that even a small, but maintained, imbalance in the outputs of the two ventricles would have catastrophic consequences.

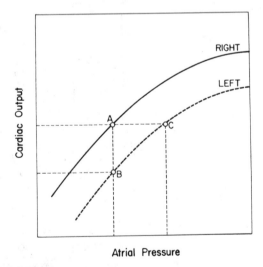

Fig. 8-15. Curves relating the outputs of right (continuous line) and left (dashed line) ventricles to mean right and left atrial pressure, respectively. At a given level of cardiac output, mean left atrial pressure (for example, point C) exceeds mean right atrial pressure (point A).

The curves relating cardiac output to mean atrial pressure for the two ventricles are not coincident; the curve for the left ventricle usually lies below that for the right, as shown in Fig. 8-15. At equal right and left atrial pressures (points A and B) right ventricular output would exceed left ventricular output. Hence venous return to the left ventricle (a function of right ventricular output) would exceed left ventricular output, and left ventricular diastolic volume and pressure would rise. By the Frank-Starling mechanism, left ventricular output would therefore increase (from B toward C.) Only when the outputs of both ventricles were identical (points A and C) would a stable equilibrium condition exist. Under such conditions, however, left atrial pressure (C) would exceed right atrial pressure (A), and this is precisely the relationship that is found in both humans and the dog. This difference in atrial pressures accounts for the observation that in *congenital atrial septal defects*, where a communication exists between the two atria, the direction of the shunt flow is ordinarily from left to right.

Homeometric autoregulation

Flow-induced regulation. In the ventricular volume tracings in Figs. 8-12 and 8-13, when venous return or arterial resistance was suddenly increased, there was a rapid increase in diastolic ventricular volume, as described previously. However, after the initial ventricular distension, some tendency for the ventricles to return toward control volumes was manifest over the next 1 or 2 minutes, even though the augmentation of venous return or arterial resistance persisted. At the time these observations were made it was not known whether these secondary changes in ventricular volume represented artifacts (caused by displacement of the heart in the cardiometer), an improvement in contractility associated with an increase in coronary blood flow, or the supervention of some secondary mechanism.

More recent experiments have been con-

ducted on hearts that have been extirpated and perfused while located entirely within a rigid plethysmograph (volume recorder). By this means, the artifacts inherent in the classical cardiometer were averted. The effects of sudden changes in the venous return on ventricular volume in such a preparation are shown in Fig. 8-16. An increase in volume is represented by an upward excursion in the figure, which is opposite to the direction of the excursion of the volume recorder in Figs. 8-12 and 8-13.

When venous return was suddenly augmented in this preparation, there was a rapid increase in the ventricular systolic and diastolic volumes. The maximum volumes were attained in approximately 30 seconds. At this point in time the stroke volume (difference between diastolic and systolic volumes) was also considerably augmented. This portion of the response is typical of heterometric autoregulation and does not differ appreciably from that displayed in Fig. 8-12. Very shortly after the maximum diastolic volume was attained, however, it is evident from the plethysmographic tracing in Fig. 8-16 that the ventricular systolic and diastolic volumes progressively diminished, whereas the enhanced stroke volume was maintained. At the new equilibrium state, which was reached after about 5 minutes, the diastolic volume was actually slightly less than the volume that existed at the lower control level of venous return. To accomplish this reduction in ventricular volume, after the maximum volume had been attained, the ventricles must have ejected during systole a stroke volume slightly greater than the volume that had entered them during the preceding diastole. Once steady state conditions had become re-established, stroke volume was, of course, equal to the volume of blood that had entered the ventricles during diastole.

The equivalent of flow-induced regulation in an isolated cardiac muscle strip is shown in Fig. 8-17. At arrow A, the initial length of an

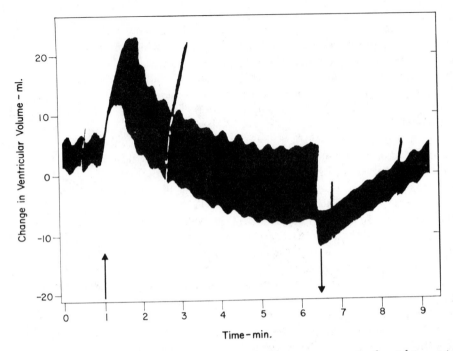

Fig. 8-16. Changes in right ventricular volume when the venous return to the right ventricle was suddenly increased (left arrow) and subsequently decreased to the control level (right arrow). While the constant increase in venous return was maintained (between the arrows), right ventricular diastolic volume (upper border of the envelope of the tracing) gradually decreased to a level slightly below the control volume. (Redrawn from Rosenblueth, A., Alanís, J., López, E., and Rubio, R.: Arch. Int. Physiol. **67**:358, 1959.)

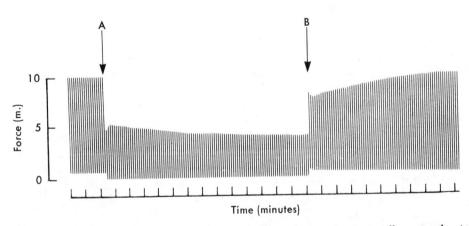

Fig. 8-17. The force developed during the isometric contraction of a cat papillary muscle. At arrow *A*, the initial length was suddenly decreased; at *B*, it was suddenly increased. Note the gradual changes in developed force, even though the length was held constant after the two abrupt changes. (Modified from Parmley, W. W., and Chuck, L.: Am. J. Physiol. **224**:1195, 1973.)

isometrically contracting papillary muscle was suddenly decreased. This is analogous to a reduction in the venous return to an intact heart. The amount of developed tension diminished promptly, in accordance with the Frank-Starling mechanism. However, over the course of the next several minutes, the developed tension continued to decrease, despite a constant initial length. A sudden increase in initial length was produced at arrow *B,* which is analogous to an augmentation of venous return to an intact heart. There was an immediate rise in developed tension, followed by a further gradual increase in the developed tension. This progressively more forceful contraction at a constant initial tension in a strip of cardiac muscle is probably equivalent to the flow-induced homeometric autoregulation displayed in the intact heart (Fig. 8-16).

Pressure-induced regulation. An increase in afterload may also evoke a type of homeometric autoregulation, which is sometimes referred to as the *Anrep effect.* In the experiment shown in Fig. 8-18, arterial and left ventricular diastolic pressures rose when peripheral resistance was increased abruptly. However the rise (phase *1*) in left ventricular diastolic pressure was only temporary. After the peak diastolic pressure was reached, the diastolic pressure diminished progressively toward the control level (phase *2*) while aortic pressure remained elevated. Finally, a new equilibrium level of ventricular diastolic pressure was reached (phase *3*); in many cases this pressure was actually at or slightly below the control value. Measurement of myocardial segment length revealed changes during diastole that paralleled the changes in ventricular diastolic pressure. With abrupt return of peripheral resistance to the control level, left ventricular diastolic pressure declined to a value appreciably below control and then gradually returned to the control level (phase *4*). Coronary blood flow was controlled in these experiments, so that the adap-

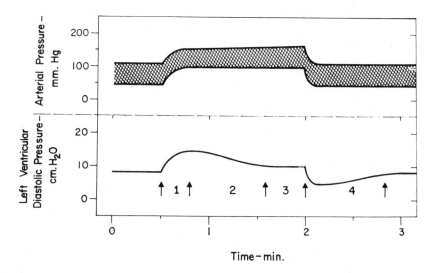

Fig. 8-18. Homeometric autoregulation in the dog heart in response to a sustained increase in peripheral resistance. Initially there was an appreciable rise in left ventricular end-diastolic pressure (phase *1*), but this subsequently returned toward the control level (phases *2* and *3*), even though peripheral resistance was still augmented. (Redrawn from Sarnoff, S. J., and Mitchell, J. H.: Am. J. Med. **34:**440, 1963.)

tation of the heart cannot be attributed to an improvement in the myocardial circulation. More recent evidence indicates that these autoregulatory changes may be caused in part by the release of myocardial catecholamines.

This evidence demonstrates that the mammalian ventricle possesses the intrinsic capability of adapting to changes in venous return and arterial resistance without a continued increase in resting fiber length. The conditions under which homeometric autoregulation, rather than maintained heterometric autoregulation, would prevail have not been delineated. Recent studies suggest that homeometric autoregulation may be relatively unimportant in the normal heart, but may be more prominent when contractility is depressed. It is likely that transient heterometric adaptation usually precedes homeometric regulation, as in the previous examples, but ventricular distension is not a necessary prelude to homeometric autoregulation. This mechanism therefore averts the mechanical disadvantage that dilation of the ventricles places on the myocardial fibers during heterometric autoregulation.

Rate-induced regulation. Approximately one century ago, Bowditch directed attention to the modifications of myocardial performance that depend on the time interval between beats. He demonstrated the occurrence of *Treppe* (the so-called *staircase phenomenon*) in the frog ventricle; that is, after a period of rest the ventricle responded to repetitive stimuli with progressively greater contractions until a plateau was attained.

The effects of contraction frequency on the tension developed in an isometrically contracting cat papillary muscle are shown in Fig. 8-19, *B*. Initially, the strip of cardiac muscle was stimulated to contract only once every 20 seconds. When the muscle was made to contract once every 0.63 second, the developed tension increased progressively over the next several beats. At the new steady state, the developed tension was more than five times as great as it was at the lower contraction frequency. A return to the slower rate had the opposite influence on developed tension, with an ultimate return to the initial value.

The effect of the interval between contrac-

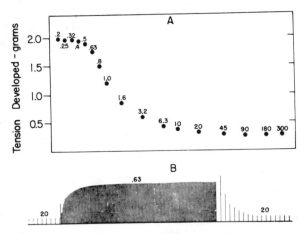

Fig. 8-19. Changes in tension development in an isolated papillary muscle from a cat as the interval between contractions is varied. The numbers in both sections of the record denote the interval (in seconds) between beats. In section *A* the points represent the steady state tensions developed at the intervals indicated. (Redrawn from Koch-Weser, J., and Blinks, J. R. Pharmacol. Rev. **15**:601, 1963.)

tions on the steady state level of developed tension is shown in panel A (Fig. 8-19) for a wide range of intervals. It is evident that, as the interval is diminished from 300 seconds down to about 10 or 20 seconds, there is little change in developed tension. As the interval is reduced further, to a value of about 0.5 second, tension increases steeply. Further reduction of the interval to 0.2 second has little additional effect on developed tension.

The progressive rise in developed tension observed in panel B as the contraction frequency is increased is ascribable to a gradual rise in intracellular calcium. During the plateau of the action potential, Ca^{++} enters the cell. As the contraction frequency is increased, there is a greater influx of Ca^{++} into the cell. The duration of the cardiac action potential diminishes as contraction frequency is raised, and so less Ca^{++} actually enters per contraction. The influx of Ca^{++} per minute equals the influx per beat times the number of beats per minute. As the frequency of contraction is increased, the increment in the number of beats per minute exceeds the decrement in Ca^{++} in-

flux per beat, and so there is a net rise in the Ca^{++} influx per minute. The intracellular content of Ca^{++} increases, and as a consequence, developed tension rises as the interval between contractions is diminished, as shown in panel A.

The role of intracellular Ca^{++} as a mediator of the force-frequency relationship has been graphically demonstrated in experiments in which aequorin was injected into frog atrial cells. Aequorin is a Ca^{++}-sensitive bioluminescent protein, and hence the light output reflects the intracellular Ca^{++} concentration. It is evident from Fig. 8-20 that, as the contraction frequency was increased from 0.125 to 0.25 Hz, and then to 0.5 Hz, the amplitudes of the contraction and of the aequorin signal displayed parallel changes. When the frequency was increased to 1.0 Hz, the contraction amplitude diminished, and again there was a concordant change in the aequorin signal. Hence, the changes in contraction amplitude evoked by an alteration in contraction frequency are probably ascribable to alterations in the intracellular content of Ca^{++}.

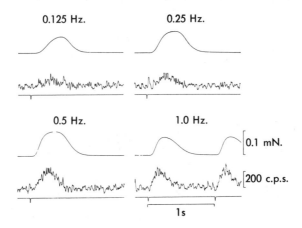

0.125 Hz.   0.25 Hz.

0.5 Hz.   1.0 Hz.
                          0.1 mN.
                          200 c.p.s.
                     1s

Fig. 8-20. The changes in the force of isometric contraction (top tracing, in mN.) and in the luminescent response (bottom tracing, in photon counts per sec.) from frog atrial myocardial fibers that were injected with aequorin, and then stimulated to contract at frequencies of 0.125, 0.25, 0.5, and 1.0 Hz. (From Allen, D. G., and Blinks, J. R.: Nature **273**:509, 1978.)

Another manifestation of the influence of the time interval between beats has been termed *postextrasystolic potentiation*. It has long been recognized that in the course of a premature ventricular systole (Fig. 2-35, *B*) the premature contraction itself is feeble, whereas the beat following the compensatory pause is considerably stronger than usual. Classically, this response has been explained exclusively on the basis of the Frank-Starling mechanism. Inadequate ventricular filling just prior to the premature beat was invoked to account for the observed ineffectual contraction during the premature systole itself. Subsequently, the exaggerated degree of filling associated with the compensatory pause served to explain the vigorous postextrasystolic contraction. Although it is true that heterometric autoregulation is involved in the usual ventricular adaptation to this arrhythmia, it is not the exclusive mechanism. For example, in the ventricular pressure curves recorded from an isovolumetric left ventricle preparation (Fig. 8-21), in which neither filling nor ejection takes place during the cardiac cycle, the premature beat *(A)* is feeble and the suceeding contraction *(B)* is supernormal. Such enhanced contractility is an example of

postextrasystolic potentiation, and it may persist for one or more additional beats (for example, contraction *C*, Fig. 8-21).

The mechanism responsible for postextrasystolic potentiation is probably related to the time course of the intracellular circulation of Ca^{++} during the contraction and relaxation process. The Ca^{++} that activates contraction is probably released from some "phasic store" in the sarcoplasmic reticulum. This phasic store probably contains only a little more Ca^{++} than is required for one beat. The Ca^{++} during excitation is then taken up at some other site in the sarcotubular network to induce relaxation. The Ca^{++} is translocated from this site back to the phasic store, where it will be released again with the next excitation. The process of translocation is relatively slow, however, and requires about 200 to 800 msec.

It is likely that the premature beat itself (beat *A*, Fig. 8-21) is feeble because it occurs before much of the Ca^{++} has been able to move from the uptake site in the sarcotubular network, where it is unavailable for interaction with the contractile proteins, to the phasic store, where the release process does occur. The postextrasystolic beat (Fig. 8-21, *B*), con-

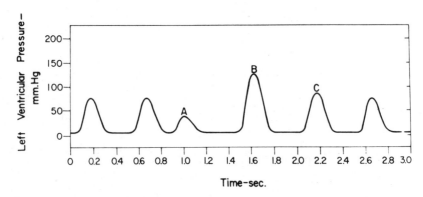

Fig. 8-21. In an isovolumetric canine left ventricle preparation, a premature ventricular systole (beat *A*) is typically feeble, whereas the postextrasystolic contraction (beat *B*) is characteristically strong, and the enhanced contractility may persist to a diminishing degree over a few beats (for example, contraction *C*).

versely, is considerably stronger than normal. A plausible reason is that after the compensatory pause (the pause between beats *A* and *B*) the phasic store will have available for release the Ca^{++} that had been taken up at the site in the sarcotubular network after two contractions—the extrasystole (beat *A*) and the normal beat that had preceded it.

EXTRINSIC REGULATION OF CARDIAC PERFORMANCE

Although the completely isolated heart is capable of a remarkable degree of adaptation to changes in venous return and peripheral resistance, in the intact animal there are various extrinsic factors that also exercise a potent regulatory influence on the myocardium. Under many natural conditions these extrinsic mechanisms may dwarf in importance the intrinsic regulatory mechanisms. These extrinsic regulatory factors may be subdivided into nervous and chemical components.

Nervous control

Sympathetic influences. The sympathetic division of the autonomic nervous system has a profound facilitatory effect on the atrial and ventricular myocardium. The concentration of norepinephrine in the various regions of the heart reflects the relative density of the sympathetic innervation to those areas. In the normal heart the norepinephrine concentration in the atria is about three times that in the ventricles. The norepinephrine concentration in the S-A and A-V nodes is no greater than that in the surrounding atrial regions. When the heart is denervated, the tissue concentrations of this neurotransmitter approach zero.

The alterations in ventricular contraction evoked by electrical stimulation of the left stellate ganglion in a canine isovolumetric left ventricle preparation are shown in Fig. 8-22. The peak pressure developed by the ventricle during systole is considerably augmented, and the maximum rate at which pressure is generated

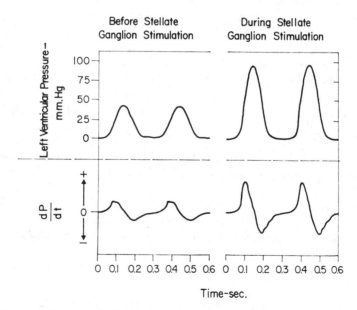

Fig. 8-22. In an isovolumetric left ventricle preparation, stimulation of cardiac sympathetic nerves evokes a substantial rise in peak left ventricular pressure and in the maximum rates of intraventricular pressure rise and fall (dP/dt).

(dP/dt) is markedly increased. Also the duration of systole is reduced and the rate of ventricular relaxation is increased during the early phases of diastole. In the dog, sympathetic nerve fibers from the left side exert a much more potent effect on left ventricular performance than do those from the right side. On the average a given stimulus applied to the left stellate ganglion produces a rise in peak left ventricular pressure that is more than twice as great as that evoked by right-sided stimulation (Fig. 8-3).

Sympathetic nervous activity facilitates myocardial performance principally by enhancing the contractility of the individual cardiac muscle cells. The manner in which this is accomplished is not fully understood, but the stimulation of adenylate cyclase appears to be a critical step in the process. The consequent elevation in the level of cyclic AMP leads to the activation of a protein kinase that is involved in the phosphorylation of proteins in the sarcolemma and sarcoplasmic reticulum. These reactions are believed to alter Ca^{++} fluxes across the sarcolemma and sarcoplasmic reticulum. Catecholamines are known to increase the permeability of the cell membrane to Ca^{++}, resulting in a greater inward Ca^{++} current during the plateau of the action potential. Such a change in Ca^{++} flux across the sarcolemma, and probably also across the sarcoplasmic reticulum, makes more Ca^{++} available for interaction with the contractile proteins, which thereby affects the enhancement of myocardial contractility.

The pronounced change in intracellular Ca^{++} concentration produced by catecholamines is clearly shown in Fig. 8-23. Under control conditions, there was only a small increase in bioluminescence during contraction in these atrial myocardial fibers that had been injected with aequorin. However, after the addition of the beta-adrenergic agonist, isoproterenol, there was a dramatic increase in the aequorin response, concomitant with the increase in contractile force.

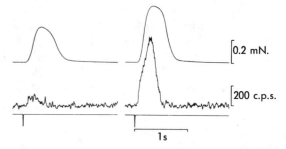

Fig. 8-23. The force of isometric contraction (top tracing, in mN.) and the luminescent response (bottom tracing, in photon counts per sec.) of frog atrial myocardial fibers that were injected with aequorin. The responses on the left were recorded before, and those on the right after, the addition of isoproterenol. (Modified from Allen, D. G., and Blinks, J. R.: Nature **273**:509, 1978.)

In addition to such direct actions, indirect mechanisms also play a contributory role. The abridgement of the duration of systole and the more rapid ventricular relaxation enhance ventricular filling. The former provides a greater fraction of the cardiac cycle duration for ventricular filling, whereas the latter permits a more rapid ventricular filling from any given atrial filling pressure. The positive inotropic influence of sympathetic activity on atrial contraction also facilitates ventricular filling. These factors assume greater importance at faster heart rates, which, of course, are also a consequence of increased sympathetic activity. Finally, contractility may be improved by virtue of the increased heart rate itself, a manifestation of rate-induced homeometric autoregulation (Fig. 8-19).

This multiplicity of facilitatory sympathetic influences on the heart of intact animals can best be appreciated in terms of families of ventricular function curves. When stepwise increases in the frequency of electrical stimulation are applied to the left stellate ganglion, the ventricular function curves are shifted progressively to the left. The changes parallel those produced by catecholamine infusions, as illus-

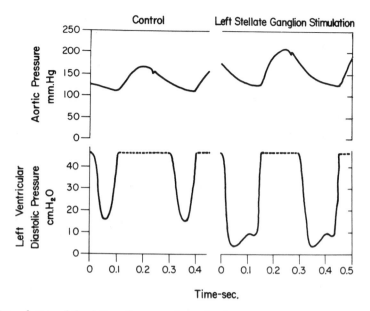

Fig. 8-24. Stimulation of the left stellate ganglion of a dog increases arterial pressure, stroke volume, and stroke work despite a concomitant reduction in ventricular end-diastolic pressure. Note also the abridgment of systole, thereby allowing more time for ventricular filling. In the ventricular pressure tracings the pen excursion is limited at 45 mm. Hg; actual ventricular pressures during systole can be estimated from the aortic pressure tracings. (Redrawn from Mitchell, J. H., Linden, R. J. and Sarnoff, S. J.: Circ. Res. **8:**1100, 1960.)

trated in Fig. 8-14. Hence, for any given left ventricular end-diastolic pressure, the ventricle is capable of performing more work as the level of sympathetic nervous activity is progressively raised. During cardiac sympathetic stimulation the shift from curve to curve usually occurs in such a manner than an increase in work is achieved despite a reduction in mean left ventricular end-diastolic pressure. An example of such a reduction in left ventricular end-diastolic pressure during stellate ganglion stimulation in a paced heart is shown in Fig. 8-24. In this experiment, stroke work increased by about 50%, despite a 7 cm. H_2O reduction in the left ventricular end-diastolic pressure. The pronounced abridgment of the duration of ventricular systole, with the consequent lengthening of the filling period, is also evident in the same tracing. The reason for the reduction in

ventricular end-diastolic pressure is explained on p. 198.

Parasympathetic influences. It is universally recognized that the vagus nerves exert profound depressant effects on the cardiac pacemaker, the atrial myocardium, and the atrioventricular conduction tissue. The ventricles have generally been considered to be essentially devoid of parasympathetic innervation. However, considerable evidence has been accumulated over the past decade that demonstrates that the vagus nerves do indeed depress ventricular contractility. Fig. 8-25 reveals that the effects of vagal stimulation on the isovolumetric left ventricle preparation are just the reverse of those induced by sympathetic stimulation (illustrated in Fig. 8-22). Vagal stimulation produces a reduction in peak left ventricular pressure, in the maximum rate of pressure

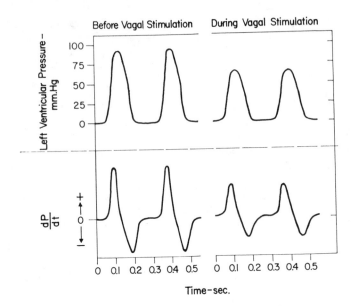

Fig. 8-25. In an isovolumetric left ventricle preparation, when the ventricle is paced at a constant frequency, vagal stimulation decreases the peak left ventricular pressure and diminishes the maximum rates of pressure rise and fall (dP/dt).

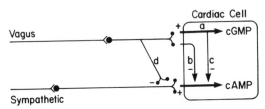

Fig. 8-26. The interneuronal and intracellular mechanisms responsible for the vagal-sympathetic interactions. (Modified from Levy, M. N. In Baan, J., Noordergraaf, A., and Raines, J., editors: Proceedings of the International Conference on Cardiovascular System Dynamics, Cambridge, Mass., 1976, The M.I.T. Press.)

development (dP/dt), and in the maximum rate of pressure decline during diastole. In pumping heart preparations the ventricular function curve is shifted to the right. No differences are detectable between the effects of the right and left vagus nerves on ventricular performance, in contrast to the distinct differences between the two sides observed in the case of the cardiac sympathetic nerves.

The depressant effect of vagal activity on the ventricular myocardium appears to be achieved by at least four mechanisms, as shown in Fig. 8-26. One process is direct, and the others involve an interaction with the sympathetic nervous system. With respect to the direct mechanism, it has been found that vagal stimulation or acetylcholine infusions raise the intracellular levels of cyclic GMP (arrow *a*). This nucleotide in turn may depress myocardial contractility through some process that has not yet been elucidated.

With respect to the mechanisms involving an interaction with the sympathetic nervous system, it has been observed that when the existing level of cardiac sympathetic nervous activity is low, the depressant effect of increased parasympathetic activity on the ventricular myocardium is relatively feeble. However, against a background of tonic sympathetic activity, the negative inotropic effect produced by vagal stimulation is considerably more prominent. A similar phenomenon occurs with respect to the chronotropic responses to autonomic neural activity. In the left half of Fig. 8-1, for example, vagal stimulation produced a much greater reduction in the heart rate during simultaneous cardiac sympathetic stimulation at 4 pulses/sec. (S = 4) than in the absence of sympathetic stimulation (S = 0).

This accentuated antagonism between the parasympathetic and sympathetic systems may be accomplished in three different ways (Fig. 8-26). First, the acetylcholine released at the vagal endings causes a direct reduction in the intracellular levels of cylic AMP (arrow *b*). As stated in the preceding ´section, cyclic AMP probably mediates the enhancement of contractility produced by sympathetic neural activity. Second, increased vagal activity raises intracellular levels of cyclic GMP, as explained previously. This nucleotide accelerates the hydrolysis of cyclic AMP (arrow *c*), thereby lowering its concentration in the myocardial cell.

The third mechanism responsible for the accentuated antagonism between the two divisions of the autonomic nervous system involves extracellular processes. As shown in Fig. 8-26, some postganglionic vagal terminals (*d*) end near the postganglionic sympathetic terminals

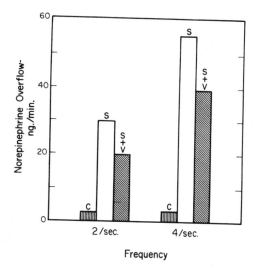

Fig. 8-27. The mean rates of overflow of norepinephrine (NE) into the coronary sinus blood in a group of seven dogs under control conditions (*C*), during cardiac sympathetic stimulation (*S*) at 2 or 4 cycles/sec., and during combined sympathetic and vagal stimulation (S + V). The combined stimulus consisted of sympathetic stimulation at 2 or 4 cycles/sec., and vagal stimulation at 15 cycles/sec. (Redrawn from Levy, M. N., and Blattberg, B.: Circ. Res. **38**:81, 1976. By permission of the American Heart Association, Inc.)

in the heart. The acetylcholine released at these vagal endings inhibits the release of norepinephrine from the sympathetic fibers. Fig. 8-27 demonstrates that stimulation of the cardiac sympathetic nerves (*S*) results in the overflow of substantial amounts of norepinephrine into the coronary sinus blood, in comparison with the rates of overflow during the control state (*C*). Concomitant vagal stimulation (S + V) causes a reduction of about 30% in the rate of overflow of norepinephrine produced by sympathetic stimulation alone at a given frequency. The amount of norepinephrine overflowing into the coronary sinus blood probably parallels the amount released at the sympathetic terminals.

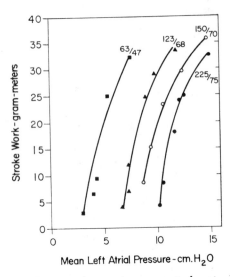

Fig. 8-28. As the pressure in the carotid sinus is progressively raised, there is a shift to the right of the ventricular function curves. The numbers at the tops of each curve represent the systolic/diastolic perfusion pressures (in mm. Hg) in the carotid sinus regions of the dog. (Redrawn from Sarnoff, S. J., Gilmore, J. P., Brockman, S. K., Mitchell, J. H., and Linden, R. J.: Circ. Res. **8:**1123, 1960.)

Baroreceptor reflex. Just as stimulation of the carotid sinus and aortic arch baroreceptors evokes marked changes in heart rate (p. 151), so also does it elicit reflex alterations of myocardial performance. Evidence of reflex alterations of ventricular contractility is presented in Fig. 8-28. Ventricular function curves were obtained at four different levels of carotid sinus perfusion pressure. With each successive rise in perfusion pressure, the ventricular function curves were displaced farther and farther to the right, denoting a progressively greater reflex depression of ventricular performance.

In normal, resting individuals and animals the tonic level of sympathetic activity is usually very low. Recent studies have shown that under such conditions, mild to moderate changes in baroreceptor activity have little reflex influence on myocardial contractility. In states of augmented sympathetic neural activity, how-

ever, the effects of the baroreceptor reflex on contractility may play an important role. In the adaptation to blood loss, for example, a reflex change in myocardial contractility may constitute an important means of compensation. Blood loss diminishes cardiac output. The associated reduction in arterial blood pressure alters the intensity of baroreceptor stimulation, thereby evoking not only an acceleration of heart rate but also an improvement of myocardial contractility. Hence this type of feedback mechanism tends to minimize the extent of the reduction in cardiac output that would be induced by the loss of a given volume of blood.

Chemoreceptor reflex. When pulmonary ventilation is controlled, carotid chemoreceptor stimulation results in profound bradycardia, often with some degree of atrioventricular conduction block. This indicates a considerable increase of vagal activity. When the atria and

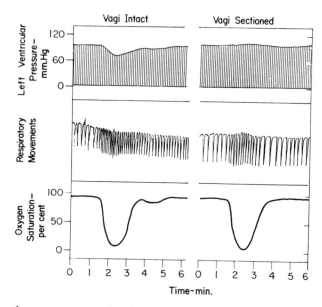

Fig. 8-29. When the oxygen saturation (bottom tracing) of the blood perfusing the carotid chemoreceptors was diminished in an isovolumetric left ventricle preparation, peak ventricular pressure decreased when the vagi were intact. After the vagi were transected, a similar degree of carotid chemoreceptor stimulation had either no effect or a slight stimulatory influence, indicating that the depression of ventricular contractility was mediated primarily by the vagi. (Redrawn from DeGeest, H., Levy, M. N., and Zieske, H.: Am. J. Physiol. **209**:564, 1965.)

ventricles are paced at a constant rate, the influence of enhanced parasympathetic activity on myocardial performance is manifested.

The ventricular response to carotid chemoreceptor stimulation is depicted in Fig. 8-29. With the vagi intact (left panel), perfusion of the isolated carotid chemoreceptor region with hypoxic or hypercapnic blood diminished the peak pressure generated by the left ventricle, whereas the frequency of respiratory muscle movements was increased. After cervical vagotomy (right panel), peak ventricular pressure was no longer diminished by carotid chemoreceptor stimulation. The chemoreceptors were still active, however, as shown by the response of the respiratory muscles. Therefore the

depression of left ventricular contractility during carotid chemoreceptor stimulation must be mediated primarily by parasympathetic pathways.

Hypoxia and ischemia of the central nervous system. When the cephalic region of an innervated, isovolumetric left ventricle preparation is perfused with hypoxic blood, a biphasic response is usually obtained from the ventricle. As shown in the left panel of Fig. 8-30, there is an initial depression of peak pressure generation by the left ventricle, followed by an enhancement of performance. The initial depression is attributable to carotid chemoreceptor excitation, for it may be abolished by bilateral denervation of the carotid sinus regions. The

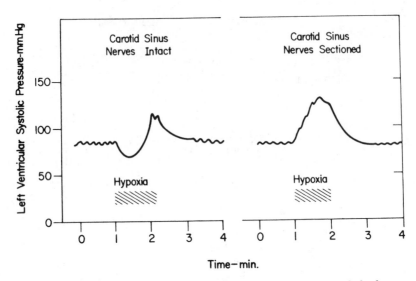

Fig. 8-30. In the innervated, isovolumetric left ventricle preparation, cephalic hypoxia characteristically evokes an initial reduction in peak left ventricular pressure, followed by a rise in pressure (left half of figure). The initial depression of ventricular performance is ascribable to carotid chemoreceptor stimulation, because after sectioning the carotid sinus nerves (right half of figure), cephalic hypoxia elicits only facilitation of ventricular contractility. The small rhythmic fluctuations in left ventricular systolic pressure occur at the frequency of the respiratory movements. (Redrawn from DeGeest, H., Levy, M. N., and Zieske, H.: Circ. Res. **17:**349, 1965.)

secondary increase in contractility is caused by the preponderant effect of hypoxia on the brain. After inactivation of the carotid chemoreceptors, acute central nervous system hypoxia elicits a pronounced enhancement of ventricular performance without any preliminary depression (right panel, Fig. 8-30).

Chemical control

Hormones

Adrenomedullary hormones. Essentially the adrenal medulla is an integral component of the autonomic nervous system. The rate of secretion of the catecholamines epinephrine and norepinephrine, by the adrenal medulla is largely regulated by the same mechanisms that control the activity of the sympathetic division of this system. The relative content of epineph-

rine and norepinephrine in the adrenal venous blood varies from species to species and in any given species probably changes under different physiological conditions.

The catecholamines exert a potent positive inotropic effect on myocardial contractility (Fig. 8-14). Since norepinephrine is the active transmitter secreted at the postganglionic sympathetic neuroeffector junctions within the heart, it may be surmised that the adrenomedullary hormones affect the myocardium in a manner virtually identical to that described on p. 169 in relation to the influence of sympathetic nervous activity. There are no important distinctions between the myocardial effects of epinephrine and norepinephrine.

Adrenocortical hormones. In adrenocortical insufficiency (*Addison's disease*), cardiovascular disturbances constitute an important clinical feature, and cardiovascular collapse characterizes the so-called addisonian crisis. A severe reduction in plasma volume is a prominent factor in producing the profound hypotension that is typical of addisonian crisis, but a cardiac factor is probably also involved. The precise influence of the adrenocortical steroids on the myocardium is controversial at present. Cardiac muscle removed from adrenalectomized animals shows a much greater tendency to fatigue than that obtained from normal animals. In some species and in certain experimental preparations the adrenocortical hormones exert a pronounced positive inotropic effect on the myocardium. Furthermore, it has recently been shown that hydrocortisone potentiates the cardiotonic effects of the catecholamines. This potentiation may be mediated in part by an inhibition of the neuronal and extraneuronal uptake mechanisms for the catecholamines.

Thyroid hormones. Cardiac activity in patients with inadequate thyroid function (*hypothyroidism*) is typically sluggish; that is, the heart rate is slow and cardiac output is diminished. The converse is true in patients with overactive thyroid glands (*hyperthyroidism*). Characteristically, such patients exhibit tachycardia, high cardiac output, palpitations, and arrhythmias (such as atrial fibrillation). In normal individuals or in experimental animals the cardiovascular manifestations of hyperthyroidism may be simulated by the administration of thyroxine.

Abundant evidence has been adduced to show that the thyroid hormones do exert a direct effect on the heart. Numerous studies on intact animals and humans and on a variety of isolated cardiac muscle preparations have demonstrated a significant enhancement of myocardial contractility. The rate of Ca^{++} uptake and the rate of ATP hydrolysis by the sarcoplasmic reticulum have been found to be increased in experimental hyperthyroidism, and the opposite effects occur in hypothyroidism. However

it is likely that the cardiovascular changes in thyroid dysfunction are also dependent on indirect mechanisms. Unquestionably, the increased cardiac output engendered by thyroid hormones (endogenous or exogenous) is evoked in part by alterations in the peripheral circulation. Thyroid hyperactivity increases the metabolic rate, and this in turn results in arteriolar vasodilation. The consequent reduction in the total peripheral resistance leads to an elevated cardiac output, as explained on p. 201. Other indirect mechanisms might involve the sympathetic nervous system. Substantial evidence has accrued that suggests that in hyperthyroidism there is either increased sympathetic neural activity or increased sensitivity of the heart to such activity. However recent evidence tends to contradict these earlier conclusions, and this aspect of the problem remains highly controversial at present.

Insulin. Insulin has a prominent, direct, positive inotropic effect on the heart of several mammalian species, including cats, rabbits, sheep, and pigs, but not dogs. Whether it also has a direct inotropic effect on the human heart remains to be established. In those species in which insulin does enhance cardiac contractility, the effect is evident even when hypoglycemia is prevented by glucose infusions and when the beta adrenergic receptors are blocked. In fact, the positive inotropic effect of insulin is potentiated by beta adrenergic receptor blockade. The enhancement of contractility cannot be explained satisfactorily by the concomitant augmentation of glucose transport into the myocardial cells. Recent studies have shown an influence of insulin on intracellular Ca^{++} binding. The inotropic effect of insulin may be mediated by inhibition of Ca^{++} binding, thereby increasing the intracellular concentration of Ca^{++} available for interaction with the myofilaments.

Glucagon. Glucagon exerts potent positive inotropic and chronotropic effects on the heart. It is unlikely that this hormone plays any significant role in the normal regulation of the cardiovascular system, but it has been used pharmacologically in the treatment of a variety of cardiac conditions. The effects of glucagon on the heart closely resemble those produced by the catecholamines, and of course certain of the metabolic effects are similar. Both glucagon and the catecholamines activate adenylate cyclase to increase the myocardial tissue levels of cyclic AMP. Therefore it has been postulated that a common mechanism is involved in eliciting the cardiotonic effects. This hypothesis is tenuous at present because it has been shown that propranolol blocks the positive inotropic effects of the catecholamines, but not of glucagon. Also it was found that the positive inotropic effect of glucagon appeared appreciably earlier than the rise in cyclic AMP content, whereas with the catecholamines these events occur in the appropriate, reverse sequence.

Anterior pituitary hormones. The cardiovascular derangements in hypopituitarism are related principally to the associated deficiencies in adrenocortical and thyroid function caused by the diminished secretion of the tropic hormones. It is also likely that growth hormone has some effect on the myocardium, at least in combination with thyroxine. In hypophysectomized experimental animals it was found that growth hormone alone had little effect on the depressed heart, whereas thyroxine by itself did restore adequate cardiac performance under basal conditions. However, under conditions of hypervolemia or increased peripheral resistance, thyroxine alone was incapable of restoring adequate cardiac function, but the combination of growth hormone and thyroxine did re-establish normal cardiac performance under such conditions of cardiac loading.

Blood gases

Oxygen. Changes in oxygen tension (Pa_{O_2}) of the blood perfusing the brain and the peripheral chemoreceptors affect the heart through nervous mechanisms, as described earlier in this chapter. The indirect effects of hypoxia on the heart are usually prepotent. Moderate degrees of hypoxia characteristically produce tachycardia, increased cardiac output, and enhanced myocardial contractility. These changes are largely abolished by beta adrenergic receptor blockade.

In addition to these preponderant indirect effects, the Pa_{O_2} of the blood perfusing the myocardium influences myocardial performance directly. The effect of hypoxia is actually biphasic, with moderate degrees being stimulatory and more severe degrees being depressant. As shown in Fig. 8-31 when the O_2 saturation is reduced to levels below 50% in isolated hearts, the effect is predominantly depressant in that peak left ventricular pressures are less than the control levels. However, with less severe degrees of hypoxia (O_2 saturation > 50%), the effects are predominantly facilitatory. Furthermore, recent studies have shown that moderate degrees of hypoxia may enhance the contractile response of the heart to circulating catecholamines.

Carbon dioxide. As with variations in Pa_{O_2}, changes in Pa_{CO_2} may affect the myocardium directly as well as through neural mechanisms. The indirect, neurally mediated effects produced by increased Pa_{CO_2} are similar to those evoked by decreased Pa_{O_2}. With respect to the direct effects on the heart, the alterations in myocardial performance elicited by changes of Pa_{CO_2} in the coronary arterial blood are illustrated in Fig. 8-32. In this experiment on an isolated, isovolumetric left ventricle preparation, the control Pa_{CO_2} was 45 mm. Hg (arrow 1). Decreasing the Pa_{CO_2} to 34 mm. Hg (arrow 2) had a pronounced stimulatory effect, whereas increasing Pa_{CO_2} to 86 mm. Hg (arrow 3) was severely depressant. In intact animals, however, systemic hypercapnia also activates the sympathoadrenal system, which tends to compensate for the direct depressant effect of the increased Pa_{CO_2} on the heart.

The changes in Pa_{CO_2} described above were achieved by varying the CO_2 content of the gas

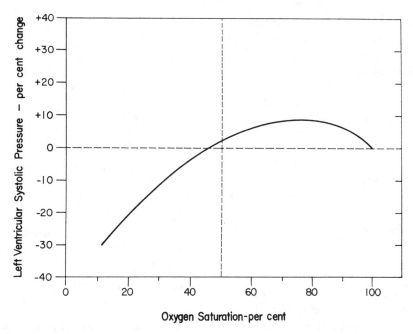

Fig. 8-31. In the isovolumetric left ventricle preparation, a reduction in the oxygen saturation of the coronary arterial blood has a mild stimulatory effect on ventricular contractility when the oxygen saturation is between 45% and 100%, but a depressant effect when the oxygen saturation falls below 45%. (Redrawn from Ng, M. L., Levy, M. N., DeGeest, H., and Zieske, H.: Am. J. Physiol. **211:**43, 1966.)

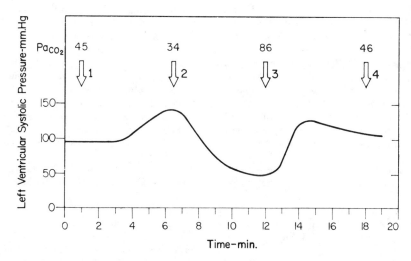

Fig. 8-32. Decrease in arterial blood carbon dioxide tension (Pa_{CO_2}) increases left ventricular systolic pressure (arrow *2*) in an isovolumetric left ventricle preparation, whereas a rise in Pa_{CO_2} (arrow *3*) has the reverse effect.

mixture in the oxygenator of the perfusion system. Therefore changes in Pa_{CO_2} of the blood were accompanied by inverse changes in blood pH. By analysis of the effects of experimental alterations of Pa_{CO_2} and pH in various combinations, it has become apparent that neither the Pa_{CO_2} nor the blood pH are actually primary determinants of myocardial behavior. It is more likely that the resultant change in intracellular pH is the critical factor. The precise mechanisms by which an intracellular acidosis depresses myocardial contractility are not known. Recent studies do suggest that a reduced intracellular pH diminishes the amount of Ca^{++} released from the sarcoplasmic reticulum in response to excitation of a myocardial cell. Furthermore, the diminished pH depresses the myofilaments directly. When they are exposed to a given concentration of Ca^{++}, the lower the prevailing pH, the less the developed tension.

BIBLIOGRAPHY
Journal articles

Berk, J. L., and Levy, M. N.: Profound reflex bradycardia produced by transient hypoxia or hypercapnia in man, Eur. Surg. Res. 9:75, 1977.

Boettcher, D. H., Vatner, S. F., Heyndrickx, G. R., and Braunwald, E.: Extent of utilization of the Frank-Starling mechanism in conscious dogs, Am. J. Physiol. 234:H338, 1978.

Coleman, T. G.: Arterial baroreflex control of heart rate in the conscious rat, Am. J. Physiol. 238:H515, 1980.

DeGeest, H., Levy, M. N., Zieske, H., and Lipman, R. I.: Depression of ventricular contractility by stimulation of the vagus nerves, Circ. Res. 17:222, 1965.

Donald, D. E.: Myocardial performance after excision of the extrinsic cardiac nerves in the dog, Circ. Res. 34:417, 1974.

Donald, D. E., and Shepherd, J. T.: Cardiac receptors: normal and disturbed function, Am. J. Cardiol. 44:873, 1979.

Donald, T. C., Peterson, D. M., Walker, A. A., and Hefner, L. L.: Afterload-induced homeometric autoregulation in isolated cardiac muscle, Am. J. Physiol. 231:545, 1976.

Downing, S. E., and Lee, J. C.: Myocardial and coronary vascular responses to insulin in the diabetic lamb, Am. J. Physiol. 237:H514, 1979.

Fabiato, A., and Fabiato F.: Effects of pH on the myofilaments and the sarcoplasmic reticulum of skinnned cells from cardiac and skeletal muscles, J. Physiol. (London) 276:233, 1978.

Frank, O.: On the dynamics of cardiac muscle (translated by Chapman, C. B., and Wasserman, E.), Am. Heart J. 58:282, 467, 1959.

Geis, G. S., and Wurster, R. D.: Cardiac responses during stimulation of the dorsal motor nucleus and nucleus ambiguus in the cat, Circ. Res. 46:606, 1980.

Higgins, C. B., Vatner, S. F., and Braunwald, E.: Parasympathetic control of the heart, Pharmacol. Rev. 25:119, 1973.

Kirchheim, H. R.: Systemic arterial baroreceptor reflexes, Physiol. Rev. 56:100, 1976.

Koch-Weser, J., and Blinks, J. R.: The influence of the interval between beats on myocardial contractility, Pharmacol. Rev. 15:601, 1963.

Kollai, M., and Koizumi, K.: Reciprocal and non-reciprocal action of the vagal and sympathetic nerves innervating the heart, J. Auton. Nerv. Syst. 1:33, 1979.

Korner, P. I.: Integrative neural cardiovascular control, Physiol. Rev. 51:312, 1971.

Kunze, D. L.: Regulation of activity of cardiac vagal motoneurons, Fed. Proc. 39:2513, 1980.

Lakatta, E. G., and Spurgeon, H. A.: Force staircase kinetics in mammalian cardiac muscle: modulation by muscle length, J. Physiol. (London) 299:337, 1980.

Levy, M. N.: Sympathetic-parasympathetic interactions in the heart, Circ. Res. 29:437, 1971.

Levy, M. N., and Blattberg, B.: Effect of vagal stimulation on the overflow of norepinephrine into the coronary sinus during cardiac sympathetic nerve stimulation in the dog, Circ. Res. 38:81, 1976.

Loewy, A. D., and McKellar, S.: Neuroanatomical basis of central cardiovascular control, Fed. Proc. 39:2495, 1980.

McAllen, R. M., and Spyer, K. M.: Two types of vagal preganglionic motoneurones projecting to the heart and lungs, J. Physiol. (London) 282:353, 1978.

Morad, M., and Goldman, Y.: Excitation-contraction coupling in heart muscle: membrane control of development of tension, Prog. Biophys. Mol. Biol. 27:257, 1973.

Noble, M. I. M.: The Frank-Starling curve, Clin. Sci. Mol. Med. 54:1, 1978.

Parmley, W. W., and Chuck, L.: Length-dependent changes in myocardial contractile state, Am. J. Physiol. 224:1195, 1973.

Patterson, S. W., Piper, H., and Starling, E. H.: The regulation of the heart beat, J. Physiol. (London) 48:465, 1914.

Sagawa, K.: The ventricular pressure-volume diagram revisited, Circ. Res. 43:677, 1978.

Sarnoff, S. J., and Berglund, E.: Ventricular function, I. Starling's law of the heart studied by means of simultaneous right and left ventricular function curves in the dog, Circulation 9:706, 1954.

Sarnoff, S. J., Mitchell, J. II., Gilmore, J. P., and Remensnyder, J. P.: Homeometric autoregulation in the heart, Circ. Res. 8:1077, 1960.

Strauer, B. E., and Scherpe, A.: Experimental hyperthyroidism, I. Hemodynamics and contractility in situ. Basic Res. Cardiol. 70:115, 1975.

Suga, H., and Sagawa, K.: Transient force responses in blood-perfused papillary muscle after step changes in load, Am. J. Physiol. 235:H267, 1978.

Thorén, P. N., Donald, D. E., and Shepherd, J. T.: Role of heart and lung receptors with nonmedullated vagal afferents in circulatory control, Circ. Res. 38(Suppl. II): II-2, 1976.

Vanhoutte, P. M., and Levy, M. N.: Prejunctional cholinergic modulation of adrenergic neurotransmission in the cardiovascular system, Am. J. Physiol. 238:H275, 1980.

Vatner, S. F., and Boettcher, D. H.: Regulation of cardiac output by stroke volume and heart rate in conscious dogs, Circ. Res. 42:557, 1978.

Vatner, S. F., Higgins, C. B., and Braunwald, E.: Sympathetic and parasympathetic components of reflex tachycardia induced by hypotension in conscious dogs with and without heart failure, Cardiovasc. Res. 8:153, 1974.

Vatner, S. F., and Rutherford, J. D.: Control of the myocardial contractile state by carotid chemo- and baroreceptor and pulmonary inflation reflexes in conscious dogs, J. Clin. Invest. 61:1593, 1978.

Books and monographs

Braunwald, E., and Ross, J., Jr.: Control of cardiac performance. In Handbook of physiology; Section 2: Cardiovascular system, The heart, Washington, D.C., 1979, American Physiological Society, vol. 1, pp. 533-580.

Brown, A. M.: Cardiac reflexes. In Handbook of physiology; Section 2: Cardiovascular system, The heart, Washington, D.C., 1979, American Physiological Society, vol. 1, pp. 677-690.

Coleridge, J. C. G., and Coleridge, H. M.: Chemoreflex regulation of the heart. In Handbook of physiology; Section 2: Cardiovascular system, The heart, Washington, D.C., 1979, American Physiological Society, vol. 1, pp. 653-676.

Downing, S. E.: Baroreceptor regulation of the heart. In Handbook of physiology; Section 2: Cardiovascular system, The heart, Washington, D.C., 1979, American Physiological Society, vol. 1, pp. 621-652.

Guz, A., Chairman: Physiological basis of Starling's law of the heart, Ciba Foundation Symposium, Amsterdam, 1974, Associated Scientific Publishers.

Hainsworth, R., Kidd, C., and Linden, R. J., editors: Cardiac receptors, Cambridge, 1979, Cambridge University Press.

Heymans, C., and Neil, E.: Reflexogenic areas of the cardiovascular system. London, 1958, J. & A. Churchill, Ltd.

Korner, P. I.: Central nervous control of autonomic cardiovascular function. In Handbook of physiology; Section 2: Cardiovascular system, The heart, Washington, D.C., 1979, American Physiological Society, vol. 1, pp. 691-740.

Langer, G. A., and Brady, A. J., editors: The mammalian myocardium, New York, 1974, John Wiley & Sons, Inc.

Levy, M. N., and Martin, P. J.: Neural control of the heart. In Handbook of physiology; Section 2: Cardiovascular system, The heart, Washington, D.C., 1979, American Physiological Society, vol. 1, pp. 581-620.

Mirsky, I., Ghista, D. N., and Sandler, H., editors: Cardiac mechanics; physiological, clinical, and mathematical considerations, New York, 1974, John Wiley & Sons, Inc.

Randall, W. C., editor: Neural regulation of the heart, New York, 1976, Oxford University Press, Inc.

Stull, J. T., and Mayer, S. E.: Biochemical mechanisms of adrenergic and cholinergic regulation of myocardial contractility. In Handbook of physiology; Section 2: Cardiovascular system, The heart, Washington, D.C., 1979, American Physiological Society, vol. 1, pp. 741-774.

Vassalle, M., editor: Cardiac physiology for the clinician, New York, 1976, Academic Press, Inc.

CHAPTER 9

CARDIAC OUTPUT AND THE VENOUS SYSTEM

Cardiac output is the volume of blood pumped by the heart per minute. Venous return is the volume returning to the heart from the veins in the same unit of time. In this chapter some of the techniques for measuring cardiac output will be presented, the interrelations between cardiac output and venous return will be discussed, and the factors that affect cardiac output and venous pressure will be described.

MEASUREMENT OF CARDIAC OUTPUT
Fick principle

In 1870, the German physiologist, Adolph Fick, contrived the first method for measuring cardiac output that would be applicable in intact animals and humans. The basis for this method has been called the "Fick principle," and it is simply an application of the law of conservation of mass. It is derived from an algebraic statement of the fact that the quantity of oxygen per minute delivered to the pulmonary capillaries via the pulmonary artery plus the quantity of oxygen per minute that enters the pulmonary capillaries from the alveoli must equal the quantity of oxygen per minute that is carried away by the pulmonary veins.

This is depicted schematically in Fig. 9-1. The rate, q_1, at which O_2 is delivered to the lungs by the pulmonary artery equals the oxygen concentration in that blood, $[O_2]_{pa}$, times

the pulmonary arterial blood flow, Q, which is, in fact, the cardiac output; that is,

$$q_1 = Q[O_2]_{pa} \qquad (1)$$

Let q_2 be the net rate of oxygen uptake by the pulmonary capillaries from the alveoli, the so-called *oxygen consumption* of the body. The rate, q_3, at which oxygen is carried away by the pulmonary veins equals the O_2 concentration in that blood, $[O_2]_{pv}$, times the total pulmonary venous blood flow, Q, which is virtually equal to the pulmonary arterial blood flow at equilibrium; that is,

$$q_3 = Q[O_2]_{pv} \qquad (2)$$

From conservation of mass,

$$q_1 + q_2 = q_3 \qquad (3)$$

Therefore,

$$Q[O_2]_{pa} + q_2 = Q[O_2]_{pv} \qquad (4)$$

Solving for cardiac output,

$$Q = q_2 /([O_2]_{pv} - [O_2]_{pa}) \qquad (5)$$

Equation (5) is the statement of the Fick principle.

In the clinical determination of cardiac output, the rate of oxygen consumption is computed from measurements of the volume and oxygen content of expired air over a given interval of time. Since the oxygen concentration

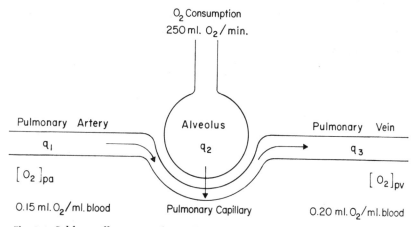

Fig. 9-1. Schlema illustrating the Fick principle for measuring cardiac output.

of peripheral arterial blood is essentially identical with that in the pulmonary veins, $[O_2]_{pv}$ is determined on a sample of peripheral arterial blood withdrawn by needle puncture. Pulmonary arterial blood actually represents mixed systemic venous blood. Samples for oxygen analysis are obtained from the pulmonary artery or right ventricle through a catheter. For many years the catheter was a relatively stiff, radiopaque tube that had to be introduced into the pulmonary artery under fluoroscopic guidance. Now, a very flexible catheter with a small balloon near the tip may be inserted into a peripheral vein. As the tube is advanced, it is carried by the flowing blood toward the heart. By simply following the pressure changes, the physician is able to know when the catheter tip has been passed through the tricuspid valve into the right ventricle, and then through the pulmonic valve into the pulmonary artery.

An example of the results ordinarily obtained in a normal, resting adult is illustrated by the values of Fig. 9-1. With an oxygen consumption of 250 ml./min., an arterial (pulmonary venous) oxygen content of 0.20 ml. O_2/ml. blood (or 20 vol.%), and a mixed venous (pulmonary arterial) oxygen content of 0.15 ml. O_2/ml.

blood (or 15 vol.%), the cardiac output would equal $250 \div (0.20 - 0.15) = 5,000$ ml./min.

The Fick principle is also used for estimating the oxygen consumption of organs in situ, when blood flow and the oxygen contents of the arterial and venous blood can be determined. Algebraic rearrangement reveals that oxygen consumption equals the blood flow times the arteriovenous oxygen concentration difference. For example, if the blood flow through one kidney is 700 ml./min., arterial oxygen content is 0.20 ml. O_2/ml. blood, and renal venous oxygen content is 0.18 ml. O_2/ml. blood, then the rate of oxygen consumption by that kidney must be $700 (0.20 - 0.18) = 14$ ml. O_2/min.

Indicator dilution techniques

The principle of the indicator dilution technique for measuring cardiac output is also based on the law of conservation of mass and is illustrated by the model in Fig. 9-2. Let a liquid flow through a tube at a rate of Q ml./sec., and let q mg. of dye be injected as a slug into the stream at point A. Let mixing occur at some point downstream. If a small sample of liquid is continually withdrawn from point B farther downstream and passed through a den-

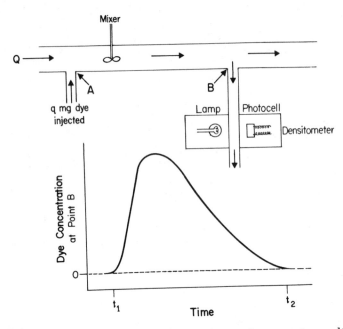

Fig. 9-2. Schema illustrating the indicator dilution technique for measuring cardiac output. In this model, in which there is no recirculation, q mg. of dye are injected instantaneously at point *A* into a stream flowing at Q ml./min. A mixed sample of the fluid flowing past point *B* is withdrawn at a constant rate through a densitometer. The resulting dye concentration curve at point *B* has the configuration shown in the lower section of the figure.

sitometer, then a curve of the dye concentration, c, may be recorded as a function of time, t, as shown in the lower half of the figure.

If there is no loss of dye between points *A* and *B*, the amount of dye, q, passing point *B*, between times t_1 and t_2 will be

$$q = \bar{c} \, Q \, (t_2 - t_1) \tag{1}$$

where $\bar{c}$ is the mean concentration of dye. The value of $\bar{c}$ may be computed by dividing the area of the dye concentration curve by the time duration $(t_2 - t_1)$ of that curve; that is

$$\bar{c} = \int_{t_1}^{t_2} c \, dt / (t_2 - t_1) \tag{2}$$

Substituting this value of $\bar{c}$ into equation (1), and solving for Q yields

$$Q = \frac{q}{\int_{t_1}^{t_2} c \, dt} \tag{3}$$

Thus, flow may be measured by dividing the amount of indicator injected upstream by the area under the downstream concentration curve.

This technique has been widely applied for the estimation of cardiac output in humans. An accurately measured quantity of some indicator (a dye or isotope that remains within the circulation) is injected rapidly as a slug into a large-central vein or into the right side of the heart through a cardiac catheter. Arterial blood is continuously drawn through a detector (densitometer or isotope rate counter), and a curve of indicator concentration is recorded as a function of time.

Because some of the indicator recirculates and reappears at the site of arterial withdrawal before the entire curve is inscribed, the concentration curve is not as simple as that shown

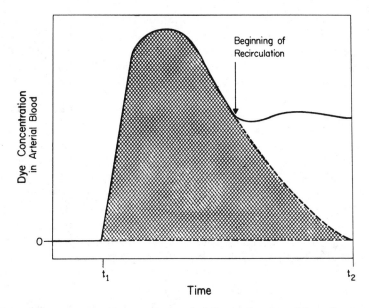

Fig. 9-3. Typical dye concentration curve recorded from a human subject. Because of recirculation of the dye, the concentration does not return to 0, as in the model in Fig. 9-2. The dashed line on the descending limb represents the semilogarithmic extrapolation of the upper portion of the descending limb, prior to the beginning of recirculation.

in the lower half of Fig. 9-2. Instead, on the downstroke of the curve there is a secondary increase in concentration (Fig. 9-3) as the recirculated dye becomes mixed with the last portions of dye still undergoing its primary passage past the site of withdrawal. To compute the area under the concentration curve, the downslope of the curve is extrapolated to zero concentration beyond the beginning of recirculation (dashed line, Fig. 9-3). The extrapolation, of course, introduces some error into the estimation of cardiac output. In model systems without recirculation it has been found that, for the descending portion of the curve, the logarithm of the concentration varies linearly with time. Usually, in the intact circulatory system this same relationship holds for the downstroke of the concentration curve up till the time of recirculation. Therefore in practice this initial

portion of the descending limb of the concentration curve is plotted on semilogarithmic paper, the curve is extrapolated to concentrations approaching zero, and the extrapolated curve is then replotted on the original linear tracing. Cardiac output equals the amount of indicator injected divided by the area of the curve to the point of recirculation plus the area under the extrapolated portion of the curve after the beginning of recirculation. The total area is represented by the shaded region in Fig. 9-3.

Over the past several years, the most popular indicator dilution technique has been that of *thermodilution*. The indicator used is cold saline. The temperature and volume of the saline are measured accurately before injection. A very flexible double-lumen catheter is introduced into a peripheral vein and advanced so that the tip lies in the pulmonary artery. A small termistor, attached to the catheter tip, permits rapid changes in temperature to be recorded. One of the openings of the catheter lies a few inches proximal to the tip. When the tip is in the pulmonary artery, the proximal opening lies in or near the right atrium. The cold saline is injected rapidly through the catheter, and it enters the circulation through this upstream opening. The resultant change in temperature downstream is recorded by the thermistor in the pulmonary artery.

The thermodiluton technique has the following distinct advantages: (1) an arterial puncture is not necessary; (2) the small volumes of saline used in each determination are innocuous, and so repeated determinations may be made; and (3) there is virtually no problem with recirculation. Temperature equilibration takes place as the cooled blood flows through the pulmonary and systemic capillary beds, before it flows by the thermistor in the pulmonary artery the second time. Therefore, the curve of temperature change resembles that shown in Fig. 9-2, and the errors associated with extrapolation are largely averted.

Circulation time

The *circulation time* is an index of the velocity of blood flow. Hence, it tends to reflect the cardiac output and has the advantage that it can be measured clinically in a simple fashion. Some indicator is introduced at one site in the circulatory system, usually in an antecubital vein, and the time required for it to appear at some other point is accurately measured. The time required for the indicator to travel between these two points varies inversely with the rate of blood flow and directly with the volume of blood between the two loci.

Two of the circulation times most commonly measured are the arm-to-tongue and arm-to-lung times. In the former an indicator such as the bile salt, sodium dehydrocholate, is rapidly injected into an arm vein. The time required for the patient to detect the bitter taste is

noted. Normally, this requires 10 to 16 seconds. In determining the arm-to-lung time, ether is injected and the time until the patient coughs and grimaces is measured. The normal time is 4 to 8 seconds.

CONTROL OF CARDIAC OUTPUT

The various factors that determine cardiac performance have been discussed in detail in Chapters 4 and 8. There are usually considered to be four factors that affect the heart directly; namely, preload, afterload, heart rate, and myocardial contractility. The last two of these factors (i.e., heart rate and contractility) are characteristics of the cardiac tissues per se, although they are subject to modulation by various neural and humoral mechanisms. The first two factors (i.e., preload and afterload), however, are dependent on the characteristics of both the heart and the vascular system. The preload and afterload are critical determinants of cardiac performance, but at the same time, the preload and afterload are determined by the cardiac and vascular components of the circulatory system.

Over the past two decades, we have become increasingly aware of the fact that changes in the peripheral circulation are often just as important in determining the level of the cardiac output as are changes in the cardiac tissues themselves. Therefore, in order to understand the regulation of cardiac output, it is important to appreciate the nature of the coupling between the heart and the vascular system. Guyton and his colleagues have made important contributions to this field. They have developed very useful graphic techniques that we shall use in modified form for analyzing the interactions between the cardiac and vascular components of the circulatory system.

The graphic analysis involves two simultaneous functional relationships between the cardiac output and the *central venous pressure* (i.e., the pressure in the right atrium and thoracic venae cavae). The curve defining one of these relationships will be called the *cardiac function curve*. It is an expression of the well-known Frank-Starling relationship and reflects the fact that the cardiac output depends, in part, on the preload (i.e., the central venous, or right atrial, pressure). The cardiac function curve is a characteristic of the heart itself and has been studied in hearts completely isolated from the rest of the circulatory system.

The second functional relationship between the central venous pressure and the cardiac output is defined by a second curve, which we shall call the *vascular function curve*. This relationship depends only on certain critical characteristics of the vascular system; namely, the peripheral resistance, the arterial and venous capacitances, and the blood volume. It is entirely independent of the characteristics of the heart, and it can be studied even if the heart were replaced by a mechanical pump.

VASCULAR FUNCTION CURVE

The vascular function curve defines the change in central venous pressure that occurs as a consequence of a change in cardiac output; that is, central venous pressure is the dependent variable (or response), and cardiac output is the independent variable (or stimulus). This

contrasts with the cardiac function curve (or Frank-Starling mechanism), for which the central venous pressure (or preload) is the independent variable, and the cardiac output is the dependent variable.

The simplified model of the circulation illustrated in Fig. 9-4 will be used to explain how the cardiac output determines the level of the central venous pressure. For the sake of simplicity the essential components of the cardiovascular system may be lumped into four elements, as illustrated in Fig. 9-4. The right and left sides of the heart as well as the pulmonary vascular bed will be considered simply as a pump-oxygenator, much as that employed during open heart surgery. The high-resistance microcirculation is designated the peripheral resistance. Finally, the entire *capacitance* of the system is subdivided into two components, the total arterial capacitance, C_a, and the total venous capacitance, C_v. As defined on p. 96, *capacitance* (C) is the increment of volume (dV) accommodated per unit change of pressure (dP); that is, $C = dV/dP$. The venous capacitance is approximately twenty times as great as the arterial capacitance. In the example to follow, the ratio of $C_v : C_a$ will be set at 19:1; this will simplify the mathematics, as will

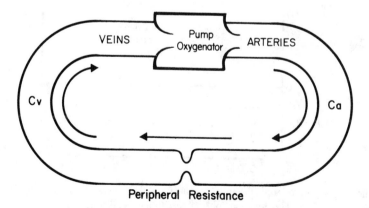

Fig. 9-4. Simplified model of the cardiovascular system, consisting of a pump-oxygenator, an arterial capacitance (C_a), a peripheral resistance, and a venous capacitance (C_v).

be evident below. Thus, for a 1 mm. Hg increment in venous pressure, nineteen times as much blood would be stored on the venous side of the circuit as would be stored on the arterial side for an equivalent rise in arterial pressure. It must be emphasized that these statements apply strictly to the usual pressure ranges that exist on both sides of the vascular circuit, because capacitance varies as a function of transmural pressure (p. 96).

With the model system at rest, pressures are the same throughout the entire circuit and there is no flow. The pressure that exists at rest is a function of only the total volume of blood contained within the system and the elastic characteristics of the walls (that is, the overall capacitance of the system). This equilibrium pressure has been termed the *mean circulatory pressure* by Guyton. At normal blood volumes

and with normal vessels, the magnitude of the mean circulatory pressure has been estimated to be about 7 mm. Hg. To the left of arrow *1* in Fig. 9-5, the arterial pressure (P_a) and the venous pressure (P_v) are both equal to 7 mm. Hg when the cardiac output (C.O.) is zero.

Let the pump-oxygenator (or simply the pump) in Fig. 9-4 start suddenly to deliver a constant flow of 1 L./min. (at arrow *1*, Fig. 9-5), and let peripheral resistance remain constant at 20 mm. Hg/L./min. Because of the arrangement of the valves, the direction of transfer will be from the venous to the arterial side of the circuit. Hence pressure will begin to fall on the venous side and rise on the arterial side. The arterial pressure, P_a, will continue to rise until a pressure of 20 mm. Hg above the venous pressure, P_v is attained. From the defini-

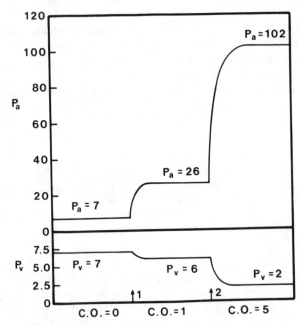

Fig. 9-5. The changes in arterial (P_a) and venous (P_v) pressures in the circulatory model shown in the preceding figure. The total peripheral resistance is 20 mm. Hg/L./min., and the ratio of C_v to C_a is 19:1. The cardiac output (C.O.) is 0 to the left of arrow *1*. It is increased to 1 L./min. at arrow *1*, and 5 L./min. at arrow *2*.

tion of peripheral resistance, as explained on p. 99:

$$R = (P_a - P_v)/Q \qquad (1)$$

$$P_a = P_v + QR = P_v + 20 \qquad (2)$$

Hence, P_a will continue to be 20 mm. Hg above P_v, as long as the pump output is maintained at 1 L./min. and the peripheral resistance remains at 20 mm. Hg/L./min. The arterial volume increment required to achieve this new, elevated level of P_a is entirely dependent on the arterial capacitance, C_a. For a rigid arterial system (low capacitance) this volume will be small; for a distensible system the volume will be large. Whatever the magnitude of this volume, however, it represents the translocation of some quantity of blood from the venous to the arterial side of the circuit. For a given total blood volume, any increment in arterial volume (ΔV_a) must be equal to the decrement in venous volume (ΔV_v); that is

$$\Delta V_a = -\Delta V_v \qquad (3)$$

From the definition of capacitance,

$$C_a = \Delta V_a / \Delta P_a \qquad (4)$$

and

$$C_v = \Delta V_v / \Delta P_v \qquad (5)$$

By substitution into equation (3),

$$\frac{\Delta P_v}{\Delta P_a} = -\frac{C_a}{C_v} \qquad (6)$$

Given that C_v is nineteen times as great as C_a, then the increment in P_a will be nineteen times as great as the decrement in P_v; that is

$$\Delta P_a = -19\Delta P_v \qquad (7)$$

Let ΔP_a represent the difference between the prevailing P_a and the mean circulatory pressure (P_{mc}); that is let

$$\Delta P_a = P_a - P_{mc} \qquad (8)$$

and let ΔP_v represent the difference between the prevailing P_v and the mean circulatory pressure:

$$\Delta P_v = P_v - P_{mc} \qquad (9)$$

Substituting these values for ΔP_a and ΔP_v into equation (7):

$$P_a - P_{mc} = -19 (P_v - P_{mc}) \qquad (10)$$

By solving equations (2) and (10) simultaneously:

$$P_a = P_{mc} + 19 \qquad (11)$$

and

$$P_v = P_{mc} - 1 \qquad (12)$$

Hence, in Fig. 9-5, P_v is shown to decrease to 6 mm. Hg from the mean circulatory pressure of 7 mm. Hg. Concomitantly, P_a increases to 26 mm. Hg, resulting in the required arteriovenous pressure gradient of 20 mm. Hg.

The flow through the peripheral resistance (Fig. 9-4) is a function of the pressure gradient ($P_a - P_v$) and the resistance. The pressure gradient across the peripheral resistance is often referred to as the *vis a tergo*, or force from behind. *It is the single most important factor responsible for the venous return and is directly ascribable to the pumping action of the heart itself.*

If the pump output is abruptly increased to a constant level of 5 L./min. (Fig. 9-5, arrow 2) and peripheral resistance remains constant at 20 mm. Hg/L./min., an additional volume of blood will again be translocated from the venous to the arterial side of the circuit. It will progressively accumulate in the arteries until P_a reaches a level of 100 mm. Hg above P_v. By substitution into equation (1),

$$P_a = P_v + QR = P_v + 100 \qquad (13)$$

By solving equations (10) and (13) simultaneously, we find that P_a rises to a value of 95 mm. Hg above P_{mc}, and P_v falls to a value 5 mm. Hg below P_{mc}. In Fig. 9-5, therefore, P_v

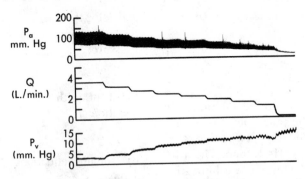

Fig. 9-6. The changes in arterial (P_a) and central venous (P_v) pressures produced by changes in systemic blood flow (Q) in a canine right-heart bypass preparation. Stepwise changes in Q were produced by altering the rate at which blood was mechanically pumped from the right atrium to the pulmonary artery. (From Levy, M. N.: Circ. Res. **44**:739, 1979. By permission of the American Heart Association, Inc.)

declines to 2 mm. Hg and P_a rises to 102 mm. Hg. The resultant pressure gradient of 100 mm. Hg is that value which will force a cardiac output of 5 L./min. through a constant peripheral resistance of 20 mm. Hg/L./min.

Experiments in animals and observations in human patients have demonstrated that alterations in cardiac output do indeed evoke the directional changes in P_a and P_v that have been predicted above for our simplified model. In the experiment on an anesthetized dog shown in Fig. 9-6, a mechanical pump was substituted for the right ventricle. As the cardiac output (Q) was diminished in a series of small steps, there were concomitant reductions in P_a and elevations of P_v. Similarly, a major coronary artery may suddenly become occluded in a human subject. The resultant *acute myocardial infarction* often leads to a substantial reduction in cardiac output, which is attended by a fall in the arterial pressure and a rise in the central venous pressure.

In animal experiments, if the output of the heart or of a substitute mechanical pump were varied over a range of values and the resultant values of P_v were recorded, a *vascular function curve* could be constructed. Such a curve is

shown in Fig. 9-7, and it is derived from the theoretical data shown in Fig. 9-5. Contrary to the usual convention, the independent variable, cardiac output in this instance, is plotted along the ordinate, and the dependent variable, venous pressure, is plotted along the abscissa. The customary arrangement of axes is reversed for reasons of convenience, which will become apparent later in this chapter. In the vascular function curve depicted in Fig. 9-7, point *A* represents the mean circulatory pressure, point *B*, the value of P_v at a cardiac output of 1 L./min., and point *C*, the value of P_v at a cardiac output of 5 L./min. From equations (1), (6), (8), and (9) above, the following equation for the linear portion of the vascular function curve can be derived:

$$P_v = - \frac{RC_a}{C_a + C_v} Q + P_{mc} \qquad (14)$$

From this equation, it is evident that at Q = 0, $P_v = P_{mc}$. Also, the slope depends only on R, C_v, and C_a.

Fig. 9-7 also shows that there is a limit to the reduction of P_v that can be produced by an increase in cardiac output. At some critical maximum value of cardiac output, sufficient

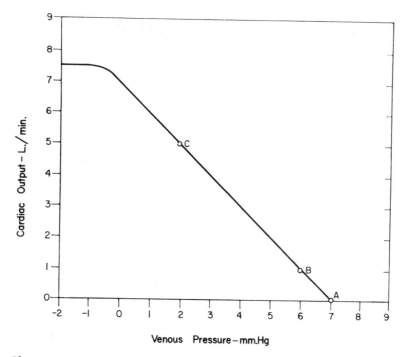

Fig. 9-7. Changes in venous pressure as cardiac output is varied over a range from 0 to over 7 L./min. Point *A* is the mean circulatory pressure, which is the equilibrium pressure throughout the cardiovascular system when cardiac output is 0. Points *B* and *C* represent the values of venous pressure at cardiac outputs of 1 and 5 L./min., respectively. Contrary to the usual convention, the independent variable (in this instance, cardiac output) is plotted along the ordinate, whereas the dependent variable (venous pressure) is plotted along the abscissa.

fluid will be translocated from the venous to the arterial side of the circuit such that P_v will drop below the ambient pressure. In a system of distensible tubes, the venous system will be collapsed by this negative transmural pressure (P_v minus ambient pressure). This will, of course, limit the maximum value of cardiac output, regardless of the capabilities of the pump.

In the simplified schema in Fig. 9-4, the venous system was considered to be without resistance. In the body, however, there is a continuous pressure gradient from the venules to the right side of the heart. In an open-chest animal, therefore, if an artificial pump is substituted for the heart and if the pump output is progressively increased, the site of collapse will be at the junction of the venae cavae with the inflow side of the pump. In the normal, closed-chest animal, venous collapse occurs at the points of entry of the veins into the chest, for reasons to be described later in this chapter.

Blood volume

The venous pressure curve is affected by variations in the total blood volume. During circulatory standstill (cardiac output zero), the mean circulatory pressure is simply a function

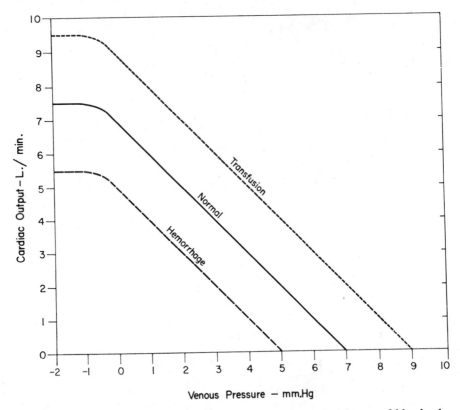

Fig. 9-8. Effects of increased blood volume (*transfusion* curve) and of decreased blood volume (*hemorrhage* curve) on the venous pressure curve. Similar shifts in the venous pressure curve are produced by increases and decreases, respectively, in venomotor tone.

of vascular capacitance and blood volume, as stated previously. Thus for a given vascular capacitance the mean circulatory pressure will be increased during *hypervolemia*, when the blood volume is expanded (as by transfusion) and decreased during *hypovolemia*, when the blood volume is diminished (as by hemorrhage). This is illustrated by the X-axis intercepts in Fig. 9-8, where the mean circulatory pressure is 5 mm. Hg with hemorrhage and 9 mm. Hg with transfusion, as compared with the value of 7 mm. Hg at the normal blood volume (*normovolemia*).

Furthermore, the differences in P_v during hypervolemia, normovolemia, and hypovole-

mia in the static system are preserved at each level of cardiac output so that the venous pressure curves parallel each other (Fig. 9-8). To illustrate, consider the example of hypervolemia, in which the mean circulatory pressure is 9 mm. Hg. In Fig. 9-5 both P_a and P_v would be 9 mm. Hg, instead of 7 mm. Hg, when the cardiac output was zero. With a sudden increase in cardiac output to 1 L./min. (at arrow *1*), if the peripheral resistance were still 20 mm. Hg/L./min., an arteriovenous pressure gradient of 20 mm. Hg would still be necessary for 1 L./min. to flow through the resistance vessels. This does not differ from the example for normovolemia. Assuming the same ratio of

C_v to C_a of 19:1, the pressure gradient would be achieved by a 1 mm. Hg decline in P_v and a 19 mm. Hg rise in P_a. Hence, a change in cardiac output from 0 to 1 L./min. would evoke the same 1 mm. Hg reduction in P_v irrespective of the blood volume, as long as the C_v/C_a ratio and the peripheral resistance were independent of blood volume. Equation (14) also discloses that the slope of the vascular function curve remains constant as long as there is no change in R, C_v, or C_a.

From Fig. 9-8 it is also apparent that the value for cardiac output at which $P_v = 0$ varies directly with the blood volume. Therefore the maximum value of cardiac output becomes progressively more limited as the total blood volume is reduced, but the pressure at which the veins collapse (sharp change in slope on the venous pressure curve) is not altered appreciably by changes in blood volume.

Venomotor tone

The venous pressure curves representing changes in venomotor tone closely resemble those for changes in blood volume. In Fig. 9-8 for example, the transfusion curve could just as well represent increased venomotor tone, whereas the hemorrhage curve could represent decreased tone. During circulatory standstill, for a given blood volume, the pressure within the vascular system will rise as the tension exerted by the smooth muscle within the vascular walls increases. Present evidence indicates that it is principally the arteriolar and venous smooth muscle that is under any appreciable nervous or humoral control. Since the fraction of the blood volume located within the arterioles is small (Fig. 1-2), only changes in venomotor tone are of any practical importance in altering the magnitude of the mean circulatory pressure. Hence mean circulatory pressure rises with increased venomotor tone and falls with diminished tone. Changes in venomotor tone may alter the elastic characteristics of the veins. The consequent change in C_v may there-

fore alter the slope of the vascular function curve, as predicted by equation (14).

Experimentally, the pressure attained shortly after abrupt circulatory standstill is usually above 7 mm. Hg, even when blood volume is normal. This is attributable to the generalized venoconstriction elicited by cerebral ischemia, activation of the chemoreceptors, and reduced stimulation of the baroreceptors. If resuscitation is not achieved, then this reflex response subsides as central nervous activity ceases, and, at normal blood volumes the mean circulatory pressure usually approaches a value close to 7 mm. Hg.

Blood reservoirs

The extent of venoconstriction is considerably greater in certain regions of the body than in others. In effect, vascular beds that undergo appreciable venoconstriction constitute blood reservoirs. The vascular bed of the skin is one of the major blood reservoirs in humans. During blood loss, profound subcutaneous venoconstriction occurs, giving rise to the characteristic pale appearance of the skin. The resultant redistribution of blood thereby liberates several hundred milliliters of blood to be perfused through more vital regions. The vascular beds of the liver, lungs, and spleen also serve as important blood reservoirs. In the dog the spleen is packed with red blood cells and is capable of constricting to a small fraction of its normal size. During hemorrhage this mechanism serves to autotransfuse blood of high erythrocyte content into the general circulation. However, in humans the volume changes of the spleen are considerably less extensive and play only a minor role in compensating for blood loss.

Peripheral resistance

The modification of the venous pressure curve induced by changes in arteriolar tone is shown in Fig. 9-9. It has been estimated that the arterioles contain only about 3% of the total

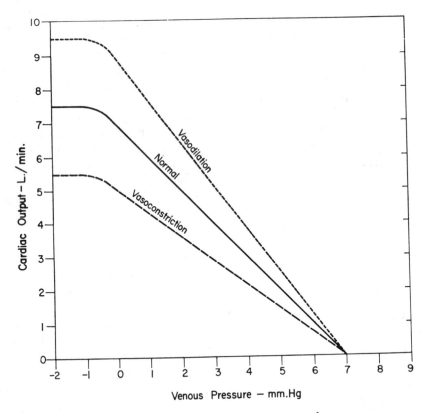

Fig. 9-9. Effects of arteriolar vasodilation and vasoconstriction on the venous pressure curve.

blood volume (Fig. 1-2). Hence changes in the contractile state of these vessels do not appreciably alter the pressure-volume relationship in the static system. Therefore the mean circulatory pressure is virtually independent of arteriolar tone, and the family of venous pressure curves representing a range of peripheral resistances will converge at a common point on the abscissa.

At any given level of cardiac output, P_v will vary inversely with the arteriolar tone, all other factors remaining constant. Vasoconstriction sufficient to double the peripheral resistance will result in a twofold rise in P_a. In the example shown in Fig. 9-5, a change in the cardiac output from 0 to 1 L./min. caused P_a to rise from 7 to 26 mm. Hg, an increment of 19 mm.

Hg. If peripheral resistance had been twice as great, the same change in cardiac output would have evoked twice as great an increment in P_a. To achieve this greater rise in P_a, twice as great an increment in blood volume would be required on the arterial side of the circulation, assuming a constant arterial capacitance. Given a constant total blood volume, this larger arterial volume signifies a corresponding reduction in venous volume. Hence the decrement in venous volume would be twice as great when the peripheral resistance is doubled. With a constant venous capacitance, a twofold reduction in venous volume would be reflected by a twofold decline in P_v. Therefore, in Fig. 9-5, an increase in cardiac output to 1 L./min. (arrow *1*) would have caused a 2 mm. Hg decrement in P_v, to a level of 5 mm. Hg, instead of the 1 mm. Hg decrement that occurred with the normal peripheral resistance. Similarly, greater increases in cardiac output would have evoked proportionately greater decrements in P_v under conditions of increased peripheral resistance than with normal levels of resistance. This inverse relationship between the peripheral resistance and the decrement in P_v, together with the failure of peripheral resistance to affect the mean circulatory pressure, accounts for the counterclockwise rotation of the venous pressure curves with increased peripheral resistance that is observed in Fig. 9-9. Conversely, arteriolar vasodilation produces a clockwise rotation from the same horizontal axis intercept. From Fig. 9-9, it is also evident that a higher maximal level of cardiac output is attainable during vasodilation than with normal or increased arteriolar tone.

INTERRELATIONSHIPS BETWEEN CARDIAC OUTPUT AND VENOUS RETURN

Cardiac output and venous return are inextricably interdependent. Clearly, except for small, transient disparities, the heart will be unable to pump any more blood than is delivered to it through the venous sytsem. Similarly, since the circulatory system is a closed circuit, the rate of venous return must equal the cardiac output over any appreciable time interval. The flow around the entire closed circuit depends upon the capability of the pump, the characteristics of the circuit, and the total volume of fluid in the system. Cardiac output and venous return are simply two terms for expressing the flow around the closed circuit. Cardiac output is the volume of blood being pumped by the heart per unit time. Venous return is the volume of blood returning to the heart per unit time. At equilibrium, these two flows are equal.

The techniques of circuit analysis will be applied in an effort to gain some insight into the control of flow around the circuit. Acute changes in cardiac contractility, peripheral resistance, or blood volume may transiently exert differential effects on cardiac output and venous return. Except for such brief disparities, however, such factors simply alter flow around the entire circuit, and it is irrelevant whether one thinks of that flow as "cardiac output" or "venous return." Not uncommonly, authors have ascribed the reduction in cardiac output during hemorrhage, for example, to a decrease in venous return. It will become clear that such an explanation is a blatant example of circular reasoning in its most literal sense. Hemorrhage reduces flow around the entire circuit, for reasons to be elucidated. To attribute the reduction in cardiac output to a curtailment of venous return is equivalent to ascribing the decrease in total flow to a decrease in total flow.

COUPLING BETWEEN THE HEART AND THE VASCULATURE

In accordance with Starling's law of the heart, or so-called heterometric autoregulation, cardiac output is intimately dependent on P_v, since right atrial pressure is virtually identical to central venous pressure. Furthermore, the right atrial pressure is approximately equal to the right ventricular end-diastolic pressure because the normal tricuspid valve constitutes a low resistance junction between the right atrium and ventricle. In the discussion to follow, graphs of cardiac ouput as a function of P_v will be called *cardiac function curves*. Supervention of other types of intrinsic regulatory mechanisms (for example, homeometric autoregulation, p. 163) will modify such curves appreciably. For completeness such changes must be taken into consideration, but for purposes of simplicity the cardiac function curves will be drawn to represent only the Frank-Starling relationship. Extrinsic regulatory influences may be expressed as shifts in such curves, as has previously been indicated (p. 162).

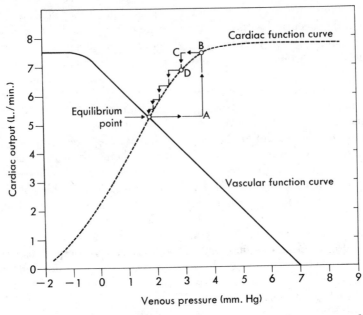

Fig. 9-10. Typical vascular and cardiac function curves plotted on the same coordinate axes. The coordinates of the equilibrium point, at the intersection of these curves, represent the stable values of cardiac output and central venous pressure at which the system tends to operate. Any perturbation (such as when venous pressure is suddenly increased to point A) institutes a sequence of changes in cardiac output and venous pressure such that these variables are returned to their equilibrium values.

A typical cardiac function curve is plotted on the same coordinates as a normal vascular function curve in Fig. 9-10. Contrary to the vascular function curve, the cardiac function curve will be plotted according to the usual convention; that is, the independent variable (P_v) will be plotted along the abscissa, and the dependent variable (cardiac output) will be plotted along the ordinate. In accordance with the Frank-Starling mechanism, the cardiac function curve reveals that a rise in P_v is associated with an increase in cardiac output. Conversely, the vascular function curve describes an inverse relationship between cardiac output and P_v. The *equilibrium point* of such a system is defined by the point of intersection of these two curves. The coordinates of this equilibrium point represent the values of cardiac output

and P_v at which such a system tends to operate. Only transient deviations from such values for cardiac output and P_v are possible, as long as the given cardiac and vascular function curves accurately describe the system.

The tendency for the cardiovascular system to operate about such an equilibrium point may best be illustrated by examining its response to a sudden perturbation. Consider the changes elicited by a sudden rise in P_v from the equilibrium point to point A in Fig. 9-10. Such a change might be induced experimentally by the rapid injection, during ventricular systole, of a given volume of blood on the venous side of the circuit, accompanied by the withdrawal of an equal volume from the arterial side so that total blood volume would remain constant. Because of the Frank-Starling mechanism, this

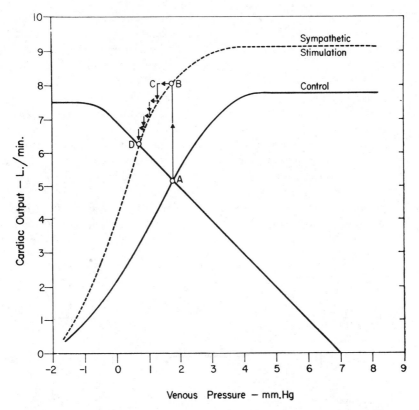

Fig. 9-11. Enhancement of myocardial contractility, as by cardiac sympathetic nerve stimulation, causes the equilibrium values of cardiac output and P_v to shift from the intersection (point *A*) of the control vascular and cardiac function curves (continuous lines) to the intersection (point *D*) of the same vascular function curve with that cardiac function curve (dashed line) representing enhanced myocardial contractility.

elevated P_v would result in an enhanced cardiac output (point *B*) during the very next ventricular systole. The increased cardiac output, in turn, would result in the net transfer of blood from the venous to the arterial side of the circuit, with a consequent reduction in P_v (to point *C*). During the next systole, cardiac output would therefore be less (point *D*), although still above the equilibrium point. This process would continue in ever-diminishing steps until the equilibrium values for P_v and cardiac output were re-established.

Enhanced myocardial contractility

The effect of alterations in ventricular contractility may be comprehended on the basis of a similar graphic representation. In Fig. 9-11 the lower cardiac function curve represents the control state, whereas the upper curve represents a condition of improved contractility, analogous to the family of ventricular function curves described on p. 161. Such enhancement of ventricular contractility might be achieved by electrical stimulation of the cardiac sympathetic nerves. Since the effects of such stimulation would be restricted to the heart, a single vascular function curve would be appropriate to both control and experimental conditions, as shown in Fig. 9-11.

During the control state the equilibrium values for cardiac output and P_v are designated by point A. With the onset of cardiac sympathetic nerve stimulation (assuming the effects to be instantaneous and constant), the prevailing level of P_v would elicit an abrupt rise in cardiac output to point B because of the enhanced contractility. However, this level of cardiac output would result in the net transfer of blood from the venous to the arterial side of the circuit, and consequently P_v will fall (to point C). Car-

diac output will therefore continue to diminish until a new equilibrium point (D) is reached, which is located at the point of intersection of the vascular function curve with the new cardiac function curve (representing the effects of increased sympathetic activity on the heart). The new equilibrium point (D) lies above and to the left of the control equilibrium point (A), revealing that sympathetic stimulation evokes a greater cardiac output at a lower level of P_v. That such a change accurately describes the actual experimental situation is illustrated by the tracings reproduced in Fig. 9-12. In this experiment the left stellate ganglion was stimulated between the two arrows at the top of Fig. 9-12.

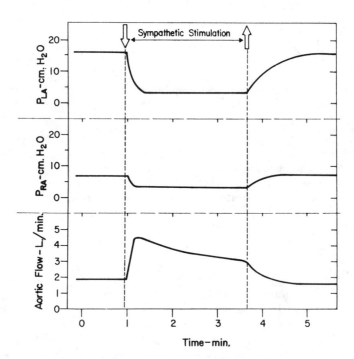

Fig. 9-12. During electrical stimulation of the left stellate ganglion (containing cardiac sympathetic nerve fibers), aortic blood flow increased while pressures in the left atrium (P_{LA}) and right atrium (P_{RA}) diminished. These data conform with the conclusions derived from Fig. 9-11, in which the equilibrium values of cardiac output and venous pressure are observed to shift from point A to point D during cardiac sympathetic nerve stimulation. (Redrawn from Sarnoff, S. J., Brockman, S. K., Gilmore, J. P., Linden, R. J., and Mitchell, J. H.: Circ. Res. **8:**1108, 1960.)

During stimulation there was a substantial rise in aortic flow, accompanied by reductions in right and left atrial pressures (P_{RA} and P_{LA}).

Blood volume

Changes in blood volume per se do not affect myocardial contractility but do influence the vascular function curve in the manner shown in Fig. 9-8. Therefore, to understand the circulatory alterations evoked by a given change in blood volume, it is necessary to plot the appropriate cardiac function curve along with the vascular function curves that represent the control and experimental states. Fig. 9-13 illustrates the response to a blood transfusion.

Since equilibrium point B, which denotes the values for cardiac output and P_v after transfusion, lies above and to the right of the control equilibrium point A, it is evident that transfusion results in an enhanced cardiac output and elevated P_v. Hemorrhage would have the opposite effect. Pure increases or decreases in venomotor tone elicit responses that are analogous to those evoked by augmentations or reductions, respectively, of the total blood volume, for reasons that were discussed on p. 193.

Heart failure

Heart failure may be acute or chronic. Acute heart failure may be caused by toxic quantities

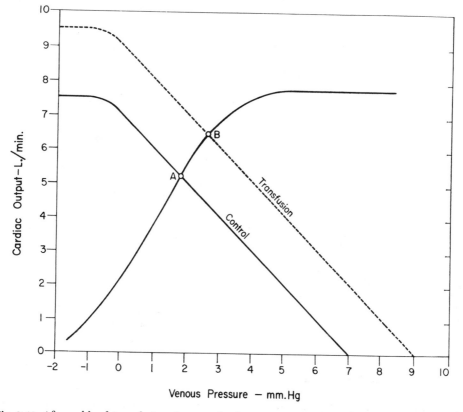

Fig. 9-13. After a blood transfusion the vascular function curve is shifted to the right (dashed curve). Therefore cardiac output and venous pressure are both increased, as denoted by the translocation of the equilibrium point from A to B.

of certain drugs and anesthetics or by certain pathological conditions, such as coronary artery occlusion. Chronic heart failure may occur in such conditions as essential hypertension or rheumatic heart disease. In these various forms of heart failure, myocardial contractility is impaired. Consequently, the cardiac function curve is shifted to the right as depicted in Fig. 9-14.

In acute heart failure there has not been sufficient time for blood volume to change. Therefore the equilibrium point will shift from the intersection of the normal curves (Fig. 9-14,

point A) to the intersection of the normal vascular function curve with a depressed cardiac function curve (point B or C).

In chronic heart failure there is not only a shift in the cardiac function curve but also an increase in blood volume caused in part by fluid retention by the kidneys. The fluid retention is related to the concomitant reduction in glomerular filtration rate and to the increased secretion of aldosterone by the adrenal cortex. The resultant hypervolemia causes the vascular function curve to be shifted to the right, as shown in Fig. 9-14. Hence, with moderate de-

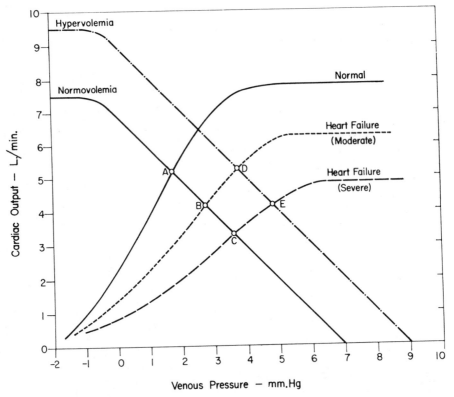

Fig. 9-14. With moderate or severe heart failure, the cardiac function curves are shifted to the right. With no change in blood volume, cardiac output decreases and venous pressure rises (from control equilibrium point A to point B or point C). With the increase in blood volume that usually occurs in heart failure, the vascular function curve is shifted to the right. Hence venous pressure may be elevated with no reduction in cardiac output (point D) or (in severe heart failure) with some diminution in cardiac output (point E).

grees of heart failure, P_v will be elevated but cardiac output will be approximately normal (point *D*). With more severe degrees of heart failure, P_v will be still higher and cardiac output will be subnormal (point *E*).

Classically heart failure has been explained on the basis of the descending limb of the Frank-Starling curve. As explained on p. 160, when myocardial fibers are stretched beyond an optimal point, then further degrees of stretch impair rather than improve performance. Beyond the optimal point, cardiac output diminishes as venous pressure is raised; this portion of a cardiac function curve beyond the optimal point is designated the descending limb. In the normal heart the descending limb does not become manifest until the ventricular end-diastolic pressure reaches very high levels, but with myocardial failure it may appear at

much lower end-diastolic pressures. Not uncommonly, patients in severe heart failure improve when blood is withdrawn and become worse when their blood volume is expanded. However, whether such clinical observations indicate that the heart is operating on the descending limb of a cardiac function curve or whether other mechanisms are responsible remains controversial.

Peripheral resistance

Predictions concerning the precise alterations evoked by changes in peripheral resistance are complex because shifts in both the cardiac and vascular function curves are involved. With increased peripheral resistance (Fig. 9-15) the vascular function curve is displaced downward but converges to the same X-axis intercept as the control curve, as described

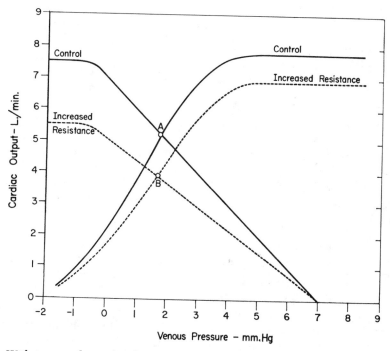

Fig. 9-15. With increased peripheral resistance, both the cardiac and vascular function curves are displaced downward.

on p. 194. The cardiac function curve is also shifted downward because at any given P_v the heart is able to pump less blood against a greater resistive load. Since both the vascular and cardiac function curves are displaced downward by an increase of peripheral resistance, it is clear that the new equilibrium point, B, will fall below the control point, A.

Whether point B will fall directly below point A, or will lie somewhat to the right or left, depends on the magnitude of the shift in each of the two curves. For example, if a given increase in peripheral resistance resulted in a relatively greater downward shift of the vascular function curve than of the cardiac function curve, then equilibrium point B would fall not only below A but also to the left of A; that is, both cardiac output and P_v would diminish. Conversely, if the cardiac function curve is displaced more than the vascular function curve, then point B will fall below and to the right of point A; that is, cardiac output would decrease but P_v would rise.

ANCILLARY FACTORS AFFECTING THE VENOUS SYSTEM AND CARDIAC OUTPUT

The interrelationships between central venous pressure and cardiac output have been discussed in terms of an oversimplified schema. The effects evoked by changes in single variables have been described. However it must be recognized that, partly because of the multitude of feedback control loops that regulate the cardiovascular system, an isolated change in a single variable rarely occurs. A change in blood volume, for example, would rapidly elicit reflex alterations in cardiac function, peripheral resistance, and venomotor tone. Aside from the complications involved in analyzing the effects of multiple variables acting simultaneously, several auxiliary factors must also be considered for a complete understanding of the control of cardiac output. Such ancillary factors may be considered to modulate the operation of the simplified schema described previously.

Gravity

Experience has shown that gravitational forces may exert dramatic effects on the cardiac output. Among soldiers standing at attention for protracted periods, particularly in warm weather, it is not unusual for some individuals to faint because of reduced cardiac output. Gravitational effects are exaggerated in cases of airplane pilots executing pullouts from dives, where a centrifugal force several times greater than the force of gravity may be exerted briefly in the footward direction; such individuals characteristically black out during the maneuver, as blood is drained from the cephalic regions and pooled in the lower parts of the body.

Specious reasoning is sometimes applied to explain the curtailment of cardiac output under such conditions. It is often argued that when an individual is oriented in a vertical, footdown position, the forces of gravity act counter to those forces that ordinarily promote venous return from the dependent regions of the body. This statement is incomplete, however, since it ignores the fact that any impedimentary gravitational force on the venous side of the circuit is exactly balanced by a facilitatory counterforce on the arterial side of the same circuit.

In this sense, therefore, the vascular system may be considered to be a U tube. To comprehend the action of gravity on flow through such a system, the models depicted in Figs. 9-16 and 9-17 will be analyzed. In Fig. 9-16 all the U tubes represent rigid cylinders of constant diameter. With both limbs of the U tube oriented horizontally (A) the flow is dependent only on the pressures at the inflow and outflow ends of the tube (P_i and P_o, respectively), the viscosity of the fluid, and the length and radius of the tube, in accordance with Poiseuille's equation (p. 58). With a constant cross section the pressure gradient will be uniform; hence

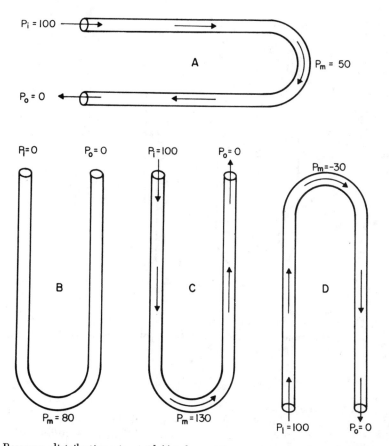

Fig. 9-16. Pressure distributions in rigid U tubes with constant internal diameters, all with the same dimensions. For a given inflow pressure (P_i = 100) and outflow pressure (P_o = 0), the pressure at the midpoint (P_m) depends on the orientation of the U tube, but the flow through the tube is independent of the orientation.

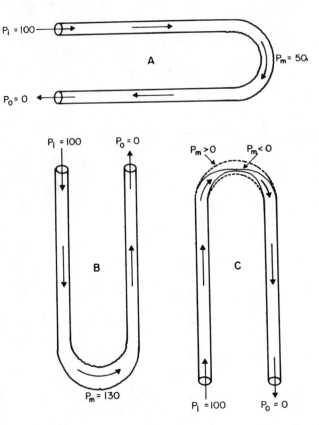

Fig. 9-17. In U tubes with a distensible section at the bend, even when inflow (P_i) and outflow (P_o) pressures are the same, the resistance to flow and the fluid volume contained within each tube vary with the orientation of the tube.

the pressure midway down the tube (P_m) will be the average of the inflow and outflow pressures.

When the U tube is oriented vertically (B to D), hydrostatic forces must be taken into consideration. In tube B both limbs are open to atmospheric pressure at the same hydrostatic level; hence there is no flow. The pressure at the midpoint of the tube will simply be ρhg; that is, it will depend on the density of the fluid, ρ, the height of the U tube, h, and the acceleration of gravity, g. In the example the length of the U tube is such that the pressure at the midpoint is 80 mm. Hg.

Now consider case C, where the U tube is oriented the same as tube B, but where a 100 mm. Hg pressure difference is applied across the two ends. The flow will precisely equal that in A, since the pressure gradient, the tube dimensions, and the fluid viscosity are all the same. Gravitational forces are precisely equal in magnitude but opposite in direction in the two limbs of the U tube. Therefore, since the flow will be the same as that in A, there will be a pressure drop of 50 mm. Hg at the midpoint because of the viscous losses resulting from flow. Furthermore, there will be an increased pressure of 80 mm. Hg at the midpoint because of gravitational effects, just as in B. The actual pressure at the midpoint of the tube C then will be the resultant of the viscous loss and hydrostatic gain, or 130 mm. Hg in this example.

In D a pressure gradient of 100 mm. Hg is applied to the same U tube, but oriented in the opposite direction. Gravitational forces will be so directed that the pressure at the midpoint will tend to be 80 mm. Hg less than that at the ends of the U tube. Viscous losses will still produce a 50 mm. Hg pressure drop at the midpoint relative to P_i. Hence, with this orientation, pressure at the midpoint of the U tube will be 30 mm. Hg below ambient pressure. Flow will, of course, be the same as in tubes A and C, for the reasons stated in relation to C.

Therefore in a system of rigid tubes gravitational effects will not alter the rate of blood flow. Since experience shows that gravity does affect the cardiovascular system, it may be suspected that the reason resides in the fact that the cardiovascular system is composed of distensible rather than rigid vessels. Therefore the pressures in a set of U tubes with distensible components (at the bends in the tubes of Fig. 9-17) will be examined. In tubes A and B the pressure distributions will resemble those in tubes A and C, respectively, of Fig. 9-16. Since the pressure is higher at the bend of tube B than at the bend of tube A, and since the segment is distensible in this region, the tubing at the bend of B will be distended more than at the bend of A. The extent of the distension will depend on the elastic characteristics of these tube segments. Since flow is directly related to the tube diameter, the flow through B will exceed the flow through A for a given pressure difference applied at the ends.

Since orienting a U tube with its bend downward actually increases rather than diminishes flow for any given pressure difference, how then is the observed impairment of cardiovascular function explained when the body is sim-

ilarly oriented? Of course the explanation resides in the fact that the cardiovascular system is a closed circuit of constant volume, whereas in the example the U tube is conceived of as an open conduit supplied by a source of unlimited volume. In the cardiovascular system, most of the distension will occur on the venous rather than the arterial side of the circuit in the dependent regions of the body because the venous capacitance is so much greater than the arterial capacitance. Such venous distension is readily observed on the back of the hands when the arms are allowed to hang down below heart level. Whatever the volume of blood involved in such venous distension (*venous pooling*), the hemodynamic effects resemble those caused by the loss of an equivalent volume of blood from the body during hemorrhage. It has been estimated that when a person shifts from a supine to a relaxed standing position, from 300 to 800 ml of blood are pooled in the legs. This results in a decrease in cardiac output of about 2 L./min., and a 40% reduction in stroke volume.

The compensatory adjustments that enable humans to adapt to the erect position are largely identical to those that permit them to adapt to blood loss. For example, the diminution in baroreceptor excitation reflexly initiates acceleration of the heart, a more vigorous cardiac contraction, and increased arteriolar and venular constriction. The baroreceptor reflex per se apparently has a much more pronounced effect on the resistance than on the capacitance vessels. Warm ambient temperatures tend to interfere with the compensatory vasomotor reactions, and the absence of muscular activity exaggerates the effects. Therefore fainting is not uncommon during prolonged standing in hot weather.

The reverse situation prevails when the U tube is rotated so that the bend is directed upward. In Fig. 9-17 the pressure at the bend of tube C would tend to be -30 mm. Hg, just as in tube D of the preceding figure. Since the ambient pressure exceeds the internal pressure, however, the tube will collapse. Flow will then cease, and there will no longer be any viscous decline of pressure associated with flow. In U tube C, when flow stops, the pressure at the top of each limb will be 80 mm. Hg less than at the bottom (the hydrostatic pressure difference). Hence, in the left, or inflow, limb the pressure will approach +20 mm. Hg. As soon as this pressure exceeds ambient pressure, the collapsed tubing will be forced open and flow will begin. With the onset of flow, however, pressure will again drop below the ambient pressure. Thus the tubing at the bend will flutter; that is, it will fluctuate between the open and closed states.

When an arm is raised, the cutaneous veins in the hand and forearm are seen to collapse, for the reasons described previously. Fluttering does not occur here because the deeper veins are protected from collapse by surrounding structures; these deeper veins therefore accommodate the flow. The situation would be analogous to adding a rigid tube (representing the deeper veins) in parallel with the collapsible tube (representing the superficial veins) at the bend of tube C in Fig. 9-17. The collapsible tube would no longer flutter, but would remain closed. All flow would occur through the rigid tube, just as in tube D in Fig. 9-16.

It may not be appreciated why the superficial veins in the neck are ordinarily partially collapsed in the normal individual in the upright position. Venous return from the head is conducted largely through the deeper cervical veins. However, when *central venous pressure* (pressure in the thoracic venae cavae and right atrium) is abnormally elevated, the superficial neck veins are distended and they do not collapse even when the subject assumes the upright position. Such cervical venous distension is an important clinical sign, often indicative of congestive heart failure.

Muscular activity and venous valves

When an individual assumes the upright position but remains at rest, the pressure rises in the veins in the dependent regions of the body. The venous pressure in the legs increases gradually and does not reach an equilibrium value until almost one minute after standing. The gradual nature of the rise in P_v is attributable to the valves located within the veins, which permit flow only in the direction toward the heart. When a person stands up, the valves prevent blood in veins at higher levels from actually falling toward the feet. Hence the column of venous blood is supported at numerous levels by these valves; temporarily the venous column consists of many separate segments. However, blood continues to enter the column at many points from the venules and smaller tributary veins, and pressure continues to rise. As soon as the pressure in one segment exceeds that in the segment just above it, the valve is forced open. Ultimately all the valves are opened, and the column is then continuous, similar to the outflow limbs of the U tubes shown in Figs. 9-16 and 9-17.

Precise measurement reveals that the final level of P_v in the feet during standing is only slightly greater than that which would have existed in a static column of blood extending from the right atrium to the feet. This indicates that the pressure drop caused by flow from foot veins to the right atrium is very small. This is one justification for lumping all the veins as a common venous capacitance in the model illustrated in Fig. 9-4.

When an individual who has been standing quietly begins to walk, the venous pressure in his legs decreases appreciably during this muscular activity. Because of the intermittent venous compression produced by the contracting muscles and because of the orientation of the venous valves, blood is forced from the veins toward the heart. Hence, muscular contraction lowers the mean venous pressure and serves as an auxiliary pump to assist venous return. Furthermore, it prevents venous pooling and lowers capillary hydrostatic pressure (which must always exceed P_v), thereby reducing the tendency toward the formation of tissue edema in the dependent regions.

Respiratory activity

During quiet, normal breathing, the periodic activity of the respiratory muscles results in rhythmic variations in vena caval flow and serves as an auxiliary pump to promote venous return. Other activities involving the muscles of respiration, such as coughing and straining at stool, usually exert more profound influences on cardiac output and venous return; substantial changes in cardiac output may also occur during artificial respiration.

The changes in blood flow in the superior vena cava associated with normal spontaneous respiration are shown in Fig. 9-18. During in-spiration the reduction in intrathoracic pressure is transmitted to the lumina of the blood vessels located wihin the thoracic cavity. This is reflected by the diminution of right atrial pressure accompanying the intrathoracic pressure change. This reduction in central venous pressure during inspiration increases the pressure gradient between extrathoracic and intrathoracic veins. The consequent acceleration of venous return to the right atrium is displayed in the figure as an increase in superior vena caval blood flow from 5.2 ml./sec. during expiration to 11 ml./sec. during inspiration.

An exaggerated reduction in intrathoracic pressure achieved by a strong inspiratory effort against a closed glottis (called the *Müller maneuver*) does not evoke a proportionately greater acceleration of venous return. This is largely caused by the tendency for many of the extrathoracic veins to collapse near their points of entry into the chest when their pressures fall

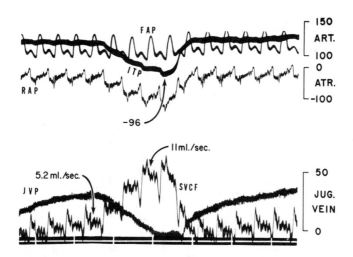

Fig. 9-18. During a normal inspiration, intrathoracic *(ITP)*, right atrial *(RAP)*, and jugular venous *(JVP)* pressures decrease, and flow in the superior vena cava *(SVCF)* increases in this case from 5.2 to 11 ml./sec. All pressures are in mm. H_2O, except for femoral arterial pressure *(FAP)*, which is in mm. Hg. (Modified from Brecher, G. A.: Venous return, New York, 1956, Grune & Stratton, Inc.)

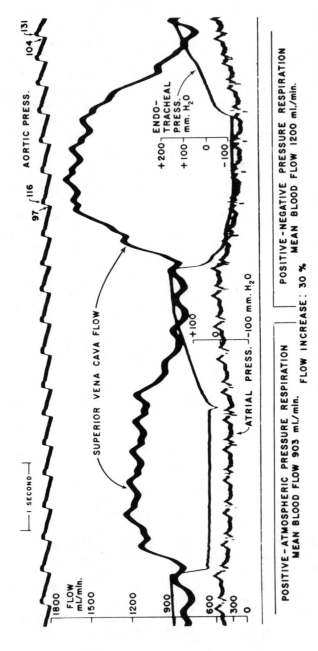

Fig. 9-19. During intermittent positive pressure respiration, the flow in the superior vena cava is approximately 30% greater when the lungs are deflated actively by applying negative endotracheal pressures (right half of the figure) than when they are allowed to deflate passively against atmospheric pressure (left of the figure). (Modified from Brecher, G. A.: Venous return, New York, 1956, Grune & Stratton, Inc.)

below the atmospheric level. As the veins collapse, flow into the chest momentarily stops. The cessation of flow results in a rise in pressure upstream, forcing the collapsed segment to open again. The process is repetitive, such that the segments of veins adjacent to the chest will actually flutter rapidly between the open and closed states.

During expiration there is, of course, a deceleration of the flow into the central veins. However it has been shown that the mean rate of venous return during normal respiration exceeds the rate in the temporary absence of respiratory activity. Hence normal inspiration apparently exerts a greater effect toward promoting venous return than does normal expiration toward impeding it. In part this must be attributable to the presence of valves in the veins of the extremities and neck, which would prevent any reversal of flow during expiration. Thus the respiratory muscles and venous valves constitute an auxiliary pump for venous return.

Sudden, sustained increases in intrathoracic pressure tend to impede venous return. Straining against a closed glottis (termed the *Valsalva maneuver*) regularly occurs during coughing, defecation, and lifting. Intrathoracic pressures in excess of 100 mm. Hg have been recorded in trumpet players, and pressure over 400 mm. Hg have been observed during paroxysms of coughing. Arterial pressure may rise considerably above the control level during the cough by virtue of a direct transmission of the greatly increased intrathoracic pressure to the lumina of the aorta and its branches within the chest. However, after cessation of the coughing the arterial blood pressure might drop transiently to very low levels, because of the severe impediment to venous return during the preceding paroxysm of coughing.

Artificial respiration

An understanding of the preceding section enables one to appreciate the circulatory effects induced by various forms of artifical respiration. When the entire subject, except for the head, is placed in an iron lung, the periodic drop in pressure below the atmospheric level expands the thorax. The effects of this device on venous return most closely simulate those of normal respiration, since lung inflation is accompanied by a decreased intrathoracic pressure. With most other forms of artificial respiration (for example, mouth-to-mouth resuscitation and various mechanical respirators), lung inflation is achieved by applying endotracheal pressures above atmospheric pressure, and expiration occurs primarily by passive recoil of the thoracic cage. Thus lung inflation is attended by aɴ appreciable rise in intrathoracic pressure. In Fig. 9-19 it is evident that superior vena caval flow decreases sharply during the phase of positive-pressure lung inflation (indicated by the progressive rise in endotracheal pressure in the central portion of Fig. 9-19). When deflation is achieved by applying a negative endotracheal pressure (indicated by the abrupt decrease in endotracheal pressure in the right half of Fig. 9-19), vena caval flow then accelerates to a considerably greater extent than when the lungs are allowed to deflate passively at atmospheric pressure (near left border of the figure).

BIBLIOGRAPHY
Journal articles

Abel, F. L., and Waldhausen, J. A.: Respiratory and cardiac effects on venous return, Am. Heart J. **78:**266, 1969.

Carneiro, J. J., and Donald, D. E.: Blood reservoir function of dog spleen, liver, and intestine, Am. J. Physiol. **232:**H67, 1977.

Gauer, O. H., and Thron, H. L.: Properties of veins in vivo: integrated effects of their smooth muscle, Physiol. Rev. **42**(supp.5):283, 1962.

Grodins, F. S.: Integrative cardiovascular physiology: a mathematical synthesis of cardiac and blood vessel hemodynamics, Q. Rev. Biol. **34:**93, 1959.

Grodins, F. S., Stuart, W. H., and Veenstra, R. L.: Performance characteristics of the right heart bypass preparation, Am. J. Physiol. **198:**552, 1960.

Lautt, W. W., and Greenway, C. V.: Hepatic venous compliance and role of liver as a blood reservoir, Am. J. Physiol. **231**:292, 1976.

Levy, M. N.: The cardiac and vascular factors that determine systemic blood flow, Circ. Res. **44**:739, 1979.

Levy, M. N.: The cardiovascular physiology of the critically ill patient, Surg. Clin. North Am. **55**:483, 1975.

Ludbrook, J.: The musculovenous pumps of the human lower limb, Am. Heart J. **71**:635, 1966.

Rothe, C. F.: Reflex control of the veins in cardiovascular function, Physiologist **22**:28, 1979.

Sagawa, K.: Critique of a large-scale organ system model: guytonian cardiovascular model, Ann. Biomed. Engin. **3**:386, 1975.

Samar, R. E., and Coleman, T. G.: Measurement of mean circulatory filling pressure and vascular capacitance in the rat, Am. J. Physiol. **234**:H94, 1978.

Shepherd, A. P., Granger, H. J., Smith, E. E., and Guyton, A. C.: Local control of tissue oxygen delivery and its contribution to the regulation of cardiac output, Am. J. Physiol. **225**:747, 1973.

Shoukas, A. A., and Sagawa, K.: Control of total systemic vascular capacity by the carotid sinus baroreceptor reflex, Circ. Res. **33**:22, 1973.

Weisel, R, D., Berger, R. L., and Hechtman, H. B.: Measurement of cardiac output by thermodilution, N. Engl. J. Med. **292**:682, 1975.

Books and monographs

Alexander, R. S.: The peripheral venous system. In Handbook of physiology; Section 2: Circulation, Washington, D. C., 1963, American Physiological Society, vol. II, pp. 1075-1098.

Bloomfield, D. A.: Dye curves: the theory and practice of indicator dilution, Baltimore, 1974, University Park Press.

Brecher, G. A.: Venous return, New York, 1956, Grune & Stratton, Inc.

Green, J. F.: Determinants of systemic blood flow; International Review of Physiology III: Cardiovascular Physiology, vol. 18, Baltimore, 1979, University Park Press.

Guyton, A. C., Jones, C. E., and Coleman, T. G.: Circulatory physiology: cardiac output and its regulation, ed. 2, Philadelphia, 1973, W. B. Saunders Co.

Shepherd, J. T., and Vanhoutte, P. M.: Veins and their control, Philadelphia, 1975, W. B. Saunders Co.

Zierler, K. L.: Circulation times and the theory of indicator-dilution methods for determining blood flow and volume. In Handbook of physiology; Section 2: Circulation, Washington, D.C., 1962, American Physiological Society, vol. I, pp. 585-615.

CHAPTER 10

CORONARY CIRCULATION

ANATOMY OF CORONARY VESSELS

The right and left coronary arteries, which arise at the root of the aorta behind the right and left cusps of the aortic valve, respectively, provide the entire blood supply to the myocardium. The right coronary artery supplies principally the right ventricle and atrium; the left coronary artery, which divides near its origin into the anterior descendens and the circumflex branches, supplies principally the left ventricle and atrium, but there is some overlap. In the dog the left coronary artery supplies about 85% of the myocardium, whereas in humans the right coronary artery is dominant in 50% of individuals, the left coronary artery is dominant in another 20%, and the flow delivered by each main artery is about equal in the remaining 30%. Coronary blood flow is measured in humans by (1) the nitrous oxide technique, as described for cerebral blood flow measurement on p. 232, (2) measurement of ^{84}rubidium (a positron emitter) uptake by the myocardium, utilizing a procedure known as coincident counting over the chest, or (3) the clearance rate of an inert radioactive gas (for example, ^{133}xenon or ^{85}krypton) injected directly into a coronary artery by means of a catheter passed from a peripheral artery (for example, brachial) into the coronary artery. With the last method the rate of washout of the injected radioactivity, as monitored by a radiation detector placed over the precordium, is proportional to the blood flow through the myocardium supplied by the injected vessel.

After passage through the capillary beds, most of the venous blood returns to the right atrium through the coronary sinus, but some reaches the right atrium by way of the anterior coronary veins. There are also vascular communications directly between the vessels of the myocardium and the cardiac chambers; these comprise the *arteriosinusoidal,* the *arterioluminal,* and the *thebesian vessels.* The arteriosinusoidal channels consist of small arteries or arterioles that lose their arterial structure as they penetrate the chamber walls and divide into irregular, endothelium-lined sinuses (50 to 250 μ). These sinuses anastomose with other sinuses and with capillaries and communicate with the cardiac chambers. The arterioluminal vessels are small arteries or arterioles that open directly into the atria and ventricles. The thebesian vessels are small veins that connect capillary beds directly with the cardiac chambers and also communicate with cardiac veins and other thebesian veins. On the basis of anatomical studies, intercommunication appears to exist among all the minute vessels of the myocardium in the form of an extensive plexus of subendocardial vessels. It has been suggested that some myocardial nutrition can be derived from the cardiac cavities through those channels. Isotope-labeled blood in the cardiac chambers does penetrate a short distance into the endocardium, but does not constitute a significant source of oxygen and nutrients to the myocardium. In the dog a major fraction of the left coronary inflow returns to the right atrium

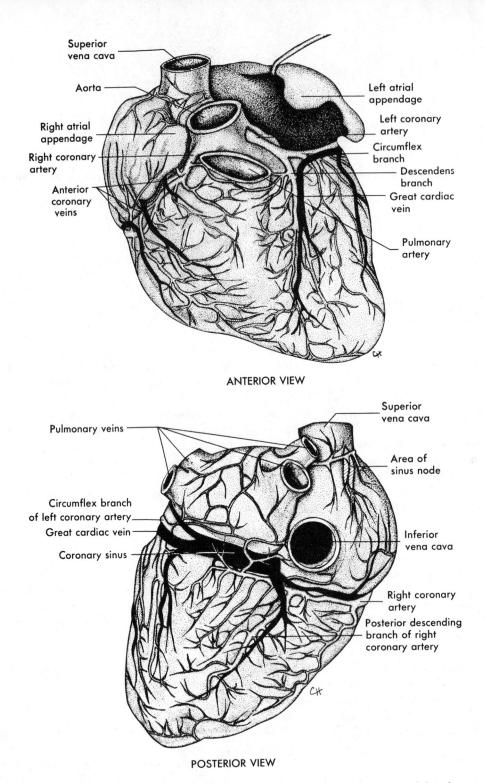

Fig. 10-1. Anterior and posterior surfaces of the heart, illustrating the location and distribution of the principal coronary vessels.

via the coronary sinus, and a small fraction supplying the interventricular septum returns directly to the right ventricular cavity. Right coronary artery drainage is primarily via the anterior cardiac veins to the right atrium. The epicardial distribution of the coronary arteries and veins is illustrated in Fig. 10-1.

FACTORS THAT INFLUENCE CORONARY BLOOD FLOW
Physical factors

The primary factor responsible for perfusion of the myocardium is the aortic pressure, which is, of course, generated by the heart itself. Changes in aortic pressure generally evoke parallel directional changes in coronary blood flow. However, alterations of cardiac work, produced by an increase or decrease in aortic pressure, have a considerable effect on coronary resistance. By means of mechanisms that have not yet been fully elucidated, increased metabolic activity of the heart results in a decrease in coronary resistance, and a reduction in cardiac metabolism produces an increase in coronary resistance. If a cannulated coronary artery is perfused by blood from a pressure-controlled reservoir, perfusion pressure can be altered without changing aortic pressure and cardiac work. Under these conditions abrupt variations in perfusion pressure produce equally abrupt changes in coronary blood flow in the same direction. However, maintenance of the perfusion pressure at the new level is associated with a return of blood flow toward the level observed prior to the induced change in perfusion pressure (Fig. 10-2). This phenomenon is an example of autoregulation of blood flow and is discussed in Chapter 7. Under normal conditions blood pressure is kept within relatively narrow limits by the baroreceptor reflex mechanisms so that changes in coronary blood flow are primarily caused by caliber changes of the coronary resistance vessels in response to metabolic demands of the heart.

In addition to providing the head of pressure to drive blood through the coronary vessels, the heart also influences its blood supply by the squeezing effect of the contracting myocardium of the blood vessels that course through it *(extravascular compression)*. This force is so great during early ventricular systole that blood flow, as measured in a large coronary artery supplying the left ventricle, is briefly reversed. Maximal left coronary inflow occurs in early diastole, when the ventricles have relaxed and extravascular compression of the coronary vessels is virtually absent. This flow pattern is seen in the phasic coronary flow curve for the left coronary artery (Fig. 10-3). After an initial reversal in early systole, left coronary blood flow follows the aortic pressure until early diastole, when it rises abruptly and then declines slowly as aortic pressure falls during the remainder of diastole. Left ventricular myocardial pressure (pressure within the wall of the left ventricle) is greatest near the endocardium and lowest near the epicardium.

However, under normal conditions this pressure gradient does not result in impairment of endocardial blood flow since the greater diastolic flow in the endocardium compensates for the greater systolic flow in the epicardium. In fact, when 10 μ diameter radioactive microspheres are injected into the coronary arteries, their distribution indicates that the blood flow to the epicardial and endocardial halves of the left ventricle are approximately equal under normal conditions. Since extravascular compression is greatest at the endocardial surface of the ventricle, equality of epicardial and endocardial blood flow must mean that the tone of the endocardial resistance vessels is less than that of the epicardial vessels.

Under abnormal conditions, when diastolic pressure in the coronary arteries is low, such as in severe hypotension, partial coronary artery occlusion, or severe aortic stenosis, the ratio of endocardial to epicardial blood flow falls below a value of 1.0. This indicates that the blood flow to the endocardial region is more

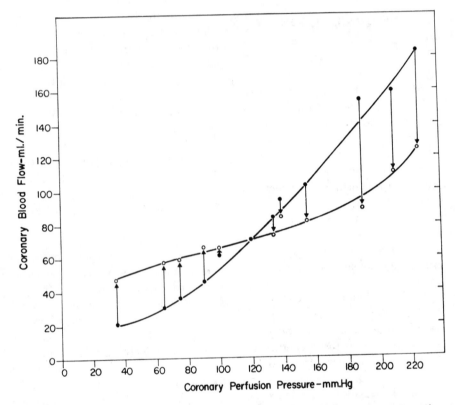

Fig. 10-2. Pressure-flow relationships in the coronary vascular bed. At constant aortic pressure, cardiac output, and heart rate, coronary artery perfusion pressure was abruptly increased or decreased from the control level indicated by the point where the two lines cross. The closed circles represent the flows that were obtained immediately after the change in perfusion pressure and the open circles the steady state flows at the new pressures. There is a tendency for flow to return toward the control level (autoregulation of blood flow), and this is most prominent over the intermediate pressure range (about 60 to 180 mm. Hg).

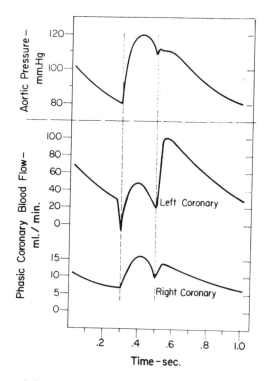

Fig. 10-3. Comparison of phasic coronary blood flow in the left and right coronary arteries.

severely impaired than that to the epicardial regions of the ventricle. This redistribution of coronary flow is also reflected in an increase in the gradient of myocardial lactic acid and adenosine concentrations from epicardium to endocardium. For this reason, the myocardial damage observed in arteriosclerotic heart disease (for example, following coronary occlusion) is greatest in the inner wall of the ventricle.

Flow in the right coronary artery shows a similar pattern (Fig. 10-3), but because of the lower pressure developed during systole by the thin right ventricle, reversal of blood flow does not occur in early systole and systolic blood flow constitutes a much greater proportion of total coronary inflow than it does in the left coronary artery. The extent to which extravas-

cular compression restricts coronary inflow can be readily seen when the heart is suddenly arrested in diastole or on the induction of ventricular fibrillation. Fig. 10-4 depicts mean left coronary flow when the vessel was perfused with blood at a constant pressure from a reservoir. At the arrow in record *A*, ventricular fibrillation was electrically induced and an immediate and substantial increase in blood flow occurred. Subsequent increase in coronary resistance over a period of many minutes reduced myocardial blood flow to below the level existing prior to the induction of ventricular fibrillation (record *B*, before stellate ganglion stimulation).

Tachycardia and bradycardia have dual effects on coronary blood flow. A change in heart

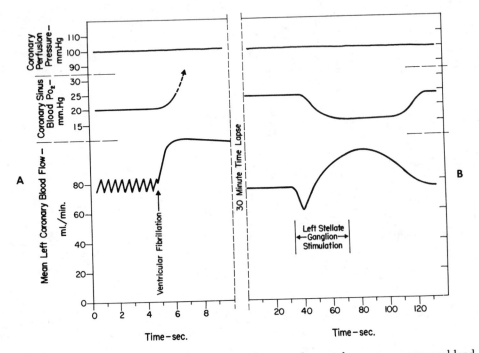

Fig. 10-4. A, Unmasking of the restricting effect of ventricular systole on mean coronary blood flow by induction of ventricular fibrillation during constant pressure perfusion of the left coronary artery. **B,** Effect of cardiac sympathetic nerve stimulation on coronary blood flow and coronary sinus blood oxygen tension in the fibrillating heart during constant pressure perfusion of the left coronary artery.

rate is accomplished chiefly by shortening or lengthening of diastole. With tachycardia the proportion of time spent in systole, and consequently the period of restricted inflow, increases. However, this mechanical reduction in mean coronary flow is overriden by the coronary dilation associated with the increased metabolic activity of the more rapidly beating heart. With bradycardia the opposite is true; restriction of coronary inflow is less (more time in diastole) but so are the metabolic (oxygen) requirements of the myocardium.

Neural and neurohumoral factors

Stimulation of the sympathetic nerves to the heart elicits a marked increase in coronary blood flow. However, the increase in flow is associated with cardiac acceleration and a more forceful systole. The stronger myocardial contractions and the tachycardia (with the consequence that a greater proportion of time is spent in systole) tend to restrict coronary flow, whereas the increase in myocardial metabolic activity, as evidenced by the rate and contractility changes, tends to evoke dilation of the coronary resistance vessels. The increase in coronary blood flow observed with cardiac sympathetic nerve stimulation is the algebraic sum of these factors. In perfused hearts in which the mechanical effect of extravascular compression is eliminated by cardiac arrest or ventricular fibrillation, an initial coronary vasoconstriction is often observed with cardiac sympathetic nerve stimulation before the vasodilation attributable to the metabolic effect comes into play (Fig. 10-4, *B*). Such observations suggest that the primary action of the sympathetic nerve fibers on the coronary resistance vessels is vasoconstriction.

In contrast to skeletal muscle (p. 134), sympathetic cholinergic innervation of the coronary vessels does not exist, and whether there are sympathetic fibers to beta adrenergic receptors on the coronary arterioles is questionable. However, experiments with the use of alpha and beta adrenergic drugs and their respective blocking agents reveal the presence of alpha (constrictor) and beta (dilator) receptors on the coronary vessels. Recent studies have indicated that the coronary resistance vessels participate in the baroreceptor and chemoreceptor reflexes and that there appears to be sympathetic constrictor tone of the coronary arterioles that can be reflexly modulated. Nevertheless coronary resistance is predominantly under local nonneural control.

Vagus nerve stimulation has little direct effect on the caliber of the coronary arterioles. In the fibrillating heart with the coronary arteries perfused at a constant pressure, or in the beating, paced heart with flow measured in late diastole when there is essentially no extravascular compression, small increments in coronary blood flow can be observed with stimulation of the peripheral ends of the vagi. In addition, activation of the carotid and aortic chemoreceptors can elicit a decrease in coronary resistance via the vagus nerves to the heart. However, it is likely that the vagi exert less effect on coronary resistance in the normal animal, because of the overriding influence of metabolic mechanisms. The failure of strong vagal stimulation to evoke a large increase in coronary blood flow is not because of insensitivity of the coronary resistance vessels to acetylcholine, since intracoronary administration of this agent elicits marked vasodilation.

Reflexes orginating in the myocardium and altering vascular resistance in peripheral systemic vessels, including the coronary vessels, have been conclusively demonstrated. However, the existence of extracardiac reflexes, with the coronary resistance vessels as the effector sites, has not been established.

Metabolic factors

One of the most striking characteristics of the coronary circulation is the close parallelism between the level of myocardial metabolic activity and the magnitude of the coronary blood flow. This relationship is also found in the denervated heart or the completely isolated heart, whether in the beating or the fibrillating state. The ventricles will continue to fibrillate for many hours when the coronary arteries are perfused with arterial blood from some external source. With the onset of ventricular fibrillation, an abrupt increase in coronary blood flow occurs because of the removal of extravascular compression (Fig. 10-4). Flow then gradually returns toward, and often decreases below, the prefibrillation level. The increase in coronary resistance, which occurs despite the elimination of extravascular compression, is a manifestation of the heart's ability to adjust its blood flow to meet its energy requirements. The fibrillating heart utilizes less oxygen than the pumping heart, and blood flow to the myocardium is reduced accordingly.

The link between cardiac metabolic rate and the coronary blood flow remains unsettled. Numerous agents, generally referred to as metabolites, have been suggested as the mediator of the vasodilation observed with increased cardiac work. Accumulation of vasoactive metabolites may also be responsible for reactive hyperemia, since the magnitude and, to a greater degree, the duration of the coronary flow following release of the briefly occluded vessel is, within certain limits, proportional to the duration of the period of occlusion. Among the substances implicated are carbon dioxide, oxygen (reduced oxygen tension), lactic acid, hydrogen ions, histamine, potassium ions, increased osmolarity, polypeptides, and adenine nucleotides. None of these has satisfied all the criteria for the physiological mediator. Although potassium release from the myocardium can account

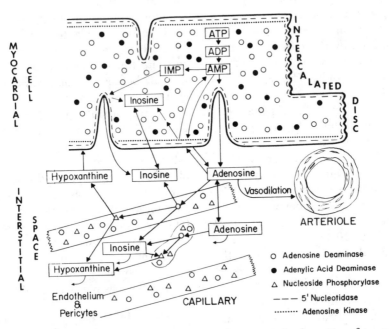

Fig. 10-5. Schematic diagram of myocardial tissue illustrating the formation, fate, and site of action of adenosine arising from intracellular ATP and the distribution of the enzymes involved in the metabolism of adenosine. (Reproduced by permission from Am. J. Physiol. **222:**550, 1972.)

for about half of the initial decrease in coronary resistance, it cannot be responsible for the increased coronary flow observed with prolonged enhancement of cardiac metabolic activity, since its release from the cardiac muscle is transitory. Recent evidence suggests that adenosine plays the role of metabolic vasodilator.

According to the adenosine hypothesis, a reduction in myocardial oxygen tension produced by low coronary blood flow, hypoxemia, or increased metabolic activity of the heart leads to the formation of adenosine, which reaches the interstitial fluid space and induces dilation of the coronary resistance vessels. The adenosine concentration of the myocardium increases in response to enhanced cardiac work in the absence of a reduced oxygen supply, and it is rapidly formed. (In a single cardiac cycle the myocardial adenosine level increases significantly during ventricular systole over that observed during diastole.) Coronary dilation results in an increase in coronary blood flow that enhances the washout of adenosine, and reduces its formation. Under normal resting conditions only small amounts of adenosine are released by the heart and exert a minimal dilating effect. The schema for adenosine regulation of coronary blood flow is illustrated in Fig. 10-5. The enzyme, 5'-nucleotidase, is located at the cell membranes and transverse tubules and dephosphorylates AMP (formed by degradation of ATP) to adenosine. The adenosine enters the interstitial space and produces arteriolar dilation. From the interstitial space the adenosine may re-enter the myocardial cell and (1) be rephosphorylated to AMP by adenosine kinase, (2) be deaminated to inosine by adenosine deaminase in the myocardial cell, or (3) enter the capillaries where it can be deaminated to inosine in the vessel wall or red cells or further degraded to hypoxanthine by the enzyme nucleoside phosphorylase, which is located in pericytes (perivascular fibroblasts), endothelial cells, or red cells. Relative to adenosine deaminase, the concentration of adenylic acid deaminase in the myocardium is low, so that only small amounts of inosinic acid (IMP) are formed.

CORONARY COLLATERAL CIRCULATION AND VASODILATORS

In the normal human heart there are virtually no functional intercoronary channels, whereas in the dog there are a few small vessels that link branches of the major coronary arteries. Abrupt occlusion of a coronary artery or one of its branches in a human or dog leads to ischemic necrosis and eventual fibrosis of the areas of myocardium supplied by the occluded vessel. However, if narrowing of a coronary artery occurs slowly and progressively over a period of days, weeks, or longer, collateral vessels develop and may furnish sufficient blood to the ischemic myocardium to prevent, or reduce the extent of, the necrosis. The development of collateral coronary vessels has been extensively studied in dogs, and the clinical picture of coronary atherosclerosis, as it occurs in humans, can be simulated by gradual narrowing of the normal dog's coronary arteries. Collateral vessels develop between branches of occluded and nonoccluded arteries. Recent work indicates that they originate from pre-existing small vessels that undergo proliferative changes of the endothelium and smooth muscle, possibly in response to wall stress and chemical agents released by the ischemic tissue.

Numerous surgical attempts have been made to enhance the development of coronary collateral vessels. However most of the techniques used do not increase the collateral circulation over and above that produced by coronary artery narrowing alone. When discrete occlusions occur in coronary arteries (even vessels as small as 1 mm. in diameter), the lesions can be bypassed with a vein graft connecting the aorta to a point on the coronary artery distal to the site of the occlusion.

A number of drugs are available that induce coronary vasodilation, and they are used in patients with coronary artery disease to relieve *angina pectoris*, the chest pain associated with myocardial ischemia. Many of these compounds are nitrites. They are not selective dilators of the coronary vessels, and the mechanism whereby they accomplish their beneficial effects has not been established. The arterioles that would dilate in response to the drugs are undoubtedly already maximally dilated by the ischemia responsible for the symptoms. It has been suggested that the relief of angina pectoris by nitrites is brought about by a reduction in cardiac work and myocardial oxygen requirement caused by the moderate hypotension these drugs produce. In short, the reduction in pressure work and oxygen requirement must be greater than the reduction in coronary blood flow and oxygen supply consequent to the lowered coronary perfusion pressure. It has also been demonstrated that nitrites dilate large coronary arteries and coronary collateral vessels, thus increasing blood flow to ischemic myocardium and alleviating precordial pain.

CARDIAC OXYGEN CONSUMPTION AND WORK

The volume of oxygen consumed by the heart is determined by the amount and the type of activity the heart performs. Under basal conditions, myocardial oxygen consumption is about 8 to 10 ml./min./100 gm. of heart. It can increase severalfold with exercise and decrease moderately under conditions such as hypotension and hypothermia. The cardiac venous blood is normally quite low in oxygen (about 5 vol.%), and the myocardium can receive little additional oxygen by further oxygen extraction from the coronary blood, a situation that also exists in contracting skeletal muscle. Therefore increased oxygen demands of the heart must be met primarily by an increase in coronary blood flow. When the heart beat is arrested, as with administration of potassium, but coronary perfusion is maintained experimentally, the oxygen consumption falls to 2 ml./min./100 gm. or

less, which is still six to seven times greater than that for resting skeletal muscle.

Left ventricular work per beat (*stroke work*, see p. 94) is generally considered to be equal to the product of the stroke volume and the mean aortic pressure against which the blood is ejected by the left ventricle. At resting levels of cardiac output the kinetic energy component is negligible (p. 53). However, at high cardiac outputs as in severe exercise, the kinetic component can account for up to 50% of total cardiac work. One can simultaneously halve the aortic pressure and double the cardiac output, or vice versa, and still arrive at the same value for cardiac work. However the oxygen requirements are greater for any given amount of cardiac work when a major fraction is so-called pressure work as opposed to volume work. An increase in cardiac output at a constant aortic pressure (volume work) is accomplished with a small increase in left ventricular oxygen consumption, whereas increased arterial pressure at constant cardiac output (pressure work) is accompanied by a large increment of myocardial oxygen consumption. Thus myocardial oxygen consumption may not correlate well with overall cardiac work. The magnitude and duration of left ventricular pressure do correlate with left ventricular oxygen consumption. The area under the systolic portion of the left ventricular pressure curve has been termed the *tension-time index*, and this index correlates reasonably well with myocardial oxygen consumption in many different physiological states. Since this index does not take into consideration the velocity of contraction, and since the velocity factor influences myocardial oxygen consumption, there are conditions (for example, exercise, sympathetic activation, epinephrine administration) in which the tension-time index fails to reflect accurately the oxygen consumption. The tension development (ventricular wall tension), the velocity of shortening, and, to a lesser extent, the degree of shortening of the myocardial fibers constitute the chief determinants of myocardial oxygen consumption. The greater energy demand of pressure work over volume work can be readily demonstrated in the heart-lung preparation and may be associated with a decrease in myocardial creatine phosphate levels. It is also of great clinical importance, especially in aortic stenosis, in which left ventricular oxygen consumption is increased because of the high intraventricular pressures developed during systole, but coronary perfusion pressure is normal or reduced because of the pressure drop across the diseased aortic valve. Since mean pulmonary artery pressure is about one-seventh that of aortic pressure, and since the outputs of the two ventricles are equal, work of the right ventricle is one-seventh that of the left ventricle.

CARDIAC EFFICIENCY

As with an engine, the efficiency of the heart is the ratio of the work accomplished to the total energy utilized. Assuming an average oxygen consumption of 9 ml./min./100 gm. for the two ventricles, a 300-gm. heart consumes 27 ml. O_2/min., which is equivalent to 130 small calories at a respiratory quotient of 0.82. Together the two ventricles do about 8 kg.-m of work per minute, which is equivalent to 18.7 small calories. Therefore the gross efficiency is 14% $\left(\dfrac{18.7}{130} \times 100 = 14\% \right)$. The net efficiency is slightly higher (18%) and is obtained by subtracting the oxygen consumption of the non-beating (asystolic) heart (about 2 ml./min./100 gm.) from the total cardiac oxygen consumption in the calculation of efficiency. It is thus evident that the efficiency of the heart as a pump is relatively low and is comparable to the efficiency of many mechanical devices used in everyday life. With exercise, efficiency improves, since mean blood pressure shows little change, whereas cardiac output and work increase considerably without a proportional increase in myocardial oxygen consumption. The energy expended in cardiac metabolism that does not contribute to the propulsion of blood through the body appears in the form of heat. The energy of the flowing blood is also dissipated as heat, chiefly in passage though the arterioles.

SUBSTRATE UTILIZATION

The heart is quite versatile in its use of substrates, and within certain limits the uptake of a particular substrate is directly proportional to its arterial concentration. The utilization of one substrate is also influenced by the presence or absence of other substrates. For example, the addition of lactate to the blood perfusing a heart metabolizing glucose leads to a reduction in glucose uptake, and vice versa. At normal blood concentrations, glucose and lactate are consumed at about equal rates, whereas pyruvate uptake is very low, but so is its arterial concentration. For glucose the threshold concentration is about 4 mM. and below this blood level no myocardial glucose uptake occurs. Insulin reduces this threshold and increases the rate of glucose uptake by the heart. A very low threshold exists for cardiac utilization of lactate; insulin does not affect its uptake by the myocardium. With hypoxia, glucose utilization is facilitated by an increase in the rate of transport across the myocardial cell wall, whereas lactate cannot be metabolized by the hypoxic heart and is in fact produced by the heart under anaerobic conditions. Associated with lactate production by the hypoxic heart is the breakdown of cardiac glycogen.

Of the total cardiac oxygen consumption, only 35% to 40% can be accounted for by the oxidation of carbohydrate. Thus the heart derives the major part of its energy from oxidation of noncarbohydrate sources. The chief noncarbohydrate fuel used by the heart is esterified and nonesterified fatty acid, which accounts for about 60% of myocardial oxygen consumption in the postabsorptive state. The various fatty acids show different thresholds for myocardial uptake but are generally utilized in direct proportion to their arterial concentration. Ketone bodies, especially acetoacetate, are readily oxidized by the heart and contribute a major source of energy in diabetic acidosis. As is true of carbohydrate substrates, utilization of a specific noncarbohydrate substrate is influenced by the presence of other substrates, both noncarbohydrate and carbohydrate. Therefore, within certain limits, the heart uses preferentially that substrate that is available in the largest concentration. Most evidence indicates that the contribution to myocardial energy expenditure provided by the oxidation of amino acids is small.

BIBLIOGRAPHY
Journal articles

Belloni, F. L.: The local control of coronary blood flow, Cardiovasc. Res. 13:63, 1979.

Berne, R. M.: Regulation of coronary blood flow, Physiol. Rev. 44:1, 1964.

Berne, R. M., De Geest, H., and Levy, M. N.: Influence of cardiac nerves on coronary resistance, Am. J. Physiol. 208:763, 1965.

Feigl, E. O.: Parasympathetic control of coronary blood flow in dogs, Circ. Res. 25:509, 1969.

Feigl, E. O.: Sympathetic control of coronary circulation, Circ. Res. 20:262, 1967.

Fox, A. C., Reed, G. E., Glassman, E., Kaltman, A. J., and Silk, B. B.: Release of adenosine from human hearts during angina induced by rapid atrial pacing, J. Clin. Invest. 53:1447, 1974.

Gregg, D. E.: The natural history of coronary collateral development, Circ. Res. 35:335, 1974.

Mommaerts, W. F. H. M.: Current problems in myocardial metabolism, Circ. Res. 34(Supp. III):2, 1974.

Murray, P. A., Belloni, F. L., and Sparks, H. V.: The role of potassium in the metabolic control of coronary vascular resistance in the dog, Circ. Res. 44:767, 1979.

Olsson, R. A., and Patterson, R. E.: Adenosine as a physiological regulator of coronary blood flow, Prog. Molec. Subcel. Biol. 4:227, 1976.

Pitt, B., Elliot, E. C., and Gregg, D. E.: Adrenergic receptor activity in the coronary arteries of the unanesthetized dog, Circ. Res. 21:75, 1967.

Rubio, R., and Berne, R. M.: Regulation of coronary blood flow, Progr. Cardiovasc. Dis. 18:105, 1975.

Thompson, C. I., Rubio, R., and Berne, R. M.: Changes in adenosine and glycogen phosphorylase activity during the cardiac cycle, Am. J. Physiol. 238:H389, 1980.

Watkinson, W. P., Foley, D. H., Rubio, R., and Berne, R. M.: Myocardial adenosine formation with increased cardiac performance in the dog, Am. J. Physiol. 236:H13, 1979.

Wearn, J. T., Mettier, S. R., Klumpp, T. G., and Zschiesche, L. J.: The nature of the vascular communications between the coronary arteries and the chambers of the heart, Am. Heart J. 9:143, 1933.

Books and monographs

Berne, R. M.: The coronary circulation. In Langer, G. A., and Brady, A. J., editors: The mammalian myocardium, New York, 1974, John Wiley & Sons, Inc., p. 251.

Berne, R. M., and Rubio, R.: Coronary circulation. In Handbook of physiology; Section 2: The cardiovascular system—the heart, Bethesda, MD, 1979, American Physiological Society, vol. I, pp. 873-952.

Buckberg, G. D., and Kattus, A. A., Jr.: Factors determining the distribution and adequacy of left ventricular myocardial blood flow. In Bloor, C. M., and Olsson, R. A., editors: Current topics in coronary research, New York, 1973, Plenum Press, pp. 95-113.

Gregg, D. E.: Coronary circulation in health and disease, Philadelphia, 1950, Lea & Febiger.

Morgan, H. E., Rannels, D. E., and McKee, E. E.: Protein metabolism of the heart. In Handbook of physiology; Section 2: The cardiovascular system—the heart, Bethesda, MD, 1979, American Physiological Society, vol. I, pp. 845-871.

Randle, P. J., and Tubbs, P. K.: Carbohydrate and fatty acid metabolism. In Handbook of physiology; Section 2: The cardiovascular system—the heart, Bethesda, MD, 1979, American Physiological Society, vol. I., pp. 805-844.

Schaper, W.: The collateral circulation of the heart, New York, 1971, North-Holland Publishing Co.

CHAPTER 11

SPECIAL CIRCULATIONS

CUTANEOUS CIRCULATION

The oxygen and nutrient requirements of the skin are relatively small and, in contrast to most other body tissues, the supply of these essential materials is not the chief governing factor in the regulation of cutaneous blood flow. The primary function of the cutaneous circulation is the maintenance of a constant body temperature. Consequently the skin shows wide fluctuations in blood flow, depending on the need for loss or conservation of body heat; mechanisms responsible for alterations in skin blood flow are primarily activated by changes in ambient and internal body temperatures.

Regulation of skin blood flow

There are essentially two types of resistance vessels in skin—arterioles, similar to those found elsewhere in the body, and *arteriovenous anastomoses*, which shunt blood from arterioles to venules and venous plexuses, and therefore bypass the capillary bed. The former are distributed over most of the body surface, whereas the latter are found primarily in the fingertips, palms of the hands, toes, soles of the feet, ears, nose, and lips. The arteriovenous anastomoses differ morphologically from the arterioles in that they are either short, straight or long, coiled vessels about 20 to 40 μ in lumen diameter, with thick muscular walls richly supplied with nerve fibers. These vessels are almost exclusively under sympathetic neural control and become maximally dilated when their nerve supply is interrupted. Conversely, reflex

stimulation of the sympathetic fibers to these vessels may produce constriction to the point of complete obliteration of the vascular lumen. Although the arteriovenous anastomoses do not exhibit *basal tone* (tonic activity of the vascular smooth muscle independent of innervation), they are nevertheless very sensitive to vasoconstrictor agents like epinephrine and norepinephrine. In fact, the absence of basal tone in these vessels and their greater sensitivity to catecholamines has been put forth as evidence against the concept that the basal tone displayed by all other resistance vessels is caused by circulating catecholamines. Furthermore, the arteriovenous anastomoses do not appear to be under metabolic control, and they fail to show reactive hyperemia or autoregulation of blood flow. Thus the regulation of blood flow though these anastomotic channels is governed principally by the nervous system in response to reflex activation by temperature receptors or from higher centers of the central nervous system.

The bulk of the skin resistance vessels exhibits some basal tone and is under dual control of the sympathetic nervous system and local regulatory factors, in much the same manner as are other vascular beds. However, in the case of skin, neural control plays a more important role than local factors. Stimulation of sympathetic nerve fibers to skin blood vessels (arteries and veins as well as arterioles) induces vasoconstriction, and severance of the sympathetic nerves, vasodilation. With chronic denervation of the cutaneous blood vessels, the degree of tone that existed prior to denervation is gradually regained over a period of several weeks. This is accomplished by an enhancement of basal tone that compensates for the degree of tone previously contributed by sympathetic nerve fiber activity. Epinephrine and norepinephrine elicit only vasoconstriction in cutaneous vessels. Whether the increased basal tone following denervation of the skin vessels is the result of their enhanced sensitivity to circulating catecholamines *(denervation hypersensitivity)* has not been established.

Parasympathetic vasodilator nerve fibers do not supply the cutaneous blood vessels. However, stimulation of the sweat glands, which are innervated by cholinergic fibers of the sympathetic nervous system, results in dilation of the skin resistance vessels. Sweat contains an enzyme that acts on a protein moiety in the tissue fluid to produce *bradykinin*, a polypeptide with potent vasodilator properties. It is thought that the bradykinin formed in the tissue acts locally to dilate the arterioles and increase blood flow to the skin. This polypeptide has also been found in the secretion of several other glandular structures and may be the mediator of the vasodilation associated with increased metabolic activity of the glands. Finally, the skin vessels of certain regions, particularly the head, neck, shoulders, and upper chest, are under the influence of the higher centers of the central nervous system. Blushing, as with embarassment or anger, and blanching, as with fear or anxiety, are examples of cerebral inhibition and stimulation, respectively, of the sympathetic nerve fibers to the affected regions. Whether blushing is caused in part by activation of the sweat glands and the local formation of bradykinin remains to be proved.

In contrast to the arteriovenous anastomoses in the skin, the cutaneous resistance vessels show autoregulation of blood flow and reactive hyperemia. If the arterial inflow to a limb is stopped with an inflated blood pressure cuff for a brief period of time, the skin shows a marked reddening below the point of vascular occlusion when the cuff is deflated. This increased cutaneous blood flow (reactive hyperemia) is also manifested by the distension of the superficial veins in the erythematous extremity. With respect to autoregulation of blood flow in the skin, the relatively low metabolic activity of the tissue favors a myogenic rather than a metabolic mechanism (p. 125).

Ambient and body temperature in regulation of skin blood flow. Since the primary function of the skin is to preserve the internal milieu and protect it from adverse changes in the environment and since ambient temperature is one of the most important external variables the body must contend with, it is not surprising that the vasculature of the skin is chiefly influenced by environmental temperature. Exposure to cold elicits a generalized cutaneous vasoconstriction that is most pronounced in the hands and feet. That this response is chiefly mediated by the nervous system is evident from the fact that when the circulation to a hand is arrested by a pressure cuff, immersion of that hand in cold water results in vasoconstriction in the skin of the other extremities that are exposed to room temperature. With the circulation to the chilled hand unoccluded, the reflex vasoconstriction is caused in part by the cooled blood returning to the general circulation and stimulating the temperature-regulating center in the anterior hypothalamus. Direct application of cold to this region of the brain produces cutaneous vasoconstriction.

The skin vessels of the cooled hand also show a direct response to cold. Moderate cooling or exposure for brief periods to severe cold (0° to 15° C.) results in constriction of the resistance and capacitance vessels including A-V anastomoses. However, prolonged exposure of the hand to severe cold has a secondary vasodilator effect. Prompt vasoconstriction and severe pain are elicited by immersion of the hand in water near 0° C., but these are soon followed by dilation of the skin vessels with reddening of the immersed part and alleviation of the pain. With continued immersion of the hand, alternating periods of constriction and dilation occur, but the skin temperature rarely drops to as low a degree as it did with the initial vasoconstriction. Prolonged severe cold, of course, results in tissue damage. The rosy faces of people working or playing in a cold environment are examples of cold vasodilation. However, the blood flow through the skin of the face may be greatly reduced despite the flushed appearance. The red color of the slowly flowing blood is in large measure the result of the reduced oxygen uptake by the cold skin and the cold-induced shift to the left of the oxyhemoglobin dissociation curve.

Direct application of heat produces not only local vasodilation of resistance and capacitance vessels as well as A-V anastomoses but also reflex dilation in other parts of the body. The local effect is independent of the vascular nerve supply, whereas the reflex vasodilation is a combination of anterior hypothalamic stimulation by the returning warmed blood and of stimulation of receptors in the heated part. However, evidence for a reflex from peripheral temperature receptors is not as definitive for warm stimulation as it is for cold stimulation.

With exposure of the extremities to cold or heat, it must be remembered that the close proximity of the major arteries and veins to each other permits considerable heat exchange (countercurrent) between artery and vein. If cold blood flows from a cooled hand toward the heart, heat is taken up by the blood in the arm veins from that in the adjacent arteries, resulting in warming of the venous blood and cooling of the arterial blood. Heat exchange is of course in the opposite direction with exposure of the extremity to heat. Thus heat conservation is enhanced and heat gain is minimized during exposure of extremities to cold and

warm environments, respectively, and the temperature of the blood returning to the core of the body is brought closer to the existing core temperature.

Skin color and special reactions of the skin vessels

The color of the skin is of course caused in large part by pigment; but in all but very dark skin, the degree of pallor or ruddiness is primarily a function of the amount of blood in the skin. With little blood in the venous plexuses the skin appears pale, whereas with moderate to large quantities of blood in the venous plexuses, the skin shows color. Whether this color is bright red, blue, or some shade between is determined by the degree of oxygenation of the blood in the subcutaneous vessels. For example, a combination of vasoconstriction and reduced hemoglobin can produce an ashen gray color of the skin, whereas a combination of venous engorgement and reduced hemoglobin can result in a dark purple hue. The skin color provides little information about the rate of cutaneous blood flow. There may coexist rapid blood flow and pale skin when the arteriovenous anastomoses are open, and slow blood flow and red skin when the extremity is exposed to cold.

White reaction and triple response. If the skin of the forearm of many individuals is lightly stroked with a blunt instrument, a white line appears at the site of the stroke within 20 seconds. The blanching becomes maximal in about 30 to 40 seconds and then gradually disappears within 3 to 5 minutes. This response is known as a *white reaction* and has been attributed by Sir Thomas Lewis to capillary contraction, since it occurs in the denervated limb and is unaffected by arrest of the limb circulation. Since all direct evidence indicates that capillaries do not contract and since skin color is primarily a result of blood content of venous plexuses, venules, and small veins, it seems logical to attribute the white reaction to venous constriction induced by mechanical stimulation.

If the skin is stroked more strongly with a sharp pointed instrument, a *triple response* is elicited. Within 3 to 15 seconds a thin *red line* appears at the site of the stroke, followed in about 15 to 30 seconds by a red blush or *flare* extending out 1 to 2 cm. from either side of the red line. This in turn is followed in 3 to 5 minutes by an elevation of the skin along the red line, with gradual fading of the red line as the elevation, a *wheal*, becomes more prominent. The red line is probably caused by dilation of the vessels because of mechanical stimulation. The flare, however, is the result of dilation of neighboring arterioles caused to relax by an *axon reflex* originating at the site of mechanical stimulation. In an axon reflex, the nerve impulse travels centripetally in the cutaneous sensory nerve fiber and then antidromically down the small branches of the afferent nerve

to adjacent arterioles to elicit vasodilation (Fig. 11-1). The flare is not affected by acute section or anesthetic block of the sensory nerve central to the point of branching, whereas it is abolished when the nerve degenerates after section. The wheal is caused by increased capillary permeability induced by the trauma. Fluid containing protein leaks out of the capillaries locally and produces edema at the site of injury. Since the triple response can be elicited by an intradermal injection of histamine, Lewis attributed the response to histamine or a histamine-like substance *(H-substance)*. Whether it is histamine, ATP, a vasoactive polypeptide like bradykinin, or some yet unidentified substance remains to be determined. In summary, slight trauma is believed to release a substance that (1) produces the red line, (2) stimulates the local sensory nerve endings, and (3) increases capillary permeability.

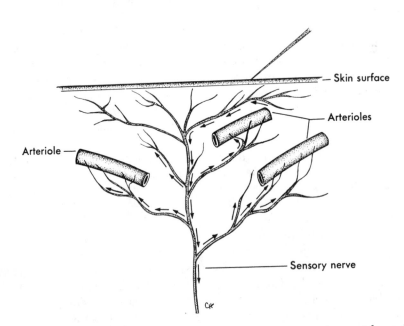

Fig. 11-1. Schematic representation of the axon reflex in response to a scratch on the skin surface with a sharp instrument. Arrows indicate the pathways of impulses in a sensory nerve from the site of stimulation to adjacent arterioles to produce local vasodilation (flare).

SKELETAL MUSCLE CIRCULATION

The rate of blood flow in skeletal muscle varies directly with the contractile activity of the tissue and the type of muscle; blood flow in red (slow-twitch, high-oxidative) muscle is greater than it is in white (fast-twitch, low-oxidative) muscle. In resting muscle the precapillary arterioles exhibit asynchronous intermittent contractions and relaxations, so that at any given moment in time a very large percentage of the capillary bed is not perfused. Consequently total blood flow through quiescent skeletal muscle is low (1.4 to 4.5 ml./min./100 g.). With exercise the resistance vessels relax and the muscle blood flow may increase manyfold (up to fifteen to twenty times the resting level), the magnitude of the increase depending largely on the severity of the exercise.

Regulation of skeletal muscle blood flow

Control of muscle circulation is achieved by neural and local factors; the relative contribution of these factors is dictated by the state of activity of the muscle. At rest neural and myogenic regulations are predominant, whereas in exercise metabolic control supervenes. As with all tissues, physical factors such as arterial pressure, tissue pressure, and blood viscosity influence muscle blood flow. However, another physical factor comes into play during exercise—the squeezing effect of the active muscle on the vessels. With intermittent contractions, inflow is restricted and venous outflow is enhanced during each brief contraction. The presence of the venous valves prevents backflow of blood in the veins between contractions, thereby aiding in the forward propulsion of the blood. With strong sustained contractions the vascular bed can be compressed to the point where blood flow actually ceases temporarily.

Neural factors. Although the resistance vessels of muscle possess a high degree of basal tone, they also display tone attributable to continuous low frequency activity in the sympathetic vasoconstrictor nerve fibers. Evidence for this sympathetic tone is the increase in blood flow observed with local anesthetic block of the sympathetic fibers.

The basal frequency of firing in the sympathetic vasoconstrictor fibers is quite low (about 1 to 2 per second), and maximal vasoconstriction is observed at frequencies as low as 8 to 10 per second. Stimulation of the sympathetic nerve fibers to skeletal muscle elicits vasoconstriction that is caused by the release of norepinephrine at the fiber endings. Intra-arterial injection of norepinephrine elicits only vasoconstriction whereas low doses of epinephrine produce vasodilation in muscle and large doses cause vasoconstriction. The vasodilator effect of epinephrine is believed to be the result of stimulation of beta adrenergic receptors on the resistance vessels or possibly to stimulation of muscle metabolism by epinephrine. However, attempts to identify the mediator of such a metabolic response have not met with success. Evidence that lactic acid serves as the mediator is controversial.

The tonic activity of the sympathetic nerves is greatly influenced by reflexes from the baroreceptors. An increase in carotid sinus pressure results in dilation of the vascular bed of the muscle, whereas a decrease in carotid sinus pressure elicits vasoconstriction (Fig. 11-2). When the existing sympathetic constrictor tone is high, as in the experiment illustrated in Fig. 11-2, the decrease in blood flow associated with common carotid artery occlusion is small, but the increase following the release of occlusion is large. The vasodilation produced by baroreceptor stimulation is caused only in part by inhibition of sympathetic vasoconstrictor activity. Usually a biphasic dilator response is obtained. A large initial transient dilation is attributed to active vasoconstrictor tone. There is some evidence to indicate that the active vasodilation is caused by histamine release from sympathetic fibers innervating the resistance vessels. This dilator effect is abolished by tissue depletion of

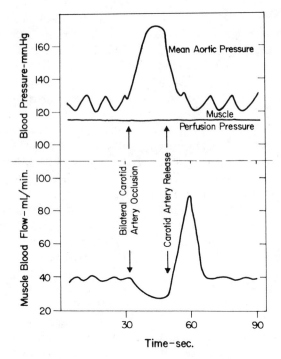

Fig. 11-2. Evidence for participation of the muscle vascular bed in vasoconstriction and vasodilation mediated by the carotid sinus baroreceptors after common carotid artery occlusion and release. In this preparation the sciatic and femoral nerves constituted the only direct connection between the hind leg muscle mass and the rest of the dog. The muscle was perfused by blood at a constant pressure that was completely independent of the animal's arterial pressure. (Redrawn from Jones, R. D., and Berne, R. M.: Am. J. Physiol. **204**:461, 1963.)

histamine or by histamine antagonists. Furthermore, histamine release into venous blood during this dilator response has been demonstrated with the aid of isotope-labeled tissue-bound histamine. Since muscle is the major body component on the basis of mass, and thereby represents the largest vascular bed, the participation of its resistance vessels in vascular reflexes plays an important role in the maintenance of a constant arterial blood pressure.

Reference has previously been made to the sympathetic cholinergic vasodilator pathway from the cortex and hypothalamus to the mus-

cle resistance vessels (p. 134). These nerve fibers are believed to induce vasodilation of the muscle vessels in anticipation of exercise; this is analogous to stepping on the accelerator prior to releasing the clutch. Whether these nerve fibers in truth serve this function and whether they exist in humans has not been established. There is some evidence that the increased muscle blood flow observed in fainting or with acute emotional upsets is in part mediated by these cholinergic fibers.

Participation of the sympathetic cholinergic vasodilator fibers in the active vasodilation induced by carotid sinus stimulation is minimal, since atropine administration reduces only slightly the degree of vasodilation obtained. From studies on the clearance of ^{131}I injected into muscle, it appears that the vasodilation produced by stimulation of sympathetic cholinergic fibers is limited to the *thoroughfare or nonnutrient* channels, since increases in blood flow through the muscle are not accompanied by increases in the rate of ^{131}I removal. Hence, activation of the sympathetic dilator nerves permits more blood to flow through open channels, thereby creating a functional shunt of blood through the muscle and bypassing the inactive capillary beds. Other procedures that increase muscle blood flow (for example, injection of dilator agents, inhibition of sympathetic vasoconstrictor tone, or increased metabolic activity of the muscle) affect nutrient blood flow (through capillary beds), since they produce proportionate increases in ^{131}I clearance.

Finally there has recently been described a slowly evoked, sustained, active vasodilation obtained with sympathetic nerve stimulation when the masking effect of the sympathetic vasoconstrictors is pharmacologically blocked. The neurohumor released by these nerve fibers is neither acetylcholine nor histamine. Its identity and the physiological role of these nerve fibers must await further investigation.

A comparison of the vasoconstrictor and vasodilator effects of the sympathetic nerves to blood vessels of muscle and skin is summarized in diagrammatic form in Fig. 11-3. Note the lower basal tone of the skin vessels, their greater constrictor response, and the absence of active cutaneous vasodilation.

Local factors. It has already been stressed that neural regulation of muscle blood flow is superseded by metabolic regulation when the muscle changes from the resting to the contracting state. However, local control is also demonstrable in innervated resting skeletal muscle of the dog's hind leg when reflex stimulation of the vasomotor nerves is minimal and local thermal or mechanical stimulation of the muscle is eliminated. Such preparations show autoregulation of blood flow (a prime example of local control) to the same degree as do actively contracting muscle and denervated muscle. If, however, vascular reflexes are activated when the vasomotor nerves to the muscle are intact or if the muscle receives mechanical or thermal stimulation (with or without the nerves intact), blood flow increases and autoregulation is either lost or diminished (Fig. 11-4). In resting muscle an autoregulating preparation is characterized by low blood flow and low venous blood oxygen saturation, whereas in contracting muscle venous blood oxygen saturation is low but the flow is high. Thus the common denominator for autoregulation of blood flow is a low venous blood oxygen level and presumably a low muscle oxygen tension. Direct measurements of Po_2 in resting skeletal muscle cells yield a range of values from 0 to 10 mm. Hg with a mean Po_2 of 4 mm. Hg. Studies on humans also reveal that at rest blood flow through muscle is low and oxygen saturation of blood draining the muscles is frequently less than 50%. These observations indicate that at rest muscle blood flow is under either local or neural control, depending on the number and the nature of the stimuli impinging on the vasomotor regions of the medulla. During exercise local metabolic factors take over blood flow regulation, regardless of the degree of sympathetic nerve activity. This concept is summarized in Fig. 11-5.

Despite the tremendous scientific effort that has gone into attempts to clarify the nature of local regulation of muscle blood flow, the mechanism still remains obscure. Some of the current hypotheses about autoregulation of blood flow and about the identity of the mediator of metabolic control of the resistance vessels are considered in Chapter 7.

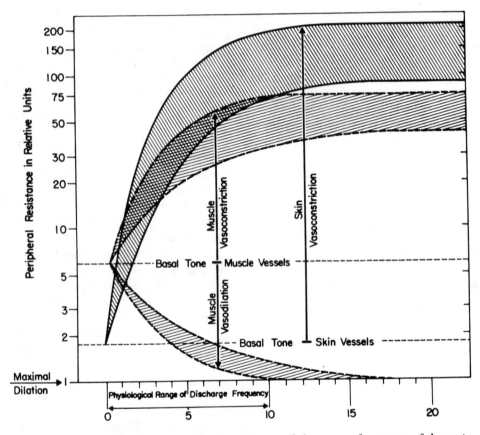

Fig. 11-3. Diagrammatic representation of basal tone and the range of response of the resistance vessels in muscle and skin to sympathetic nerve stimulation. Peripheral resistance plotted on a logarithmic scale. (Redrawn from Celander, O., and Folkow, B.: Acta Physiol. Scand. **29:**241, 1953.)

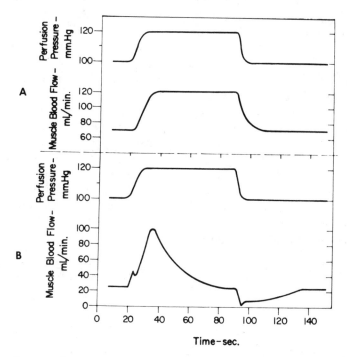

Fig. 11-4. A, Absence of autoregulation of blood flow. Skeletal muscle blood flow follows changes in perfusion pressure. **B,** Autoregulation of skeletal muscle blood flow. Blood flow returns to control level despite maintenance of elevated perfusion pressure.

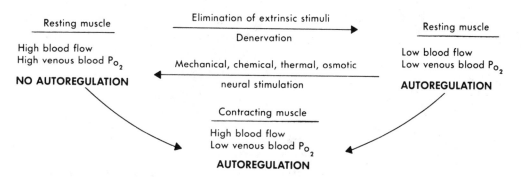

Fig. 11-5. Schema depicting the characteristics of autoregulating and nonautoregulating skeletal muscle vascular beds and the factors responsible for change from one state to the other.

CEREBRAL CIRCULATION

Blood reaches the brain through the internal carotid and the vertebral arteries. The latter join to form the basilar artery, which, in conjunction with branches of the internal carotid arteries, forms the *circle of Willis*. A unique feature of the cerebral circulation is that it all lies within a rigid structure, the cranium. Since intracranial contents are incompressible, any increase in arterial inflow, as with arteriolar dilation, must be associated with a comparable increase in venous outflow. The volume of blood and of extravascular fluid can show considerable variations in most tissues, whereas in brain the volume of blood and extravascular fluid is relatively constant; changes in either of these fluid volumes must be accompanied by a reciprocal change in the other. In contrast to most other organs, the rate of total cerebral blood flow is held within a relatively narrow range and in humans averages 55 ml./min./100 gm. of brain.

Estimation of cerebral blood flow

Total cerebral blood flow is usually measured in humans by the nitrous oxide (N_2O) method, which is based on the Fick principle (p. 182). The subject breathes a gas mixture of 15% N_2O, 21% O_2, and 64% N_2 for a period of 10 minutes, which is sufficient time to permit equilibration of the N_2O between the brain tissue and the blood leaving the brain. Simultaneous samples of arterial (any artery) blood and mixed cerebral venous (internal jugular vein) blood are taken at the start of N_2O inhalation and at 1-minute intervals throughout the 10-minute period of N_2O administration. From these data the cerebral blood flow can be calculated by the Fick equation.

Cerebral blood flow =

$$\frac{\text{Amount of } N_2O \text{ taken up by brain during time } (t_2 - t_1)}{\text{A-V difference of } N_2O \text{ across brain during time } (t_2 - t_1)}$$

Since the arterial and venous concentrations are continuously changing with time, it is nec-

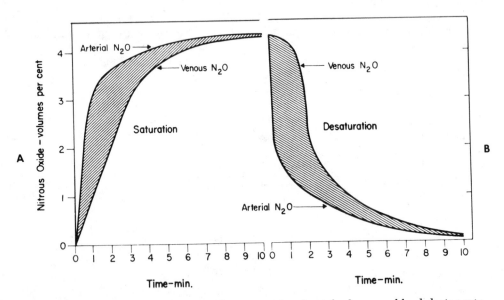

Fig. 11-6. Concentrations of nitrous oxide in arterial and cerebral venous blood during saturation, **A**, and desaturation, **B**. The shaded areas represent the arteriovenous differences of nitrous oxide during the 10-minute period of N_2O inhalation and the 10 minutes after discontinuance of the N_2O administration.

essary to get the true A-V difference during the period of N_2O inhalation by integration of the A-V difference over the 10-minute period. This is represented in Fig. 11-6, A, by the shaded area between the arterial and venous N_2O concentration curves constructed from the blood concentrations observed at successive 1-minute intervals during N_2O administration. The amount of N_2O removed by the brain as well as the concentration of N_2O in the brain tissue are unknown. Since the partition coefficient between brain and blood is about 1 and since equilibrium of N_2O between the brain and the blood leaving the brain is reached by the end of 10 minutes, the concentration of N_2O in the brain tissue closely approximates that of the cerebral venous blood in the 10-minute sample. The total weight of the brain is not known, and so for convenience the concentration in brain tissue is multiplied by 100 to express the cerebral blood flow (CBF) in ml./min./100 gm. of brain tissue. The equation is:

$$CBF = \frac{V_{10} \cdot S \cdot 100}{\int_0^{10} (A\text{-}V)dt}$$

V_{10} = Venous concentration of N_2O at equilibrium (at 10 min.)

S = Partition coefficient of N_2O between brain and blood = 1

A-V = Arteriovenous difference of N_2O

Cerebral blood flow can also be calculated from the desaturation A-V curves, which are constructed from N_2O concentrations of simultaneously drawn arterial and venous blood samples taken each minute for 10 minutes, starting when equilibrium is reached between brain tissue and cerebral venous blood (Fig. 11-6, B). In this procedure the subject breathes the N_2O mixture for 10 minutes, and sampling starts at the moment N_2O inhalation is stopped. The only difference from the preceding equation is that the denominator becomes

$$\int_{10}^{20} (V\text{-}A)dt$$

The nitrous oxide method is also used for determination of coronary blood flow in humans. The procedure is the same except that the venous blood samples are taken from the coronary sinus through a cardiac catheter. Since the coronary sinus drains the left ventricle, blood flow is expressed as ml./min./100 gm. of left ventricle (Chapter 10).

Cerebral blood flow and its distribution to different areas of the brain can be measured in animals by injection into the internal carotid artery of microspheres (about 15 μ) labeled with radioactive substances. The microspheres become lodged in the arterioles and capillaries, the brain tissue is sampled, and the radioactivity of the tissue is determined. Blood flow to each tissue sample is proportional to the radioactivity in that sample. By the use of microspheres labeled with different radioactive isotopes, several measurements of cerebral blood flow can be made with a gamma counter that can measure each isotope independently of the other isotopes in the sample. This method is also used for measurements of blood flow in other tissues, such as the myocardium. One can also measure cerebral blood flow in animals with the use of ^{14}C-antipyrine, which is taken up by the brain in proportion to the blood flow. The brain is then sliced and the radioactivity of the slice is determined by radioautography. The advantage of these methods over the N_2O method or the direct measurement of venous outflow from the brain is that blood flow to different regions of the brain (*regional blood flow*) can be determined. The obvious disadvantage is that the animals must be sacrificed to obtain the samples of brain tissue.

Recently, the development of multiple collimated scintillation detectors built into a helmet that fits over the cranium has made possible the measurement of regional blood flow (cortical blood flow) in animals and humans. An inert radioactive gas (such as [133]xenon) is injected into an internal carotid artery, and from its rate of washout from the brain, regional cerebral blood flow can be determined. The radioactive gas may also be given by inhalation, but more sophisticated techniques are required to eliminate noncerebral blood flow and to distinguish between blood flow to cortical (gray matter) and deep cerebral (white matter) tissue.

Regulation of cerebral blood flow

Of the various body tissues, the brain is the least tolerant of ischemia. Interruption of cerebral blood flow for as short a time as 5 seconds results in loss of consciousness, and ischemia lasting just a few minutes results in irreversible tissue damage. Fortunately, regulation of the cerebral circulation is primarily under direction of the brain itself. Local regulatory mechanisms and reflexes originating in the brain tend to maintain cerebral circulation relatively constant in the presence of possible adverse extrinsic effects such as sympathetic vasomotor nerve activity, circulating humoral vasoactive agents, and changes in arterial blood pressure. Under certain conditions the brain also regulates its blood flow by initiating changes in systemic blood pressure. For example, elevation of intracranial pressure results in an increase in systemic blood pressure. This response, first described by Cushing, is apparently caused by ischemic stimulation of vasomotor regions of the medulla. It aids in maintaining cerebral blood flow in such conditions as expanding intracranial tumors.

Neural factors. The cerebral vessels receive innervation from the cervical sympathetic nerve fibers that accompany the internal carotid and vertebral arteries into the cranial cavity. Until recently it was generally accepted that stimulation of the sympathetic nerves to the cerebral vessels elicited, at most, minimal vasoconstriction. However, the role of the sympathetic nerves in the regulation of the cerebral circulation has become quite controversial because of reports of severe constriction with either sympathetic nerve stimulation or the application of catecholamines to the pial arterioles and arteries. Despite these new findings of marked cerebral vasoconstriction with sympathetic nerve stimulation, the prevalent belief is that relative to other vascular beds sympathetic control of the cerebral vessels is weak and that the contractile state of the cerebral vascular smooth muscle is primarily dependent upon local metabolic factors. There are no known sympathetic vasodilator nerves to the cerebral vessels, but the vessels do receive parasympathetic fibers from the facial nerve, which produce a slight vasodilation on stimulation.

Local factors. Until quite recently it was thought that cerebral blood flow was relatively constant despite wide variations in cortical activity. However, with the advent of methods for measurement of regional cortical blood flow it became apparent that there is a tight coupling between regional metabolic activity and regional blood flow. For example, movement of one hand results in increased blood flow only in the projected hand area of the contralateral sensory-motor and premotor cortex, whereas stimulation of the retina with flashes of light increases blood flow only in the visual cortex. The reason this parallelism between cerebral metabolism and blood flow went unrecognized is that changes in the regional distribution of cerebral blood flow were masked when only total cerebral blood flow was measured. It has also been recently demonstrated that glucose uptake is closely coupled to regional cortical neuronal activity. For example, when the retina is stimulated by light, uptake of ^{14}C-2-deoxyglucose is enhanced in the visual cortex. This analogue of glucose is taken up and phosphorylated by cerebral neurons but cannot be metabolized further. The magnitude of its uptake is determined from radioautographs of slices of the brain. The mediator of the link between cerebral metabolism and blood flow has not been established, but there are currently three principal candidates—namely, pH, potassium, and adenosine.

It is well known that the cerebral vessels are very sensitive to carbon dioxide tension. Increases in arterial blood carbon dioxide tension (Pa_{CO_2}) elicit marked cerebral vasodilation; inhalation of 7% CO_2 results in a twofold increment in cerebral blood flow. By the same token decreases in Pa_{CO_2}, such as produced by hyperventilation, produce a decrease in cerebral blood flow. Carbon dioxide produces changes in arteriolar resistance by altering perivascular (and probably intracellular vascular smooth muscle) pH. By independently changing P_{CO_2} and bicarbonate concentration, it has been demonstrated that pial vessel diameter (and presumably blood flow) and pH are inversely related, regardless of the level of the P_{CO_2}. Carbon dioxide can diffuse to the vascular smooth muscle from the brain tissue or from the lumen of the vessels, whereas hydrogen ions in the blood are prevented from reaching the arteriolar smooth muscle by the "blood-brain barrier." Hence, the cerebral vessels dilate when the hydrogen ion concentration of the cerebrospinal fluid is increased but show only minimal dilation in response to an increase of the hydrogen ion concentration of the arterial blood. Despite the responsiveness of the cerebral vessels to pH changes, the precise role of hydrogen ions in the regulation of cerebral blood flow remains obscure. The initiation of increases in cerebral blood flow produced by seizures has been reported to be associated with transient increases rather than decreases in perivascular pH. Also, the intracellular and extracellular decreases in pH that occur with electrical stimulation of the brain or hypoxia often occur after cerebral blood flow has increased in response to the stimulus. Furthermore, with prolonged hypocapnia, cerebrospinal fluid pH may return to control levels

in the face of a persistent reduction in cerebral blood flow.

With respect to potassium ions, such stimuli as hypoxia, electrical stimulation of the brain, and seizures elicit rapid increases in cerebral blood flow and are associated with increases in perivascular K^+. The increments in K^+ are similar to those that produce pial arteriolar dilation when applied topically to these vessels. However, the increase in K^+ may not be sustained throughout the period of stimulation. Hence, only the initial increment in cerebral blood flow may be attributed to the release of K^+.

Adenosine levels of the brain increase with ischemia, hypoxemia, hypotension, hypocapnia, electrical stimulation of the brain, or induced seizures. When it is applied topically, adenosine is a potent dilator of the pial arterioles. In short, any intervention that either reduces the oxygen supply to the brain or increases the oxygen need of the brain results in rapid (within 5 sec.) formation of adenosine in the cerebral tissue. Unlike pH or K^+, the adenosine concentration of the brain increases with initiation of the stimulus and remains elevated throughout the period of oxygen imbalance. The adenosine released into the cerebrospinal fluid during conditions associated with inadequate brain oxygen supply is available to the brain tissue for reincorporation into cerebral tissue adenine nucleotides.

In all likelihood, all three factors,—pH, K^+, and adenosine—act in concert to adjust the cerebral blood flow to the metabolic activity of the brain but how these factors interact in accomplishing this regulation of cerebral blood flow remains to be elucidated.

The cerebral circulation shows reactive hyperemia and excellent autoregulation between pressures of about 60 and 160 mm. Hg. Mean arterial pressures below 60 mm. Hg result in reduced cerebral blood flow and syncope, whereas mean pressures above 160 may lead to increased permeability of the blood-brain barrier and cerebral edema. Autoregulation of cerebral blood flow is abolished by hypercapnia or any other potent vasodilator, and none of the candidates for metabolic regulation of cerebral blood flow has been shown to be responsible for this phenomenon. To what extent autoregulation of the cerebral vessels is attributable to a myogenic mechanism has not been established. However, the bulk of evidence favors a metabolic rather than a myogenic mechanism for autoregulation of cerebral blood flow.

PULMONARY CIRCULATION

The pulmonary and systemic vascular beds are in series with each other. Therefore, under steady state conditions, the total pulmonary and systemic blood flows are virtually identical. Despite this similarity in the rate of blood flow, however, there are enormous anatomical, hemodynamic, and physiological differences between these two sections of the cardiovascular system.

Functional anatomy

Pulmonary vasculature. The pulmonary vascular system is a low-resistance network of highly distensible vessels. The main pulmonary artery is much shorter than the aorta. The walls of the pulmonary artery and its branches are much thinner than the walls of the aorta, and they contain less smooth muscle and elastin. Contrary to systemic arterioles, which have very thick walls comprised mainly of circularly arranged smooth muscle, the pulmonary arterioles are very thin and contain little smooth muscle. The pulmonary arterioles certainly do not have the same capacity for vasoconstriction as do their counterparts in the systemic circulation. The pulmonary venules and veins are also very thin and possess little smooth muscle.

The pulmonary capillaries differ markedly from the systemic capillaries. Whereas the systemic capillaries are usually arranged as a network of tubular vessels with some interconnections, the pulmonary capillaries instead are sandwiched between adjacent capillaries in

such a manner that the blood flows in thin sheets. This provides for the maximum exposure of the capillary blood to the alveolar gases. The total surface area for exchange between alveoli and blood has been estimated to be about 50 to 70 square meters. Only thin layers of vascular and alveolar endothelium separate the blood and alveolar gas. The thickness of the sheets of blood between adjacent alveoli depends on the intravascular pressure and the intra-alveolar pressure. Ordinarily, the width of an interalveolar sheet of blood is about equal to the diameter of a red blood cell. During pulmonary vascular congestion, as when the left atrial pressure becomes elevated, the width of the sheet may increase several-fold. Conversely, when the local alveolar pressure exceeds the adjacent capillary pressure, the capillaries may collapse and there may be no blood flow to those alveoli. Hydrostatic factors play a crucial role in this phenomenon, particularly with respect to the distribution of blood flow to the various regions of the lungs, as described below.

Bronchial vasculature. The bronchial arteries are branches of the thoracic aorta. These arteries and their branches, down to and including the arterioles, have the structural characteristics of most systemic arteries, i.e., much thicker walls and more smooth muscles than the pulmonary arterial vessels of equivalent caliber. The bronchial vessels supply blood to the tracheobronchial tree, down to the level of the terminal bronchioles.

The bronchial veins drain partly into the pulmonary venous system and partly into the azygos veins, which are a part of the systemic venous system. There is some disagreement about the fraction of the bronchial blood flow that drains into each of these venous systems. The bronchial circulation normally constitutes about 1% of the cardiac output. Therefore, the fraction of the bronchial blood flow that returns to the left rather than to the right atrium constitutes at most 1% of the venous return to the heart. This small quantity of bronchial venous blood, plus a small amount of coronary venous blood that drains directly into the left atrium or left ventricle, "contaminates" the pulmonary venous blood, which is ordinarily fully saturated with oxygen. Hence, the aortic blood is very slightly desaturated. This small quantity of venous drainage to the left side of the heart also accounts for the fact that, even under true equilibrium conditions, the output of the left ventricle slightly exceeds that of the right ventricle.

In certain pathological states, the bronchial circulation may become substantial, and there may be considerable admixture of blood between the systemic and pulmonary circuits. Sustained abridgment of the pulmonary arterial blood supply to a lung, after pulmonary embolism, for example, usually results in a large increase in precapillary (arterial) communications between vessels of the systemic and pulmonary circuits. Conversely, inflammatory and degenerative disease of the lung are often associated with an increased bronchial blood flow and significant admixtures between the two systems at the postcapillary (venous) level.

Pulmonary hemodynamics

Pressures in the pulmonary circulation. In normal individuals, the average systolic and diastolic pressures in the pulmonary artery are about 22 and 10 mm. Hg, respectively, and the mean pressure is about 15 mm. Hg. These pressures are much lower than those in the aorta and reflect the much lower resistance of the pulmonary than of the systemic vascular bed. The mean pressure in the left atrium is normally about 5 to 8 mm. Hg, and so the total pulmonary arteriovenous pressure gradient is only about 10 mm. Hg. The mean hydrostatic pressure in the pulmonary capillaries lies between the pulmonary arterial and pulmonary venous values, but somewhat closer to the latter. Mean pulmonary capillary pressure is therefore about 10 mm. Hg, and there is a small pressure oscillation during each cardiac cycle.

The mean left atrial pressure is an index of the left ventricular filling pressure. In patients suspected of being in left heart failure, it would be desirable to measure the left atrial pressure, but this is difficult technically. It is relatively simple, however, to position a flexible, balloon-tipped catheter in the pulmonary artery. It has been found that a reasonably accurate estimate of mean left atrial pressure may be obtained by advancing the catheter and wedging the tip into a small branch of the pulmonary artery. The pressure thus obtained probably reflects the pressure in the smaller pulmonary veins, which is ordinarily 1 or 2 mm. Hg greater than that in the left atrium.

Pulmonary blood flow. At equilibrium, the pulmonary and systemic blood flows are equal, except for the small disparity noted above. The pulmonary arteriovenous pressure gradient is only about 10 mm. Hg, which is about 10% of that for the systemic circulation. Hence, the pulmonary vascular resistance, when computed as the ratio of pressure drop to flow, is about 10% of the systemic vascular resistance.

Because of the relatively low pressures in the pulmonary blood vessels and because of their great distensibility and collapsibility, gravitational effects produce certain important influences on the regional distribution of blood flow in the lungs. Three distinct flow regimes may be found at different hydrostatic levels in the lung, as illustrated in Fig. 11-7. Consider that the pulmonary artery delivers blood at a steady pressure of 15 mm. Hg, and that the pulmonary venous pressure remains constant at 5 mm. Hg. In those pulmonary arterial and venous branches that are 13 cm. below (zone C) the hydrostatic levels of the main pulmonary vessels, the respective pressures will be 10 mm. Hg (equivalent to 13 cm. blood) greater than those in the main vessels, by virtue of gravitational effects. Conversely, in pulmonary arterial and venous branches that are 13 cm.

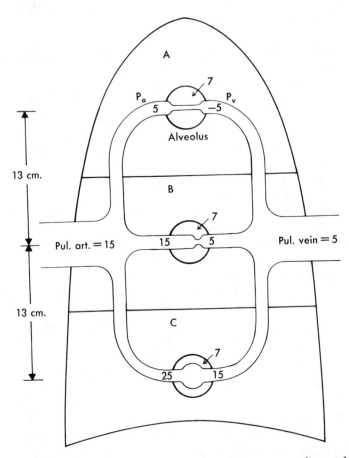

Fig. 11-7. Schematic representation of the three types of flow regimes that might exist in the pulmonary circulation. In zone A, alveolar pressure exceeds intravascular pressures. Pulmonary capillaries in this zone will not be perfused. In zone B, alveolar pressure is intermediate between pulmonary arterial and venous pressures. Pulmonary capillaries will flutter between the open and closed states. In zone C, intravascular pressures exceed alveolar pressure. The pulmonary capillaries are always open, but the flow resistances in individual vessels vary with the hydrostatic level of the vessel.

above (zone A) the main vessels, the respective pressures will be 10 mm. Hg less than those in the main vessels. At the same hydrostatic level (zone B) as the main vessels, the respective pressures in the branches will be approximately equal to those in the main vessels.

Consider that the alveolar pressure equals 7 mm. Hg in all alveoli. Such an alveolar pressure might exist in an individual on a positive pressure respirator. In zone A, the alveolar pressure would exceed the local arterial and venous pressures. The pulmonary capillary pressures lie between those of the arteries and veins, and therefore alveolar pressure would also exceed capillary pressure. Therefore, capillaries lying between adjacent alveoli would be collapsed. There would be no perfusion of those alveoli, and hence no gas exchange.

Under ordinary conditions, the mean pressure in the alveoli is atmospheric. Therefore, the conditions depicted in zone A do not prevail in any region of the lungs. In hypovolemic states, however, the mean pulmonary artery pressure is often very low. Under such conditions, the hydrostatic pressures in the vessels in the apices of the lungs might be subatmospheric. The atmospheric pressure in the alveoli would then compress the capillaries in the apical zone, such that there might be virtually no pulmonary blood flow to that region. However, the bronchial circulation, which operates at much higher pressures, would be relatively unaffected.

In zone B, the alveolar pressure lies between the local arterial and venous pressures (Fig. 11-7). A capillary in that region will flutter between the open and closed state. When the capillary is open, blood will flow through it, and the pressure will decrease progressively in the capillary from the arterial to the venous end. The pressure at the venous end will be less than the alveolar pressure, and therefore the capillary will quickly collapse at that end. With the cessation of flow, the pressures at a given hydrostatic level will equalize. Thus, the

capillary pressure will quickly rise to the level of that in the local small arteries, which exceeds the prevailing alveolar pressure. Hence, the capillary will be forced open. With the restitution of flow, however, the viscous forces will result in a pressure drop along the length of the capillary. As the pressure at the venous end drops below the ambient alveolar pressure, the capillary will again be forced shut. The greater the alveolar pressure, the greater the fraction of the total time that the capillary will be in the closed state. Conversely, the lower the alveolar pressure, the greater the fraction of the total time that the capillary will be in the open state.

The critical pressure gradient for flow in zone B is not the arteriovenous difference, as it is for most vessels in the body, but it is the arterioalveolar pressure difference. As long as the venous pressure is less than alveolar pressure, the actual level of the venous pressure has virtually no influence on the flow. Such a flow regime has been called a "waterfall effect," because the height of a waterfall has no influence on the rate of fluid flow.

In zone C, the arterial and venous pressures both exceed the alveolar pressure. Hence, the pressure everywhere along the capillary exceeds the alveolar pressure, and the capillary remains permanently in the open state. In this zone, the flow is determined by the arteriovenous pressure gradient, and the resistance may be calculated by the usual formula. The large and small pulmonary vessels, including the capillaries, are very distensible, as noted above. The pressure difference that determines the caliber of a distensible tube is the so-called transmural pressure; i.e., the difference between the internal and external pressures. Because of gravitational effects, the internal pressures in all vessels progressively increase in the downward direction. In an erect individual, therefore, the intravascular pressures in the lungs increase in the direction from apex to base. Hence, the transmural pressures increase accordingly, and the pulmonary vessels are more and more distended in the direction from apex to base. Because resistance to flow varies inversely with vessel caliber, there is a progressive reduction in resistance in zone C, and hence a progressive increase in flow, in the apex to base direction. Such predicted changes in flow have been verified in human subjects.

Regulation of the pulmonary circulation

The total volume of blood pumped per minute by the heart (i.e., the cardiac output) passes through the pulmonary circulation. Therefore, the various cardiac and vascular factors that determine cardiac output in general also determine the total pulmonary blood flow. These factors have been discussed in detail in Chapter 9.

The pulmonary blood vessels are innervated by the autonomic nervous system. Although there is a relatively small amount of smooth muscle in the small pulmonary vessels, the pulmonary circulation is a low-pressure system. Therefore, relatively small changes in vascular smooth muscle tone can evoke appreciable percentage changes in vascular resistance. It has been demonstrated that baroreceptor stimulation can evoke a reflex dilation of the pulmonary resistance vessels. Conversely, peripheral chemoreceptor stimulation has the opposite reflex effect on pulmonary vascular resistance. The importance of such neural regulation of the pulmonary vessels under natural conditions remains to be established, however.

Hypoxia is probably the most important factor that influences pulmonary vasomotor tone. Acute and chronic hypoxia both lead to an increase in total pulmonary vascular resistance. Regional reductions in alveolar oxygen tension evoke constriction of the nearby arterioles. This is probably an important mechanism in maintaining an optimal ventilation-perfusion ratio. The oxygen tension in poorly ventilated alveoli will fall toward the level of the P_{O_2} in the pulmonary arterial blood. Blood flowing by such alveoli will not be well oxygenated, and therefore it will tend to lower the oxygen tension of the blood returning to the left atrium. Arteriolar vasoconstriction tends to lower the blood flow in the region of such poorly ventilated alveoli, thereby reducing the contamination of the pulmonary venous blood with poorly oxygenated blood from inadequately ventilated regions of the lung. Thus, this mechanism shunts the pulmonary blood flow from the more poorly ventilated regions to the better ventilated regions of the lungs, thereby tending to attain full oxygen saturation of the systemic arterial blood. The means whereby hypoxia produces increments in pulmonary vascular resistance is still obscure, despite considerable research efforts to reveal the mechanism.

RENAL CIRCULATION
Anatomy

The primary branches of the *renal artery* divide into a number of *interlobar arteries*. These interlobar branches (Fig. 11-8) proceed radially from the hilus toward the corticomedullary junction, between adjacent medullary pyramids. As an interlobar artery approaches the corticomedullary junction (horizontal dashed line in Fig. 11-8), it gives rise to a number of branches, the so-called *arcuate arteries*, which travel in various directions over the bases of the adjacent medullary pyramids, in the zone between the renal cortex and medulla. There is no continuity, however, between the arcuate branches that arise from adjacent interlobar arteries. Because of this absence of an adequate collateral circulation, occlusion of an interlobar artery leads to the destruction (infarction) of a pyramidal-shaped region of the kidney.

From the arcuate arteries, a number of *interlobular branches* arise and travel toward the capsular surface of the kidney (Fig. 11-8). The *afferent arterioles* to the *glomeruli* are branches of these interlobular arteries, and the

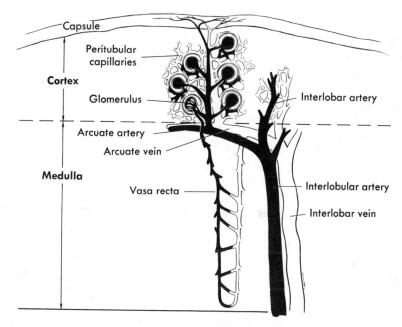

Fig. 11-8. Anatomy of the vasculature to one lobe of the kidney.

terminal branches of the interlobular arteries supply the capillary bed of the renal capsule. There are approximately 1 million glomeruli in each human kidney. The afferent arteriole to each glomerulus divides into a number of vessels which form discrete capillary loops (Fig. 11-9). The proximal and distal limbs of each loop are interconnected by a number of smaller capillaries. The distal limbs of each capillary loop within a glomerulus rejoin to form the *efferent arteriole,* the diameter of which is usually less than that of the afferent arteriole. The entire glomerular capillary tuft is enveloped by *Bowman's capsule,* which collects the glomerular filtrate.

The efferent arterioles break down into another capillary network, the *peritubular capillaries.* The architecture of the peritubular capillary network varies, depending upon whether

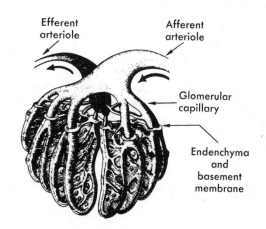

Fig. 11-9. Reconstruction of a human glomerulus. (Modified from Elias, H., Hossmann, A., Barth, I.B., and Solmor, A.: J. Urol. **83:**790, 1960.)

the efferent arteriole arises from glomeruli close to the medullary border (the so-called *juxtamedullary glomeruli*) or from glomeruli located in the remaining, more peripheral regions of the renal cortex (Fig. 11-8).

The capillaries that originate from the regular cortical glomeruli surround the relatively short renal tubules that are located almost entirely in the cortex itself. The capillary networks of neighboring cortical nephrons freely communicate with one another. Some of the capillaries that arise from the efferent arterioles of juxtamedullary glomeruli also form freely anastomosing networks about the neighboring renal tubules. However, most of the capillaries that originate from such glomeruli form long hairpin loops (the *vasa recta*) that accompany the loops of Henle deep into the renal medulla, sometimes to the tips of the renal papillae. The vasa recta participate in the countercurrent exchange system that is involved in the renal mechanism for concentrating the urine.

The renal venous system, in general, parallels the arterial distribution to the renal tissues (Fig. 11-8).

Renal hemodynamics

Resistance of the vascular segments. There is only a slight pressure drop in the interlobar, arcuate, and interlobular arteries; the main preglomerular resistance resides in the afferent arterioles. The pressure in the glomerular capillaries is normally about 50 to 60 mm. Hg. Thus, the net balance of forces across the capillary wall favors the outward filtration of plasma water along the entire length of the capillary loop. Furthermore, the glomerular capillary walls have a much higher filtration coefficient than do the walls of most other capillaries in the body. Hence, about 20% of the plasma water that enters the glomerular capillaries is filtered into Bowman's capsule. The greatest resistance in the renal circulation is at the level of the efferent arterioles, such that the pressure in the peritubular capillaries is

only about 10 to 20 mm. Hg under normal conditions. Such pressure levels favor the net reabsorption of the very large quantities of fluid that pass from the renal tubular lumina to the interstitial spaces of the kidney. The peritubular capillaries also are considerably more permeable than most of the other capillaries in the body.

Renal blood flow. The weight of the kidneys comprises only about 0.5% of the total body weight, and yet the kidneys receive about 20% of the cardiac output. Most of this rich blood supply perfuses the renal cortex; the inner medulla and papillae receive only about one-tenth the blood flow per unit weight that perfuses the cortex. Nevertheless, even this relatively poorly perfused renal tissue receives as much blood per unit weight as does the brain, for example.

The kidney has one of the highest metabolic rates per unit weight in the body. The large renal blood flow is ascribable only slightly to the great metabolic rate, however. The kidney ordinarily extracts less than 10% of the oxygen present in the renal arterial blood. In terms of metabolic requirements, therefore, the renal blood flow is at least ten times greater than that needed to deliver the required oxygen and metabolic substrates. The excessively high renal blood flow mainly subserves the function of delivering large volumes of blood to the glomeruli for the process of ultrafiltration.

Regulation of the renal circulation

Autoregulation. Renal blood flow tends to remain constant, despite any wide fluctuations that may occur in the arterial perfusion pressure. In the experiment shown in Fig. 11-10, for example, the arterial pressure was suddenly raised from 140 to 190 mm. Hg, and then held at this higher level. This stepwise change in pressure was attended by a rapid increase in renal blood flow, from 135 to 155 ml./min. This rise in renal blood flow was transitory, however, and it returned close to the control level

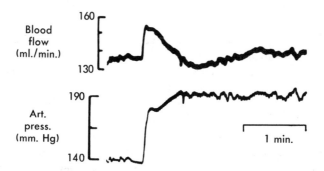

Fig. 11-10. Changes in renal blood flow in an isolated dog kidney evoked by a sudden increase in arterial perfusion pressure from 140 to 190 mm. Hg. The kidney was perfused from a peripheral artery of another dog. (Modified from Semple, S. J. G., and deWardener, H. E.: Circ. Res. 7:643, 1959. With permission of the American Heart Association.)

in less than 1 minute. Over a pressure range from about 75 to 275 mm. Hg, the steady-state level of renal blood flow is relatively insensitive to changes in arterial pressure. Below a pressure of 75 mm. Hg, however, renal blood flow varies proportionately with perfusion pressure.

This *autoregulatory* tendency for the steady-state level of the renal blood flow to remain constant, despite fluctuations in the perfusion pressure, is a process intrinsic to the kidney itself; it has been demonstrated even in isolated kidney preparations. Concomitant measurement of glomerular filtration rate (GFR) reveals that there is an equally pronounced tendency for GFR also to be autoregulated. It is apparent, therefore, that the resistance change that is induced by an alteration in perfusion pressure must occur predominantly at the level of the afferent arteriole. As the perfusion pressure is raised, for example, a subsequent afferent arteriolar constriction would tend both to limit the increase in renal blood flow and also to restrict the rise in glomerular capillary pressure and the concomitant increment in GFR.

The process of renal autoregulation has been studied extensively, but the mechanism remains controversial. A number of mechanisms

have been proposed, including the tissue pressure, myogenic, and metabolic hypotheses that have been discussed in Chapter 7, for the vascular system as a whole. None of these mechanisms appears to account entirely for the autoregulatory process in the kidneys, although each mechanism may play some contributory role. Considerable attention has also been directed toward the *juxtaglomerular apparatus,* which is involved in the release of *renin* by the kidneys. Renin is an enzyme that acts on a substrate circulating in the blood to release a peptide, *angiotensin,* which is a potent vasoconstrictor. It has been proposed that, with an increase in renal perfusion pressure, there is an initial increase in GFR. The consequent increase in the rate of fluid flow in the renal tubules is somehow sensed by the juxtaglomerular apparatus, which responds by releasing renin. The angiotensin that is then generated serves to constrict the afferent arterioles, which thereby attenuates the increases in GFR and renal blood flow.

Neural regulation. The principal innervation of the renal vasculature is by the sympathetic nervous system. Stimulation of the renal sympathetic nerves elicits a reduction in renal blood flow and a relatively smaller diminution in GFR. The neural activity constricts the afferent and efferent arterioles and the proximal segments of the vasa recta. Presumably the postglomerular constriction exceeds the preglomerular constriction, which accounts for the relatively greater reduction of renal blood flow than of GFR.

In resting subjects, the basal level of renal sympathetic tone is low; abolition of that tone has negligible effects on renal blood flow. The arterial baroreceptor reflexes appear to have only a small influence on the renal vasculature. Relatively large changes in baroreceptor activity in either direction produce only slight, inverse changes in renal blood flow. Much larger reflex effects on the renal circulation are induced by activation of receptors on the low-pressure side of the circulation. A reduction in left atrial pressure, for example, evokes a large increase in renal nerve activity and in renal vascular resistance. Emotional reactions, such as anxiety, fear, and rage, also abridge renal blood flow dramatically.

SPLANCHNIC CIRCULATION

The splanchnic circulation consists of the blood supply to the gastrointestinal (GI) tract, liver, spleen, and pancreas. There are several features that distinguish the splanchnic circulation, the most noteworthy of which is that there are two large capillary beds partially in series with one another. The small splanchnic arterial branches supply the capillary beds in the GI tract, spleen, and pancreas. From these capillary beds, the venous blood ultimately flows into the portal vein, which normally provides most of the blood supply to the liver. However, there is also an hepatic arterial blood supply. In the next two subsections, we shall deal almost exclusively with the intestinal and hepatic circulations; the blood supply to the spleen and pancreas will not be discussed.

Intestinal circulation

Anatomy. The GI tract is supplied by three large vessels, namely the celiac, superior mesenteric, and inferior mesenteric arteries. The superior mesenteric artery is the largest of all the branches of the aorta and carries over 10% of the cardiac output. Small mesenteric arteries form an extensive vascular network in the submucosa (Fig. 11-11). Branches of these small arteries penetrate the longitudinal and circular muscle layers and give rise to third- and fourth-order arterioles. Some third-order arterioles in the submucosa penetrate to the tip of the villus as its main arteriole.

The direction of the blood flow in the capillaries and venules in a villus is opposite to that in the main arteriole. This arrangement presents the anatomical potential for a countercurrent exchange system. Experiments in certain species suggest that there is an effective countercurrent multiplier in the villus that is involved in the absorption of sodium and water. There is probably also a countercurrent exchange of O_2 from arterioles to venules. At low flow rates especially, a substantial fraction of the O_2 may be shunted from arterioles to venules near the base of the villus, thereby curtailing the supply of O_2 to the mucosal cells at the tip of the villus. When intestinal blood flow is reduced, the shunting of oxygen is exaggerated; this could lead to extensive necrosis of the intestinal villi.

Neural regulation. The neural control of the mesenteric circulation is almost exclusively by the sympathetic nervous system. Increased sympathetic activity elicits a pronounced constriction of the mesenteric arterioles, precapillary sphincters, and capacitance vessels. These responses are mediated by alpha receptors, which are prepotent in the mesenteric circulation; however, beta receptors are also present. Infusion of a beta receptor agonist, such as isoproterenol, causes vasodilation.

In spontaneously occurring fighting behav-

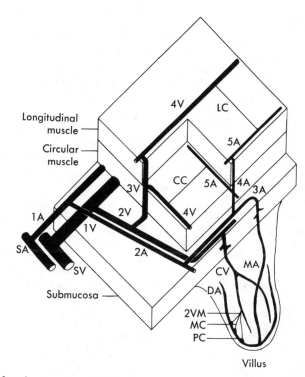

Fig. 11-11. The distribution of small blood vessels in the rat intestinal wall. *SA*, small artery; *SV*, small vein; *1A, 2A, . . ., 5A*, first-, second-, . . ., fifth-order arterioles; *1V, . . ., 4V*, first-, . . ., fourth-order venules; *CC* and *LC*, capillaries in circular and longitudinal muscle layers; *MA* and *CV*, main arteriole and collecting venule of a villus; *DA*, distributing arteriole; *2VM*, second-order mucosal venule; *PC*, precapillary sphincter; *MC*, mucosal capillary. (From Gore, R. W., and Bohlen, H. G.: Am. J. Physiol. **233**:H685, 1977.)

ior, or in response to artificial stimulation of the hypothalamic "defense" area, pronounced vasoconstriction occurs in the mesenteric vascular bed. This serves to promote a shift of blood flow from the temporarily less important intestinal circulation to the more crucial skeletal muscles, heart, and brain.

Autoregulation. Autoregulation exists in the intestinal circulation, but it is not as well developed as it is in certain other vascular beds, such as those in the brain and kidney. However, the oxygen consumption of the small intestine is more rigorously controlled. The oxygen uptake of the small intestine has been found to remain constant when the arterial perfusion pressure was varied over the range of 30 to 125 mm. Hg. The principal mechanism responsible for autoregulation appears to be metabolic, although a myogenic mechanism probably also participates. A four-fold rise in the adenosine concentration in the mesenteric venous blood was observed after a brief arterial occlusion. Adenosine is a potent vasodilator in the mesenteric vascular bed, and may be the principal metabolic mediator of autoregulation. However, potassium, and altered osmolality might also contribute to the overall response.

Functional hyperemia. The ingestion of food

leads to an appreciable increase in intestinal blood flow. Several mechanisms contribute to this hyperemia, but the secretion of certain gastrointestinal hormones is probably involved. Gastrin and cholecystokinin have been shown to augment intestinal blood flow, and they are, of course, secreted in response to food ingestion.

The absorption of food is also closely correlated with the rate of intestinal blood flow. Undigested food has no vasoactive influence, whereas several products of digestion are potent vasodilators. Among the various constituents of chyme, the principal mediators of the mesenteric hyperemia seem to be glucose and long-chain fatty acids.

Hepatic circulation

Anatomy. The blood flow to the liver normally is about 25% of the cardiac output. It is derived from two sources, the portal vein and the hepatic artery. Ordinarily, about three-fourths of the blood flow is delivered by the portal vein. The portal venous blood has already passed through the gastrointestinal capillary bed, and therefore much of the O_2 has already been extracted. The remaining one-fourth of the blood supply is delivered by the hepatic artery. Ordinarily, this blood is fully saturated with O_2. Hence, about three-fourths of the O_2 utilized by the liver is derived from the hepatic arterial blood, even though the blood flow rate in the hepatic artery is only about one-fourth that in the portal vein.

The small branches of the portal vein and hepatic artery give rise to terminal portal venules and hepatic arterioles (Fig. 11-12). These terminal vessels enter the hepatic acinus (the functional unit of the liver) at its center. Blood flows from these terminal vessels into the sinusoids, which constitute the capillary network of the liver. The sinusoids radiate toward the periphery of the acinus, where they connect with the terminal hepatic venules. Blood from these terminal venules drains into pro-

gressively larger branches of the hepatic veins, which are tributaries of the inferior vena cava.

Hemodynamics. The mean blood pressure in the portal vein is about 10 mm. Hg, and that in the hepatic artery is about 90 mm. Hg. The resistance of the vessels upstream to the hepatic sinusoids is considerably greater than that of the downstream vessels. Consequently, the pressure in the sinusoids is only 2 or 3 mm. Hg greater than that in the hepatic veins and inferior vena cava. The ratio of pre- to postsinusoidal resistance in the liver is much greater than is the ratio of pre- to postcapillary resistance for almost any other vascular bed. Hence, drugs and other interventions that tend to alter the presinusoidal resistance have only a negligible effect on the pressure in the sinusoids, and consequently such changes in presinusoidal resistance have very little effect on the fluid exchanges across the sinusoidal wall. Conversely, changes in hepatic venous (and in central venous) pressure are transmitted almost quantitatively to the hepatic sinusoids. Such venous pressure changes, therefore, have a profound effect on the transsinusoidal exchange of fluids. When central venous pressure is elevated, as in congestive heart failure, there may be a large transudation of plasma water from the liver into the peritoneal cavity, leading to the development of *ascites*.

Regulation of flow. There is a reciprocal relation between the rates of blood flow in the portal venous and hepatic arterial systems. When blood flow is curtailed in one system, the flow increases in the other system. However, the resultant increase in flow in one system usually does not fully compensate for the initiating reduction in flow in the other system.

The portal venous system does not display any significant autoregulation. As the portal venous pressure and flow are raised, resistance either remains constant or it decreases. The hepatic arterial system does display some capacity to autoregulate. However, the tendency

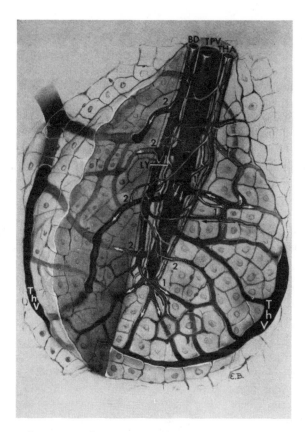

Fig. 11-12. Microcirculation to an hepatic acinus. *THA*, terminal hepatic arteriole; *TPV*, terminal portal venule; *BD*, bile ductule; *ThV*, terminal hepatic venule; *LY*, lymphatic. The hepatic arterioles empty either directly *(1)* or via the peribiliary plexus *(2)* into the sinusoids that run from the terminal portal venule to the terminal hepatic venules. (From Rappaport, A. M.: Microvasc. Res. **6:**212, 1973.)

is prominent only in denervated preparations, for some unknown reason.

Despite the weak autoregulation of hepatic blood flow, the liver does show a remarkable capacity to maintain a constant O_2 consumption. This is achieved by a very efficient mechanism for extraction of O_2 from the hepatic blood supply. As the rate of O_2 delivery to the liver is varied, the liver compensates by an appropriate change in the fraction of O_2 extracted from each unit volume of blood. This extraction process is facilitated by the distinct separation of the presinusoidal vessels at the acinar center from the postsinusoidal vessels at the periphery of the acinus (Fig. 11-12). There is little opportunity for a countercurrent exchange of O_2, contrary to the condition that exists in an intestinal villus, for example (Fig. 11-11).

The sympathetic nerves are capable of constricting the presinusoidal resistance vessels in the portal venous and hepatic arterial systems. Neural effects on the capacitance vessels are probably of greater importance, however. The effects are mediated mainly via alpha receptors.

Capacitance vessels. The liver contains about 15% of the total blood volume of the body. Under appropriate conditions, such as in response to hemorrhage, about half of this volume of blood can be rapidly expelled. Hence, the liver constitutes an important blood reservoir. It is probably the most important blood reservoir in humans. In certain other species, such as the dog, the spleen is an important blood reservoir. Smooth muscle in the capsule and trabeculae of the spleen contract in response to increased sympathetic neural activity, such as occurs during exercise or hemorrhage. However, this mechanism does not exist in humans.

FETAL CIRCULATION

The circulation of the fetus shows a number of differences from that of the postnatal infant. The fetal lungs are functionally inactive, and the fetus is completely dependent on the placenta for oxygen and nutrient supply. Oxygenated fetal blood from the placenta passes through the umbilical vein to the liver. A major fraction passes through the liver, and a small fraction bypasses the liver to the inferior vena cava through the *ductus venosus* (Fig. 11-13). In the inferior vena cava, blood from the ductus venosus joins blood returning from the lower trunk and extremities and this combined stream is in turn joined by blood from the liver through the hepatic veins. The streams of blood tend to maintain their identity in the inferior vena cava and are divided into two streams of unequal size by the edge of the interatrial septum *(crista dividens)*. The larger stream, which is primarily blood from the umbilical vein, is shunted through the *foramen ovale,* which lies between the inferior vena cava and the left atrium (*inset* Fig. 11-13), to the left atrium. The other stream passes into the right atrium, where it is joined by superior vena caval blood returning from the upper

parts of the body and by blood from the myocardium. In contrast to the adult, in whom the right and left ventricles pump in series, in the fetus the ventricles operate essentially in parallel. Because of the large pulmonary resistance, less than one-third of the right ventricular output goes through the lungs. The remainder passes through the *ductus arteriosus* from the pulmonary artery to the aorta at a point distal to the origins of the arteries to the head and upper extremities. Flow from pulmonary artery to aorta occurs because pulmonary artery pressure is about 5 mm. Hg higher than aortic pressure in the fetus. The large volume of blood coming through the foramen ovale into the left atrium is joined by blood returning from the lungs and is pumped out by the left ventricle into the aorta. About one-third of the aortic blood goes to the head, upper thorax, and arms and the remaining two-thirds to the rest of the body and the placenta. The amount of blood pumped by the left ventricle is about 20% greater than that pumped by the right ventricle, and the major fraction of the blood that passes down the descending aorta flows by way of the two umbilical arteries to the placenta.

In Fig. 11-13 the distribution of fetal blood flow is given in percentage of the combined right and left ventricular outputs. Note that over half of the combined cardiac output is returned directly to the placenta without passing through any capillary bed. Also indicated in Fig. 11-13 are the oxygen saturations of the blood (numbers in parentheses) at various points of the fetal circulation. Fetal blood leaving the placenta is 80% saturated, but the saturation of the blood passing through the foramen ovale is reduced to 67% by some mixing with desaturated blood returning from the lower part of the body and the liver. Addition of the desaturated blood from the lungs reduces the oxygen saturation of left ventricular blood to 62%, which is the level of saturation of the blood reaching the head and upper ex-

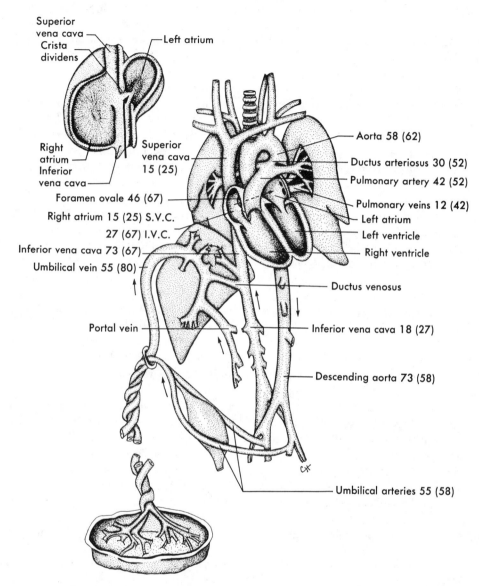

Fig. 11-13. Schematic diagram of the fetal circulation. The numbers without parentheses represent the distribution of cardiac output in percentage of the sum of the right and left ventricular outputs, and the numbers within parentheses represent the percentage of oxygen saturation of the blood flowing in the indicated blood vessel. The insert at the upper left illustrates the direction of flow of a major portion of the inferior vena cava blood through the foramen ovale to the left atrium. (Values for percentage distribution of blood flow and oxygen saturations are from Dawes, G. S., Mott, J. C., and Widdicombe, J. G.: J. Physiol. **126:**563, 1954.)

tremities. The blood in the right ventricle, a mixture of desaturated superior vena caval blood, coronary venous blood, and inferior vena caval blood, is only 52% saturated with oxygen. When the major portion of this blood traverses the ductus arteriosus and joins that pumped out by the left ventricle, the resultant oxygen saturation of blood traveling to the lower part of the body and back to the placenta is 58% saturated. Thus it is apparent that the tissues receiving blood of the highest oxygen saturation are the liver, the heart, and the upper parts of the body, including the head.

At the placenta the chorionic villi dip into the maternal sinuses, and oxygen, carbon dioxide, nutrients, and metabolic waste products exchange across the membranes. The barrier to exchange is quite large, and equilibrium of oxygen tension between the two circulations is not reached at normal rates of blood flow. Therefore the oxygen tension of the fetal blood leaving the placenta is very low. Were it not for the fact that fetal hemoglobin has a greater affinity for oxygen than does adult hemoglobin, the fetus would not receive an adequate oxygen supply. The fetal oxyhemoglobin dissociation curve is shifted to the left so that at equal pressures of oxygen fetal blood will carry significantly more oxygen than will maternal blood. If the mother is subjected to hypoxia, the reduced blood oxygen tension is reflected in the fetus by tachycardia and an increase in blood flow through the umbilical vessels. If the hypoxia persists or if flow through the umbilical vessels is impaired, fetal distress occurs and is first manifested as bradycardia. In early fetal life the high cardiac glycogen levels that prevail (which gradually decrease to adult levels by term) may protect the heart from acute periods of hypoxia.

Circulatory changes that occur at birth

The umbilical vessels have thick muscular walls that are very reactive to trauma, tension, sympathomimetic amines, bradykinin, angiotensin, and changes in oxygen tension. In animals in which the umbilical cord is not tied, hemorrhage of the newborn is prevented by constriction of these large vessels in response to one or more of these stimuli. Closure of the umbilical vessels produces an increase in total peripheral resistance and of blood pressure. When blood flow through the umbilical vein ceases, the ductus venosus, a thick-walled vessel with a muscular sphincter, closes. What initiates closure of the ductus venosus is still unknown. The asphyxia, which starts with constriction or clamping of the umbilical vessels, plus the cooling of the body activate the respiratory center. With the filling of the lungs with air, pulmonary vascular resistance decreases to about one-tenth of the value existing prior to lung expansion. This resistance change is not caused by the presence of oxygen in the lungs, since the change is just as great if the lungs are filled with nitrogen. However, filling the lungs with liquid does not reduce pulmonary vascular resistance.

The left atrial pressure is raised above that in the inferior vena cava and right atrium by (1) the decrease in pulmonary resistance, with the resulting large flow of blood through the lungs to the left atrium, (2) the reduction of flow to the right atrium caused by occlusion of the umbilical vein, and (3) the increased resistance to left ventricular output produced by occlusion of the umbilical arteries. This reversal of the pressure gradient across the atria abruptly closes the valve over the foramen ovale, and fusion of the septal leaflets occurs over a period of several days.

With the decrease in pulmonary vascular resistance, the pressure in the pulmonary artery falls to about one-half its previous level (to about 35 mm. Hg), and this change in pressure, coupled with a slight increase in aortic pressure, reverses the flow of blood through the ductus arteriosus. However, within several minutes the large ductus arteriosus begins to constrict, and this constriction produces turbulent flow, which is manifest as a murmur in the newborn. Constriction of the ductus arteriosus is progressive and usually is complete within 1 to 2 days after birth. Closure of the ductus arteriosus appears to be initiated by the high oxygen tension of the arterial blood passing through it, since pulmonary ventilation with oxygen or with air low in oxygen induces, respectively, closure and opening of this shunt vessel. Whether oxygen acts directly on the ductus or through the release of a vasoconstrictor substance is not known. Similarly, in a heart-lung preparation made from a newborn lamb, the ductus arteriosus may be made to close with high Pa_{O_2} and to open with low Pa_{O_2}. However, recent studies on fetal lambs suggest that bradykinin formed from kininogen in the lungs when they fill with air, contributes to the closure of the ductus arteriosus. The bradykinin may act directly or by inducing a release of catecholamines from the adrenal medulla. In addition, the prostaglandins may play a role in the closure of the ductus arteriosus. Permanent closure usually takes several weeks.

At birth the walls of the two ventricles are approximately of the same thickness, with possibly slight preponderance of the right ventricle. There is also present in the newborn thickening of the muscular layers of the pulmonary arterioles, which is apparently responsible in part for the high pulmonary vascular resistance of the fetus. After birth the thickness of the walls of the right ventricle diminishes, as does the muscle layer of the pulmonary arterioles, whereas the left ventricular walls increase in thickness. These changes are progressive over a period of weeks after birth.

Failure of the foramen ovale or ductus arteriosus to close after birth is occasionally observed and constitutes some of the more common congenital cardiac abnormalities that are now amenable to surgical correction.

BIBLIOGRAPHY
Journal articles

Bergofsky, E. H.: Mechanisms underlying vasomotor regulation of regional pulmonary blood flow in normal and disease states, Am. J. Med. **57**:378, 1974.

Betz, E.: Cerebral blood flow: its measurement and regulation, Physiol. Rev. **52**:595, 1972.

Carneiro, J. J., and Donald, D. E.: Blood reservoir function of dog spleen, liver and intestine, Am. J. Physiol. **1**:H67, 1977.

Celander, O., and Folkow, B.: A comparison of sympathetic vasomotor fiber control of the vessels within the skin and the muscles, Acta Physiol. Scand. **29**:241, 1953.

Fox, R. H., and Hilton, S. M.: Bradykinin formation in human skin as a factor in heat vasodilatation, J. Physiol. (London) **142**:219, 1958.

Granger, D. N., and Kvietys, P. R.: The splanchnic circulation; intrinsic regulation, Ann. Rev. Physiol. **43**:409, 1981.

Granger, D. N., Richardson, P. D. I., Kvietys, P. R., and Mortillaro, N. A.: Intestinal blood flow, Gastroenterology **78**:837, 1980.

Greenway, C. V., and Stark, R. D.: Hepatic vascular bed, Physiol. Rev. **51**:23, 1971.

Hertzman, A.: Vasomotor regulation of cutaneous circulation, Physiol. Rev. **39**:280, 1959.

Heymann, M. A., Iwamoto, H. S., and Rudolf, A. M.: Factors affecting changes in the neonatal systemic circulation, Ann. Rev. Physiol. **43**:371, 1981.

Heymann, M. A., and Rudolph, A. M.: Control of the ductus arteriosus, Physiol. Rev. **55**:62, 1975.

Hyman, C., Rosell, S., Rosen, A., Sonnenschein, R. R., and Uvnäs, B.: Effects of alterations of total muscular blood flow on local tissue clearance of radio-iodide in the cat, Acta Physiol. Scand. **46**:358, 1959.

Jones, R. D., and Berne, R. M.: Intrinsic regulation of skeletal muscle blood flow, Circ. Res. **14**:126, 1964.

Kontos, H. A.: Regulation of the cerebral circulation, Ann. Rev. Physiol. **43**:397, 1981.

Lanciault, G., and Jacobsen, E. D.: Gastrointestinal circulation, Gastroenterology **71**:851, 1976.

Lautt, W. W.: Hepatic vasculature: a conceptual review, Gastroenterology **73**:1163, 1977.

Lautt, W. W.: Hepatic nerves: a review of their functions and effects, Canad. J. Physiol. Pharmacol. **58**:105, 1980.

Melmon, K. L. Cline, M. J., Hughes, T., and Nies, A. S.: Kinins: possible mediators of neonatal circulatory changes in man, J. Clin. Invest. **47**:1295, 1968.

Narar, L. G.: Renal autoregulation: perspectives from whole kidney and single nephron studies, Am. J. Physiol. **234**:F357, 1978.

Olsson, R. A.: Local factors regulating cardiac and skeletal muscle blood flow, Ann. Rev. Physiol. **43**:385, 1981.

Rappaport, A. M.: The microcirculatory hepatic unit, Microvasc. Res. **6**:212, 1973.

Rappaport, A. M., and Schneiderman, J. H.: The function of the hepatic artery, Rev. Physiol. Biochem. Pharmacol. **76**:129, 1976.

Rubio, R., Berne, R. M., Bockman, E. L., and Curnish, R. R.: Relationship between adenosine concentration and oxygen supply in rat brain, Am. J. Physiol. **22**:1896, 1975.

Uvnäs, B.: Sympathetic vasodilator system and blood flow, Physiol. Rev. **40** (Supp. 4):69, 1960.

West, J. B.: Blood flow to the lung and gas exchange, Anesthesiology **41**:124, 1974.

Books and monographs

Berne, R. M., Winn, H. R., and Rubio, R.: Metabolic regulation of cerebral blood flow. In Mechanisms of vasodilation—2nd Symposium, New York, 1981, Raven Press.

Dawes, G. S.: Foetal and neonatal physiology, Chicago, 1968, Year Book Medical Publishers, Inc.

Hughes, J. M. B.: Pulmonary circulation and fluid balance. In Widdicombe, J. G., editor: International review of physiology; Respiratory physiology II, vol. 14, Baltimore, 1977, University Park Press, p. 136.

Johnson, P. C., editor: Peripheral circulation, New York, 1978, John Wiley & Sons, Inc.

Lewis, T.: Blood vessels of the human skin and their responses, London, 1927, Shaw & Son, Ltd.

Longo, L. D., and Reneau, D. D., editors: Fetal and newborn cardiovascular physiology; vol. 1, Developmental aspects, New York, 1978, Garland STPM Press.

Owman, C., and Edvinsson, L., editors: Neurogenic control of the brain circulation, Oxford, 1977, Pergamon Press.

Parker, J. C., Guyton, A. C., and Taylor, A. E.: Pulmonary transcapillary exchange and pulmonary edema. In Guyton, A. C., and Young, D. B., editors: International review of physiology; Cardiovascular physiology III, vol. 18, Baltimore, 1979, University Park Press, p. 262.

Zelis, R.: The peripheral circulations, New York, 1975, Grune & Stratton, Inc.

CHAPTER 12

INTERPLAY OF CENTRAL AND PERIPHERAL FACTORS IN THE CONTROL OF THE CIRCULATION

The primary function of the circulatory system is to deliver the supplies needed for tissue metabolism and growth and to remove the products of metabolism. To explain how the heart and blood vessels serve this function, it has been necessary to analyze the system morphologically and functionally and to discuss the mechanisms of action of the component parts in their contribution to the maintenance of adequate tissue perfusion under different physiological conditions. Once the functions of the various components are understood, it is essential that their interrelationships in the overall role of the circulatory system be considered. Tissue perfusion is dependent on arterial pressure and local vascular resistance, and arterial pressure in turn is dependent on cardiac output and total peripheral resistance. Arterial pressure is maintained within a relatively narrow range in the normal individual, a feat that is accomplished by reciprocal changes in cardiac output and total peripheral resistance. However, cardiac output and peripheral resistance are each influenced by a number of factors and it is the interplay among these factors that determines the level of these two variables. The autonomic nervous system and the baroreceptors play the key role in the regulation of blood pressure. However, from the long-range point of view the control of fluid balance by the kidney, by the adrenal cortex, and by the central nervous system, with maintenance of a constant blood volume, is of the greatest importance.

In a well-regulated system one way to study the extent and the sensitivity of the regulatory mechanism is to disturb the system and observe its response to restore the preexisting equilibrium state. With respect to the circulatory system, disturbances in the form of physical exercise and hemorrhage will be used to illustrate the effects of the various factors that go into its regulation.

EXERCISE

Over the years ideas about the cardiovascular changes that take place in exercise have undergone considerable revision, and the adjustments that were once explained in a simple, seemingly logical sequence are now known to be incorrect. It was formerly thought that the chain of events was initiated in the active muscles, where vasodilation led to a decrease in blood pressure, which activated the baroreceptors to induce an increase in heart rate and myocardial contractility and an enhancement of peripheral vasoconstriction in the inactive tissues. Evidence to substantiate the baroreceptor mechanism is lacking; in fact, the baroreceptor reflex is diminished in exercise.

253

Furthermore, attempts to demonstrate that either peripheral, neural, or humoral input alone triggers the increase in cardiac output have not been supported by experimental data. With regard to humoral stimuli, receptors that are known to respond to reduced Po_2 and increased Pco_2 are all located on the arterial side of the circulation and the Pa_{O_2} and Pa_{CO_2} are within normal limits during exercise. Furthermore, in exercise experiments in which venous blood from the active muscles was prevented from returning to the central veins, subjects still displayed an increase in heart rate and cardiac output, indicating the lack of need of a humoral stimulus to initiate the cardiovascular adjustments in exercise. With respect to the role of the nervous system, it has been observed that reflexes originating in the contracting muscles result in activation of the sympathetic nerves to the heart and peripheral blood vessels. Hence, neural input is in large part responsible for the cardiovascular changes observed with exercise. The earliest circulatory adjustments, particularly those occurring prior to the onset of exercise, apparently take origin from the cerebral cortex.

Preparation for exercise

In humans or in trained animals, anticipation of and preparation for physical activity evoke primarily inhibition of vagal nerve activity to the heart and generalized sympathetic discharge. The concerted effects of inhibition of parasympathetic and activation of sympathetic areas of the medulla on the heart are an increase in heart rate and in myocardial contractility (a greater ventricular emptying at the same or at reduced ventricular filling pressure). The tachycardia and the enhanced contractility result in a rapid transfer of blood from the venae cavae and the pulmonary vessels to the aorta. This increase in cardiac output would be only transient were it not for the fact that the enhanced output of the pump also increases blood flow back to the heart (*vis a tergo*).

At the same time that cardiac stimulation occurs, the sympathetic nervous system also elicits vascular resistance changes in the periphery. In cats and dogs and possibly in humans, the cholinergic sympathetic vasodilator system is activated and produces dilation of the larger resistance vessels (arterioles and metarterioles) in the muscles. In the skin, kidneys, and splanchnic regions, the sympathetic vasoconstrictor fibers increase vascular resistance, which in effect diverts blood away from these areas. The increased renal and splanchnic vascular resistance persists throughout the period of exercise. The renal vasoconstriction is probably an autoregulatory response, since it occurs even in the denervated kidney. Blood flow to the renal and splanchnic regions remains at rest levels during exercise. Hence the increase in vascular resistance is solely a result of the rise in blood pressure. Only with impaired cardiac function, as in heart failure, does the blood flow to the renal and splanchnic regions decrease with exercise. Under conditions of poor cardiac performance, this reduction in blood flow is caused by intense sympathetically mediated vasoconstriction.

Moderate exercise

Local factors in muscle. With the onset of exercise, several other circulatory adjustments come into play; the major one involves the vasculature of the active muscles. The local formation of vasoactive metabolites induces marked dilation of the resistance vessels, which progresses rapidly to maximal degrees at moderate levels of exercise. Potassium is one of the vasodilator substances released by contracting muscle, and it is at least in part responsible for the initial decrease in vascular resistance in the active muscles. Other contributing factors may be an increase in interstitial fluid osmolarity during the initial phase of exercise, and the release of adenosine during sustained exercise. Afferent impulses from receptors in the muscle travel to the medulla oblongata and an increase in sympathetic nerve activity ensues. The result is an increase in heart rate, in myocardial contractility, and in peripheral resistance in nonactive tissues. The local accumulation of metabolites induces relaxation of the terminal arterioles. Hence, blood flow through the muscle may increase to fifteen to twenty times the resting level, although the increment in blood volume in the muscle increases only about 50%. This metabolic vasodilation of the precapillary vessels in the active muscles occurs very soon after the onset of exercise, and the decrease in total peripheral resistance enables the heart to pump more blood at a lesser load and more efficiently (less pressure work; see p. 220) than if peripheral resistance were unchanged. Since only a small percentage of the capillaries are perfused at rest, whereas in exercise all or nearly all of the capillaries contain flowing blood (*capillary recruitment*), the surface available for exchange of gases, water, and solutes is increased many-fold in exercise. Furthermore, the hydrostatic pressure in the capillaries is increased by virtue of the relaxation of the resistance vessels, and there is a net movement of water and solutes into the muscle tissue. Tissue pressure rises and remains elevated during exercise as fluid continues to move out of the capillaries and is carried away by the lymphatics. Lymph flow is increased as a result of the increase in capillary hydrostatic pressure and the massaging effect of the contracting muscles on the valve-containing lymphatic vessels.

The contracting muscle avidly extracts needed oxygen from the perfusing blood (increased arteriovenous O_2 difference, Fig. 12-1), and the venous blood leaving the active muscles has a low oxygen content (about 5 vol.%). The removal of oxygen is facilitated by the nature of oxyhemoglobin dissociation. The reduction in pH caused by the high concentration of CO_2 and the formation of lactic acid and the increase in temperature in the contracting muscle contribute to shifting the oxyhemoglo-

bin dissociation curve to the right, so that at any given partial pressure of oxygen, less oxygen is held by the hemoglobin in the red cells, and consequently there is a more effective oxygen removal from the blood. Oxygen consumption may increase as much as sixtyfold with only a fifteenfold increase in muscle blood flow. Some investigators believe that the muscle myoglobin may serve as an important oxygen store in exercise. Myoglobin will release attached oxygen only at very low partial pressures that are probably reached in actively contracting muscle. However, relative to the rate of oxygen consumption of the active muscle, the amount of oxygen bound to myoglobin is negligible.

Arterial pressure. If the exercise involves a large proportion of the body musculature, such as in running or swimming, the reduction in total vascular resistance can be considerable (Fig. 12-1). Nevertheless arterial pressure starts to rise with the onset of exercise, and the increase in blood pressure roughly parallels the severity of the exercise performed (Fig. 12-1). Therefore the increase in cardiac output is proportionally greater than the decrease in total peripheral resistance. The vasoconstriction produced in the inactive tissues by the sympathetic nervous system (and to some extent by the release of catecholamines from the adrenal medulla) is important for the maintenance of normal or increased blood pressure, since sympathectomy or drug-induced block of the adrenergic sympathetic nerve fibers results in a decrease in arterial pressure (*hypotension*) during exercise. In general, systolic pressure increases more than diastolic pressure, resulting in an increase in pulse pressure (Fig. 12-1). The latter may be in part attributable to a greater stroke volume and in part to a more rapid ejection of blood by the left ventricle with less peripheral runoff during the brief ventricular ejection period.

As body temperature rises, the skin vessels dilate in response to thermal stimulation of the

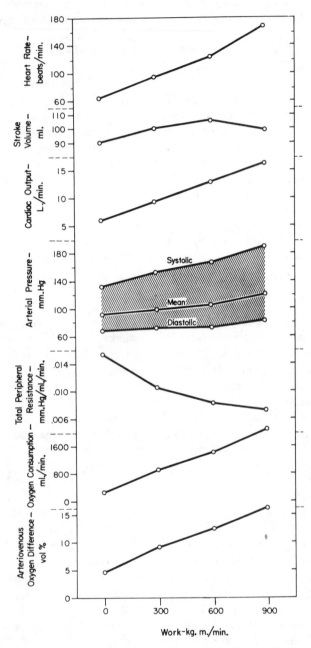

Fig. 12-1. Effect of different levels of exercise on several cardiovascular variables. (Data from Carlsten, A., and Grimby, G.: The circulatory response to muscular exercise in man, Springfield, Ill., 1966, Charles C Thomas, Publisher.)

heat-regulating center in the hypothalamus, and total peripheral resistance decreases further. This would result in a decline in blood pressure were it not for the increasing cardiac output and constriction of arterioles in the renal, splanchnic, and other inactive tissues. Coronary blood flow increases as heart work increases, but flow to the brain remains constant throughout the period of exercise.

Heart rate and stroke volume—cardiac output

The enhanced sympathetic drive and the reduced parasympathetic inhibition of the sinoatrial node continue during exercise, and consequently tachycardia persists. If the work load is moderate and constant, the heart rate will reach a certain level and remain there throughout the period of exercise. However, if the work load increases, a concomitant increase in heart rate occurs until a plateau is reached in severe exercise at about 180 beats per minute (Fig. 12-1). In contrast to the large increment in heart rate, the increase in stroke volume is only about 10% to 35% (Fig. 12-1), the larger values occurring in trained individuals (in very well-trained distance runners, whose cardiac outputs can reach six to seven times the resting level, stroke volume reaches about twice the resting value). Thus it is apparent that the increase in cardiac output observed with exercise is achieved principally by an increase in heart rate. If the baroreceptors are denervated, the cardiac output and heart rate responses to exercise are sluggish when compared to the changes in animals with normally innervated baroreceptors. However, in the absence of autonomic innervation of the heart, as produced experimentally in dogs by total cardiac denervation, exercise still elicits an increment in cardiac output comparable to that observed in normal animals, but chiefly by means of an elevated stroke volume. However, if a beta adrenergic receptor blocking agent is given to dogs with denervated hearts, exercise

performance is impaired. The beta adrenergic receptor blocker apparently prevents the cardiac acceleration and enhanced contractility caused by increased amounts of circulating catecholamines and hence limits the increase in cardiac output necessary for maximal exercise performance.

Venous return. In addition to the contribution made by sympathetically mediated constriction of the capacitance vessels in both exercising and nonexercising parts of the body, venous return is aided by the working skeletal muscles and the muscles of respiration. As pointed out in Chapter 9, the intermittently contracting muscles compress the vessels that course through them and, in the case of veins with their valves oriented toward the heart, pump blood back toward the right atrium. The flow of venous blood to the heart is also aided by the increase in the pressure gradient developed by the more negative intrathoracic pressure produced by deeper and more frequent respirations. In humans, with the exception of the skin, lungs, and liver, there is little evidence that blood reservoirs contribute much to the circulating blood volume. In fact, blood volume is usually slightly reduced in exercise, as evidenced by a rise in the hematocrit ratio, because of loss of water externally by sweating and enhanced ventilation, and internally into the contracting muscle. The fluid loss into contracting muscle reaches a plateau as interstitial fluid pressure rises and opposes the increased hydrostatic pressure in the capillaries of the active muscle. The fluid loss is partially offset by movement of fluid from the splanchnic regions and inactive muscle into the bloodstream. This influx of fluid occurs as a result of a decrease of hydrostatic pressure in the capillaries of these tissues and of an increase in plasma osmolarity due to movement of osmotically active particles into the blood from the contracting muscle. In addition, reduced urine formation by the kidneys helps to conserve body water.

The large volume of blood returning to the

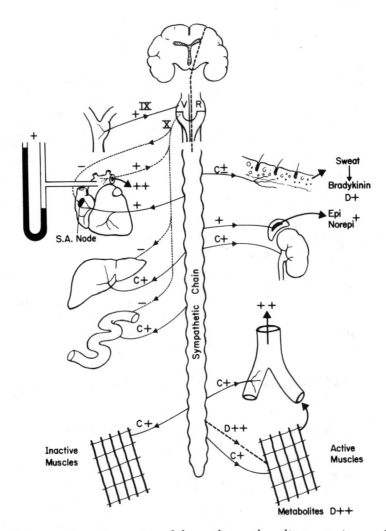

Fig. 12-2. Diagrammatic representation of the cardiovascular adjustments in exercise. *VR*, Vasomotor region; *C*, vasoconstrictor activity; *D*, vasodilator activity; *IX*, glossopharyngeal nerve; *X*, vagus nerve; ---, sympathetic cholinergic system; +, increased activity; -, decreased activity.

heart is so rapidly pumped through the lungs and out into the aorta that central venous pressure remains essentially constant. Thus the Frank-Starling mechanism of a greater initial fiber length does not account for the greater stroke volume in moderate exercise. Chest x-ray films of individuals at rest and during exercise reveal a decrease in heart size in exercise, a finding contrary to old beliefs but in harmony with the observations of a constant ventricular filling pressure. However, in maximal or near maximal exercise, right atrial pressure and end-diastolic ventricular volume do increase. Thus, the Frank-Starling mechanism contributes to the enhanced stroke volume in very vigorous exercise. In trained individuals the resting heart rate is slow and stroke volume is large. With exercise the cardiac output of the trained person can reach much greater values than can that of the untrained person as a result of greater increments in stroke volume. Heart rates will reach similar peak values with maximal exercise stress in trained and untrained subjects. Hence, there is greater oxygen delivery to the active muscles and a greater oxygen consumption in the trained individual. In additon to an enhanced maximal oxygen consumption, physical conditioning is associated with an increase in capillary density (number of capillaries per unit cross-sectional area), in the concentration of certain mitochondrial oxidative enzymes (e.g., cytochrome oxidase and succinate dehydrogenase), and probably in ATPase activity, myoglobin, and enzymes involved in lipid metabolism. Deconditioning, as occurs with complete bed rest, results in a decrease in capillary density and oxidative enzyme activity. A summary of the neural and local effects of exercise on the cardiovascular system is depicted in Fig. 12-2.

Severe exercise

In severe exercise taken to the point of exhaustion, the compensatory mechanisms begin to fail. Heart rate attains a maximum level of about 180 beats per minute, and stroke volume reaches a plateau and often decreases. Dehydration occurs. Sympathetic vasoconstrictor activity supersedes the vasodilator influence on the cutaneous vessels and has the hemodynamic effect of a slight increase in effective blood volume. However, cutaneous vasoconstriction also results in a decrease in the rate of heat loss. Body temperature is normally elevated in exercise, and reduction in heat loss through cutaneous vasoconstriction can, under these conditions, lead to very high body temperatures with associated feelings of acute distress. Since the arterial blood oxygen concentration remains at normal levels throughout exercise, the limiting factor in the performance of severe exercise is believed to be the functional capacity of the cardiovascular system. The oxygen delivery to the tissue is the cardiac output times the arterial blood oxygen content. The venous blood oxygen reaches very low levels in active muscle (large A-V O_2) and the heart rate tends to reach a plateau at about 180 beats per minute. Therefore, the limiting factor in whole body exercise is the extent to which the heart can increase stroke volume, and thereby increase oxygen delivery to the active muscles. In exercise involving only a small group of muscles, such as in finger exercise, the limiting factor is the tissue utilization of oxygen and not the supply of oxygen.

The tissue and blood pH decrease, as a result of increased lactic acid and CO_2 production, and the reduced pH is probably the key factor in determining the maximal amount of exercise a given individual can tolerate because of muscle pain, subjective feeling of exhaustion, and inability or loss of will to continue.

Postexercise recovery

When exercise stops, there is an abrupt decrease in heart rate and cardiac output—the sympathetic drive to the heart is essentially removed. In contrast, total peripheral resistance remains low for some time after the exercise is ended, presumably because of the accumulation of vasodilator metabolites in the muscles during the exercise period. As a result of the reduced cardiac output and persistence of vasodilation in the muscles, arterial pressure falls, often below pre-exercise levels, for brief periods. Blood pressure then stabilizes at normal levels as a result of activation of the baroreceptor reflexes.

HEMORRHAGE

In an individual who has lost a large quantity of blood, the principal findings, as might be anticipated, are related to the cardiovascular system. The arterial systolic, diastolic, and pulse pressures are diminished and the pulse is rapid and feeble. The cutaneous veins are collapsed and fill slowly when compressed centrally. The skin is pale, moist, and slightly cyanotic. Respiration is rapid, but the depth of respiration may be shallow or deep.

Course of arterial blood pressure changes

Cardiac output decreases as a result of blood loss for the reasons described on p. 199. This reduction in cardiac output produces the characteristic fall in arterial blood pressure. The changes in mean arterial pressure that ensue after an acute hemorrhage in experimental animals are illustrated in Fig. 12-3. If sufficient blood is withdrawn rapidly to bring mean arterial pressure to 50 mm. Hg, it is found that in almost all animals there is a tendency for the pressure to rise spontaneously toward control levels over the subsequent 20 or 30 minutes. In some animals (curve A, Fig. 12-3) this trend continues, with the result that normal pressures are regained within a few hours. Con-

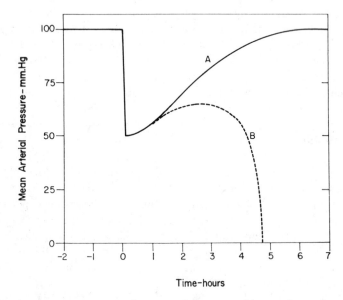

Fig. 12-3. The changes in mean arterial pressure after a rapid hemorrhage. At time zero, the animal is bled rapidly to a mean arterial pressure of 50 mm. Hg. After a period in which there is a return of pressure toward the control level, some animals will continue to improve until the control pressure is attained (curve A). However, in other animals the pressure will begin to decline until death ensues (curve B).

versely, in other animals (curve *B*), after the initial rise of pressure above 50 mm. Hg to some peak value (usually significantly below the control level), the arterial pressure begins to decline and continues to fall at an accelerating rate until death ensues.

In other experimental studies of the response to hemorrhage, animals are bled to a given hypotensive level, for example, to 35 mm. Hg, and then held at that level for a fixed period of time. This is usually accomplished by connecting a peripheral artery to a reservoir elevated to an appropriate height above the animal; the volume of blood in the reservoir is continuously monitored. The results of such a procedure are shown in Fig. 12-4. The arterial blood runs rapidly into the reservoir until the pressures become equilibrated, and then continues to flow into the reservoir at a progressively slower rate for about 2 hours. This gradual increase in shed blood volume is a manifestation of the same compensatory mechanisms that produced the rise in arterial blood pressure after hemorrhage in the experiment depicted in Fig. 12-3. However, under the experimental conditions portrayed in Fig. 12-4, as the arterial pressure tends to rise to a level higher than the hydrostatic pressure in the blood reservoir, blood flows from the cannulated vessel into the reservoir.

After the peak shed volume is attained about 2 hours after the beginning of hemorrhage, blood begins to flow in the opposite direction—from the reservoir to the animal. This is a manifestation of the tendency for arterial pressure to diminish below the established level, thereby resulting in an uptake of blood from the reservoir. It has been found that once 40% to 50% of the maximum shed volume has spontaneously returned to the ani-

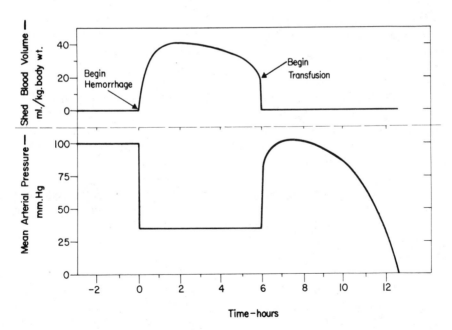

Fig. 12-4. Changes in shed blood volume during a 6-hour period of hemorrhage sufficient to hold mean arterial pressure at 35 mm. Hg, and the changes in mean arterial pressure after transfusion of the shed blood.

mal, rapid reinfusion of the remainder of the shed blood results only in a transient improvement in arterial pressure. There follows an accelerating decline of arterial pressure, eventuating in death (Fig. 12-4). This progressive deterioration of cardiovascular function is termed *shock*. At some point in time the deterioration appears to become irreversible; that is, a lethal outcome can be retarded only temporarily by any known therapy, including massive transfusions of donor blood.

Compensatory mechanisms

The changes in arterial pressure immediately after an acute blood loss (Fig. 12-3) and in shed blood volume during the initial stages of sustained hemorrhage (Fig. 12-4) indicate that certain compensatory mechanisms that combat the effects of blood loss must be operating. Any mechanism that tends to return the arterial pressure toward normal in response to the reduction in pressure produced by hemorrhage may be designated a *negative feedback mechanism*. It is termed "negative" because the secondary change in pressure is opposite in direction to the initiating change. In response to hemorrhage the following negative feedback mechanisms are evoked: (1) the baroreceptor reflexes, (2) the chemoreceptor reflexes, (3) cerebral ischemia responses, (4) reabsorption of tissue fluids, (5) release of endogenous vasoconstrictor substances, and (6) renal conservation of salt and water.

Baroreceptor reflexes. The reduction in mean arterial pressure and in pulse pressure during hemorrhage results in diminished stimulation of the baroreceptors located in the carotid sinuses and aortic arch. As a consequence multiple cardiovascular responses are evoked, all of which tend to return the arterial pressure toward its normal level. Reduction of vagal tone and enhancement of sympathetic tone result in tachycardia and a positive inotropic effect on the atrial and ventricular myocardium. The increased sympathetic discharge also produces

generalized venoconstriction, which has the same hemodynamic consequences as a transfusion of blood (p. 193). Venoconstriction is probably not elicited exclusively by the baroreceptor reflexes, since experimental evidence indicates that these reflexes exert a rather feeble influence on venomotor tone. Contraction of certain blood reservoirs in response to sympathetic activation provides an autotransfusion of blood into the circulating bloodstream. In the dog considerable quantities of blood are mobilized by contraction of the spleen. In humans the spleen does not have the same relative importance as a blood reservoir. Instead, the cutaneous, pulmonary, and hepatic vasculatures probably constitute the principal blood reservoir sites.

Generalized arteriolar vasoconstriction is a prominent component of the response to the diminished baroreceptor stimulation during hemorrhage. The reflex increase in total peripheral resistance minimizes the extent of the fall in arterial pressure resulting from the reduction of cardiac output. Fig. 12-5 shows the changes in mean aortic pressure in a group of dogs in response to an 8% blood loss. With both vagi cut and only the carotid sinus baroreceptors operative (left panel), this degree of hemorrhage produced a 14% reduction in mean aortic pressure. This did not differ significantly from the pressure reduction (12%) that occurred with all baroreceptor reflexes intact. When the carotid sinuses were denervated and the aortic baroreceptor reflexes were intact, the same percentage reduction in blood volume caused mean aortic pressure to decline by 38% (middle panel). Hence, it is apparent that the carotid sinus baroreceptors are more efficacious than the aortic baroreceptors in attenuating the fall in pressure. The aortic baroreceptors did play some role, however, because when both sets of afferent baroreceptor pathways were interrupted, an 8% blood loss produced a 48% decline in arterial pressure.

Although the arteriolar vasoconstriction is

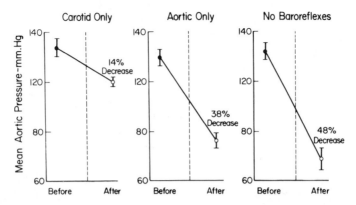

Fig. 12-5. The changes in mean aortic pressure in response to an 8% blood loss in a group of eight dogs. In the left panel the carotid sinus baroreceptor reflexes were intact and the aortic reflexes were interrupted; in the middle panel the aortic reflexes were intact and the carotid sinus reflexes were interrupted; in the right panel all sinoaortic reflexes were abrogated. (From Shepherd, J. T.: Circulation **50**:418, 1974. By permission of the American Heart Association, Inc.; derived from the data of Edis, A. J.: Am. J. Physiol. **221**:1352, 1971.)

widespread during hemorrhage, it is by no means uniform in intensity. Vasoconstriction is most severe in the cutaneous, skeletal muscle, and splanchnic vascular beds and is slight or absent in the cerebral and coronary circulations. In many instances the vascular resistance of the coronary circulation is found to be diminished. Thus a drastic redistribution of the reduced cardiac output in favor of flow through the brain and the heart occurs.

The severe cutaneous vasoconstriction accounts for the characteristic pale, cold skin of patients suffering from blood loss. Warming the skin of such patients improves their appearance considerably, much to the satisfaction of well-meaning individuals rendering first aid. However, it also inactivates an effective, natural compensatory mechanism—to the detriment of the patient.

In the early stages of mild to moderate hemorrhage, the changes in renal resistance are usually slight. The tendency for increased sympathetic activity to produce renal vasoconstriction is counteracted by intrinsic autoregulatory mechanisms. With more prolonged and more severe hemorrhages, however, intense renal vasoconstriction does occur. The reductions in renal circulation are most intense in the outer layers of the renal cortex. The inner zones of the cortex and outer zones of the medulla tend to be spared.

The severe renal and splanchnic vasoconstriction during hemorrhage undoubtedly produces more favorable conditions for the heart and the brain. However, if such constriction persists too long, it may have serious consequences. Not infrequently patients have survived the acute hypotensive period only to succumb several days later from kidney failure resulting from renal ischemia during hypotension. Intestinal ischemia also may have dire effects. In the dog, for example, intestinal bleeding and extensive sloughing of the mucosa occur after only a few hours of hemorrhagic hypotension. Furthermore, the low splanchnic flow produces swelling of the centrilobular cells in the liver. The resultant obstruction of the hepatic sinusoids produces an elevation of the portal venous pressure, which intensifies the intestinal blood loss. Fortunately the pathological changes in the liver and intestine are usually much less severe in humans.

Chemoreceptor reflexes. Once mean arterial pressure has dropped to about 60 mm. Hg, further reductions of pressure do not evoke any additional responses through the baroreceptor reflexes. However, extremely low levels of arterial pressure result in peripheral chemoreceptor stimulation because of anoxia of the chemoreceptor tissue consequent to inadequate local blood flow. Chemoreceptor excitation enhances the already existent peripheral vasoconstriction associated with baroreceptor reflexes. Also respiratory stimulation assists venous return by the auxiliary pumping mechanism described on p. 207.

Cerebral ischemia. At very low levels of arterial pressure, below 40 mm. Hg, extensive sympathetic nervous discharge occurs in response to inadequate cerebral blood flow. The intensity of the nervous discharge is severalfold greater than the maximum that occurs when the baroreceptors cease to be stimulated. Therefore severe vasoconstriction and facilitation of myocardial contractility occur. With more severe degrees of cerebral ischemia, however, the vagal centers become active and pronounced bradycardia may ensue, which would tend to aggravate the hypotension.

Reabsorption of tissue fluids. As a consequence of arterial hypotension, arteriolar constriction, and reduced venous pressure, the hydrostatic pressure in the capillaries is low during and after hemorrhage. The balance of forces, therefore, is disturbed in the direction of net reabsorption of interstitial fluid into the vascular compartment. An example of the rapidity of this response is shown in Fig. 12-6. In a group of cats a single hemorrhage of 45% of the estimated blood volume was carried out over a 20 to 30 minute period, beginning at time zero on the graph. The mean arterial blood pressure declined rapidly to about 45 mm. Hg during the bleeding. The pressure then returned rapidly, but only temporarily, to near the control level. The colloid osmotic pressure declined markedly during the bleeding and continued to decrease at a more gradual rate for several hours. The reduction in plasma colloid osmotic pressure is a reflection of the extent of dilution

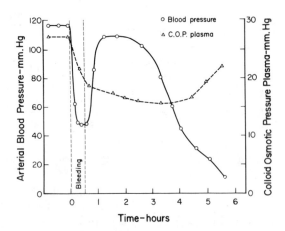

Fig. 12-6. The changes in arterial blood pressure and plasma colloid osmotic pressure in response to withdrawal of 45% of the estimated blood volume over a 30-minute period, beginning at time zero. The data are the average values for twenty-three cats. (Redrawn from Zweifach, B. W.: Anesthesiology **41:**157, 1974.)

of the blood by the ingress of tissue fluids into the vascular compartment.

Considerable quantities of fluid may thus be drawn into the circulation during hemorrhage. Values in the region of 0.25 ml. fluid reabsorbed per minute per kilogram body weight have been reported. Thus approximately 1 liter of fluid per hour might be autoinfused into the circulatory system of an average individual from his own interstitial spaces after an acute blood loss. This mechanism therefore tends to restore the depleted blood volume, and it is manifested as a reduction in hematocrit ratio soon after the onset of hemorrhage.

A slower mechanism probably also comes into play, involving the translocation of considerable quantities of fluid from intracellular to extracellular spaces. This fluid exchange is probably mediated in part by the increased secretion of cortisol by the adrenal cortex in response to hemorrhage. This hormone does cause a shift of fluid from the cells to the extracellular compartment, and it appears to be essential for a full restoration of the plasma volume after hemorrhage.

Endogenous vasoconstrictors. The *catecholamines*, epinephrine and norepinephrine, are released from the adrenal medulla in response to the same stimuli that evoke widespread sympathetic nervous discharge. For all the reasons cited previously, therefore, blood levels of catecholamines are high during and after hemorrhage. In experiments in which animals were bled to an arterial pressure level of 40 mm. Hg, it was found that there was a fiftyfold increase in blood levels of epinephrine and a tenfold increase in levels of norepinephrine. The epinephrine comes almost exclusively from the adrenal medulla, whereas the norepinephrine is derived both from that source and from the peripheral sympathetic nerve endings. These humoral substances reinforce the effects of sympathetic nervous activity listed previously.

Vasopressin, which is a potent vasoconstrictor, is actively secreted by the posterior pituitary gland in response to hemorrhage. Removal of about 20% of the blood volume in experimental animals causes an increase in the rate of vasopressin secretion to about 40 times the normal rate. The stimuli responsible for the accelerated release are the diminished pressures on both the arterial and the venous sides of the vascular system. It has been shown that the sinoaortic baroreceptors and receptors in the left atrium are involved in the regulation of vasopressin secretion.

The diminished renal perfusion during hemorrhagic hypotension leads to the secretion of *renin* from the juxtaglomerular apparatus. This enzyme acts on a plasma protein, *angiotensinogen*, to form *angiotensin*, a very powerful vasoactive substance.

Renal conservation of water. Fluid and electrolytes are conserved by the kidneys during hemorrhage for several reasons, including the effect of the increased secretion of vasopressin (antidiuretic hormone) noted previously. The lower arterial pressure per se leads to a diminished glomerular filtration rate, with a consequent reduction in the rate of excretion of water and electrolytes. Also the diminished renal blood flow results in elevated blood levels of angiotensin as described above. This polypeptide accelerates the release of *aldosterone* from the adrenal cortex. Aldosterone in turn stimulates sodium reabsorption by the renal tubules, and the sodium that is actively reabsorbed is accompanied by water. Therefore this constitutes a mechanism for the conservation of extracellular fluid.

Decompensatory mechanisms

In contrast to the negative feedback mechanisms just described, there are also latent *positive feedback mechanisms* that are evoked as a result of hemorrhage. Such mechanisms tend to exaggerate any change that occurs. Specifically, positive feedback mechanisms aggravate the hypotension induced by blood loss and tend to initiate *vicious cycles,* which may lead to death. The operation of positive feedback mechanisms is manifest in curve *B* of Fig. 12-3, and in the uptake of blood from the reservoir in Fig. 12-4.

Whether a positive feedback mechanism of itself will lead to a vicious cycle depends on the *gain* of that mechanism. For a positive feedback system, gain may be defined as the ratio of the magnitude of the secondary change evoked by the mechanism in question to the magnitude of the initiating change itself. A gain greater than 1 would lead to a vicious cycle; a gain less than 1 would not. For example, consider a positive feedback mechanism with a gain of 2. If, for any reason, mean arterial pressure decreased by 10 mm. Hg, the positive feedback mechanism would then evoke a secondary reduction of pressure of 20 mm. Hg, which in turn would cause a further decrement of 40 mm. Hg; that is, each change would induce a subsequent change that was twice as great. Hence mean arterial pressure would decline at an ever-increasing rate until death supervened, much as is depicted by curve *B* in Fig. 12-3.

Conversely, a positive feedback mechanism with a gain of 0.5 would indeed exaggerate any change in mean arterial pressure but would not necessarily lead to death. For example, if arterial pressure suddenly decreased by 10 mm. Hg, the positive feedback mechanism would initiate a secondary, additional fall of 5 mm. Hg. This in turn would provoke a further decrease of 2.5 mm. Hg, and the process would continue in ever-diminishing steps, with the arterial pressure approaching an equilibrium value asymptotically.

Some of the more important positive feedback mechanisms include (1) cardiac failure, (2) acidosis, (3) inadequate cerebral blood flow, (4) aberrations of blood clotting, and (5) depression of the reticuloendothelial system.

Cardiac failure. Considerable controversy exists at present concerning the role of cardiac failure in the progression of shock during hemorrhage. All investigators agree that the heart fails terminally, but opinions differ concerning the relative importance of cardiac failure during earlier stages of hemorrhagic hypotension. Shifts to the right in ventricular function curves (Fig. 12-7) constitute experimental evidence of a progressive depression of myocardial contractility during hemorrhage.

The hypotension induced by hemorrhage reduces the rate of coronary blood flow and therefore tends to depress ventricular function. The consequent reduction in cardiac output leads to a further decline in arterial pressure, a classical example of a positive feedback mechanism. Furthermore, the reduced tissue blood flow leads to an accumulation of vasodilator metabolites, which serves to decrease peripheral resistance and therefore to aggravate the fall in arterial pressure.

In addition to the curtailment of coronary blood flow, other mechanisms contribute to the development of cardiac failure during hemorrhagic hypotension. As described in the following section, acidosis develops in the course of hemorrhagic shock. This may depress the myocardium directly, and also indirectly by impairing the responsiveness of the heart to sympathetic stimulation and to circulating catecholamines.

Certain substances that impair cardiac function have been isolated from the blood of animals in various types of experimental shock. Such substances include the amino acid, leucine, and a small, unidentified peptide. They are apparently released from the pancreas and other splanchnic viscera as a consequence of the curtailed blood flow. The role of such *myocardial depressant factors* in the pathogenesis of cardiac failure during shock is controversial at present.

Subendocardial necrosis and hemorrhage are frequently found in the course of hemorrhagic shock. Such pathological changes probably result from a combination of inadequate coronary

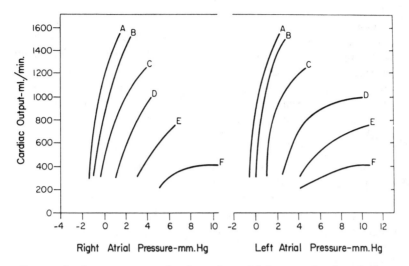

Fig. 12-7. Ventricular function curves for the right and left ventricles during the course of hemorrhagic shock. Curves *A* represent the control function curve; curves *B*, 117 min.; curves *C*, 247 min.; curves *D*, 280 min.; curves *E*, 295 min.; and curves *F*, 310 min. after the initial hemorrhage. (Redrawn from Crowell, J. W., and Guyton, A. C.: Am. J. Physiol. **203**:248, 1962.)

perfusion and excessive sympathoadrenal activity. These anatomical alterations undoubtedly accentuate the functional impairment of the heart.

Acidosis. Because of inadequate blood flow during hemorrhage, the metabolism of all cells in the body is affected. The resultant stagnant anoxia leads to increased production of lactic acid and other acid metabolites by the tissues. Furthermore, impaired kidney function prevents adequate excretion of H^+, with the result that generalized metabolic acidosis ensues. The resulting depressant effect of acidosis on the heart causes a further reduction in tissue perfusion, with an intensification of the metabolic acidosis. Acidosis also diminishes the reactivity of the resistance vessels to neurally released and circulating catecholamines, thereby intensifying the hypotension.

Inadequate cerebral blood flow. The cerebral ischemic response was shown to result in pro-

nounced sympathetic nervous stimulation of the heart, arterioles, and veins. With very severe degrees of hypotension, however, the cardiac and vasomotor centers eventually become depressed because of inadequate cerebral perfusion. The resultant loss of sympathetic tone then causes a reduction in cardiac output and peripheral resistance. The resulting reduction in mean arterial pressure intensifies the severity of the inadequate cerebral perfusion.

Aberrations of blood clotting. The alterations of blood clotting after hemorrhage are typically biphasic—an initial phase of hypercoagulability followed by a secondary phase of hypocoagulability and fibrinolysis. In the initial phase intravascular clots, or *thrombi*, have been detected within a few minutes of the onset of severe hemorrhage, and intravascular coagulation may be extensive throughout the minute blood vessels. The mortality rate from certain standard shock-provoking procedures has been reduced considerably by the administration of anticoagulants such as heparin.

In the later stages of hemorrhagic hypotension, the clotting time is prolonged and fibrinolysis is prominent. It was mentioned previously that in the dog hemorrhage into the intestinal lumen is common after several hours of hemorrhagic hypotension. Blood loss into the intestinal lumen would, of course, aggravate the effects of the original hemorrhage.

Reticuloendothelial depression. During the course of hemorrhagic hypotension, reticuloendothelial system (RES) function becomes depressed. The phagocytic activity of the RES is modulated by an opsonic protein. It has been found that the opsonic activity in plasma diminishes during the course of shock. This probably accounts in part for the depression of RES function. As a consequence, the antibacterial and antitoxic defense mechanisms of the body are impaired. Endotoxins from the normal bacterial flora of the intestine constantly enter the intestinal circulation. Ordinarily they are inactivated by the RES, principally in the liver. When the RES is severely depressed, these endotoxins invade the general circulation. Endotoxins produce a form of shock that resembles in many respects that produced by hemorrhage. Therefore depression of the RES leads to an intensification of the hemodynamic changes caused by blood loss. Sterilization of the intestine by means of antibiotics significantly reduces the mortality from certain standard shock-provoking procedures including hemorrhage.

Interactions of positive and negative feedback mechanisms

It is obvious from the preceding discussion that a multitude of circulatory and metabolic derangements are produced by hemorrhage. Some of these changes are compensatory; others are decompensatory. Some of these feedback mechanisms possess a relatively high gain; others, a relatively low gain. Furthermore,

with regard to any specific mechanism, the magnitude of the gain varies with the severity of the hemorrhage. For example, with only a slight loss of blood, mean arterial pressure will still be within the range of normal and the gain of the baroreceptor reflexes is appreciable. With greater losses of blood, when mean arterial pressure is below about 60 mm. Hg (that is, below the threshold for the baroreceptors), then further reductions of pressure will have no additional influence through the baroreceptor reflexes. Hence below this critical pressure level the gain of the baroreceptor reflexes will be zero or near zero.

As a general rule, with minor degrees of blood loss, the gains of the negative feedback mechanisms are relatively high, whereas those of the positive feedback mechanisms are low. The converse is true with more severe hemorrhages. The gains of the various mechanisms are additive algebraically. Therefore, whether a vicious cycle develops depends on whether the sum of the various gains exceeds $+1$. Total gains in excess of $+1$ are, of course, more likely to occur with severe losses of blood. Therefore to avert a vicious cycle, serious hemorrhages must be treated quickly and intensively, preferably by whole blood transfusions, before the process has become irreversible.

BIBLIOGRAPHY
Journal articles

Abel, F. L., and Kessler, D. P.: Myocardial performance in hemorrhagic shock in the dog and primate, Circ. Res. **32:**492, 1973.

Asmussen, E., and Nielsen, M.: Cardiac output during muscular work and its regulation, Physiol. Rev. **35:**778, 1955.

Barron, W., and Coote, J. H.: The contribution of articular receptors to cardiovascular reflexes elicited by passive limb movement, J. Physiol. **235:**423, 1973.

Bevegård, B. S., and Shepherd, J. T.: Regulation of the circulation during exercise in man, Physiol. Rev. **47:**178, 1967.

Chapman, C. B., editor: Symposium on physiology of muscular exercise, Circ. Res. **20** (Supp. 1):1-226, 1967.

Chien, S.: Role of the sympathetic nervous system in hemorrhage, Physiol. Rev. **47:**214, 1967.

Chien, S., and Usami, S.: Rate and mechanism of release of antidiuretic hormone after hemorrhage, Circ. Shock 1:71, 1974.

Clausen, J. P.: Circulatory adjustments to dynamic exercise and effect of physical training in normal subjects and in patients with coronary artery disease, Prog. Cardiovasc. Dis. 18:459, 1976.

Coote, J. H., Hilton, S. M., and Perez-Gonzalez, J. F.: The reflex nature of the pressor response to muscular exercise, J. Physiol. 215:789, 1971.

Donald, D. E., Ferguson, D. A., and Milburn, S. E.: Effect of beta-adrenergic receptor blockade on racing performance of greyhounds with normal and with denervated hearts, Circ. Res. 22:127, 1968.

Fredholm, B. B., Farnebo, L. O., and Hamberger, B.: Plasma catecholamines, cyclic AMP and metabolic substrates in hemorrhagic shock of the rat. The effect of adrenal demedullation and 6-OH-dopamine treatment, Acta Physiol. Scand. 105:481, 1979.

Goldfarb, R. D.: Characteristics of shock-induced circulating cardiodepressant substances: a brief review, Circ. Shock, Supp. 1, p. 23, 1979.

Kaijser, L.: Limiting factors for aerobic muscle performance, Acta Physiol. Scand., Supp. 346, pp. 1-96, 1970.

Lefer, A. M.: Properties of cardioinhibitory factors produced in shock, Fed. Proc. 37:2734, 1978.

Liang, C., and Hood, W. B., Jr.: Afferent neural pathway in the regulation of cardiopulmonary responses to tissue hypermetabolism, Circ. Res. 38:209, 1976.

Loegering, D. J.: Humoral factor depletion and reticuloendothelial depression during hemorrhagic shock, Am. J. Physiol. 232:H283, 1977.

McCloskey, D. I., and Mitchell, J. H.: Reflex cardiovascular and respiratory responses originating in exercising muscles, J. Physiol. 224:173, 1972.

Pinardi, G., Talmaciu, R. K., Santiago, E., and Cubeddu, L. X.: Contribution of adrenal medulla, spleen and lymph, to the plasma levels of dopamine β-hydroxylase and catecholamines induced by hemorrhagic hypotension in dogs, J. Pharmacol. Exp. Ther. 209:176, 1979.

Pirkle, J. C., Jr., and Gann, D. S.: Restitution of blood volume after hemorrhage: mathematical description, Am. J. Physiol. 228:821, 1975.

Rowell, L. B.: Human cardiovascular adjustments to exercise and thermal stress, Physiol. Rev. 54:75, 1974.

Saltin, B., and Rowell, L. B.: Functional adaptations to physical activity and inactivity, Fed. Proc. 39:1506, 1980.

Scheuer, J., Penpargkul, S., and Bhan, A. K.: Experimental observations on the effects of physical training upon intrinsic cardiac physiology and biochemistry, Am. J. Cardiol. 33:744, 1974.

Selkurt, E. E.: Current status of renal circulation and related nephron function in hemorrhage and experimental shock. I. Vascular mechanisms, Circ. Shock 1:3, 1974.

Smith, E. E., Guyton, A. C., Manning, R. D., and White, R. J.: Integrated mechanisms of cardiovascular response and control during exercise in the normal human, Prog. Cardiovasc. Dis. 18:421, 1976.

Sparks, H. V.: Mechanism of vasodilation during and after ischemic exercise, Fed Proc. 39:1487, 1980.

Vatner, S. F.: Effects of hemorrhage on regional blood flow distribution in dogs and primates, J. Clin. Invest. 54:225, 1974.

Vatner, S. F., and Pagani, M.: Cardiovascular adjustments to exercise: hemodynamics and mechanisms, Prog. Cardiovasc. Dis. 19:91, 1976.

Zweifach, B. W., and Fronek, A.: The interplay of central and peripheral factors in irreversible hemorrhagic shock, Prog. Cardiovasc. Dis. 18:147, 1975.

Books and monographs

Carlsten A., and Grimby, G.: The circulatory response to muscular exercise in man, Springfield, Ill., 1966, Charles C Thomas, Publisher.

Lefer, A. M., Saba, T., and Mela, L. M., editors: Advances in shock research, vol. 1, New York, 1979, Alan R. Liss, Inc.

Lefer, A. M., and Schumer, W., editors: Metabolic and cardiac alterations in shock and trauma, New York, 1979, Alan R. Liss, Inc.

Schumer, W., Spitzer, J. J., and Marshall B. E., editors: Advances in shock research, vol. 2, New York, 1979, Alan R. Liss, Inc.

Wiggers, C. J.: Physiology of shock, New York, 1950, Commonwealth Fund.

INDEX

W. SUSSEX INSTITUTE
OF
HIGHER EDUCATION
LIBRARY